DISPERSION IN HETEROGENEOUS GEOLOGICAL FORMATIONS

Edited by

BRIAN BERKOWITZ

Weizmann Institute of Science, Israel

Reprinted from *Transport in Porous Media*
Volume 42, Nos. 1–2 (2001)

KLUWER ACADEMIC PUBLISHERS
DORDRECHT/BOSTON/LONDON

A C.I.P. catalogue record for this book is available from the Library of Congress.

ISBN 0-7923-6779-0

Published by Kluwer Academic Publishers,
P.O. Box 17, 3300 AA Dordrecht, The Netherlands.

Sold and distributed in the U.S.A. and Canada
by Kluwer Academic Publishers,
101 Philip Drive, Norwell, MA 02061, U.S.A.

In all other countries, sold and distributed
by Kluwer Academic Publishers,
P.O. Box 322, 3300 AH Dordrecht, The Netherlands.

Printed on acid-free paper.

2-0602-150 ts

Printed in the Netherlands.

Table of Contents

Transport in Porous Media **42:** 1–2, 2001.

Dispersion in Heterogeneous Geological Formations: Preface (Transport in Porous Media – Special Issue)

BRIAN BERKOWITZ
Department of Environmental Sciences and Energy Research, Weizmann Institute of Science, Rehovot 76100, Israel, e-mail: brian.berkowitz@weizmann.ac.il

In spite of many years of intensive study, our current abilities to actually quantify and predict contaminant migration in natural geological formations remain severely limited. While transport theories that treat 'homogeneous' porous media are mathematically convenient, such homogeneity rarely, if ever, exists in the field. The heterogeneity of natural geological formations at a wide range of scales necessitates consideration of more sophisticated transport theories.

Over the last two decades, deterministic frameworks have given way to a variety of stochastic approaches, involving, for example, Monte Carlo simulations, random walks, spectral analysis, stochastic partial differential equations and fractal theory, to name only several. The evolution of theories for dispersion in heterogeneous porous media has thus escalated to the point that a review of this subject seems timely. While conceptual and mathematical developments were crucial with the introduction of these new approaches, there are now too many publications appearing which might be classified as 'more of the same'. These papers contain theoretical abstractions without regard to real systems, or incremental improvements to existing theories which are known not to be applicable – with the result being that the literature is expanding without concurrent improvements in actual predictive tools. A case in point (at the risk of engendering the wrath of colleagues) might be the persistent search for 'macrodispersion' parameters in the context of the advection–dispersion equation... while it is now generally accepted that the length scale necessary to reach this asymptotic limit is impractical in real geological systems.

Considering the mixture of conceptual advances and the pressing need for critical assessment of them, the scope of this project was to bring together, in a single volume, a series of articles representing a broad spectrum of approaches for characterization and quantification of contaminant dispersion in heterogeneous porous media. The authors were asked for contributions that would be as accessible as possible to a wide readership with diverse backgrounds. Thus, the authors were requested to keep the display of mathematics to a minimum, in order to emphasize the explanation of methods and discussion of their application. Each of the contributing authors has a different expertise and point of view; some of these viewpoints are complementary, while others are a source of controversy. Thoughtful consideration of these articles will, hopefully, allow the reader to discern the advantages, disadvantages and limitations of each of the methods or approaches.

It is the hope of the editor that this issue will stimulate further research in this area. Innovation and a willingness to depart from 'conventional wisdom' are desperately needed. Nature is not always accommodating, and the hunt for a 'comprehensive' dispersion theory may be daunting, yet such a theory – one which would permit genuine prediction of contaminant migration in natural geological formations – is of critical importance.

Transport in Porous Media **42:** 3–35, 2001.

On Perturbative Expansions to the Stochastic Flow Problem

F. ALEJANDRO BONILLA[1] and JOHN H. CUSHMAN[2]
[1]*Center for Applied Math and Department of Civil and Environmental Engineering*
[2]*Departments of Mathematics and Agronomy, Center for Applied Mathematics, Math Sciences Building, Purdue University, West Lafayette, IN. 47907, U.S.A.*

(Received: 31 July 1999; in final form: 30 December 1999)

Abstract. When analyzing stochastic steady flow, the hydraulic conductivity naturally appears logarithmically. Often the log conductivity is represented as the sum of an average plus a stochastic fluctuation. To make the problem tractable, the log conductivity fluctuation, f, about the mean log conductivity, $\ln K_G$, is assumed to have finite variance, σ_f^2. Historically, perturbation schemes have involved the assumption that $\sigma_f^2 < 1$. Here it is shown that σ_f may not be the most judicious choice of perturbation parameters for steady flow. Instead, we posit that the variance of the gradient of the conductivity fluctuation, $\sigma_{\nabla f}^2$, is a more appropriate choice. By solving the problem with this parameter and studying the solution, this conjecture can be refined and an even more appropriate perturbation parameter, ε, defined. Since the processes f and ∇f can often be considered independent, further assumptions on ∇f are necessary. In particular, when the two point correlation function for the conductivity is assumed to be exponential or Gaussian, it is possible to estimate the magnitude of $\sigma_{\nabla f}$ in terms of σ_f and various length scales. The ratio of the integral scale in the main direction of flow (λ_x) to the total domain length (L^*), $\rho_x^2 = \lambda_x/L^*$, plays an important role in the convergence of the perturbation scheme. For ρ_x smaller than a critical value ρ_c, $\rho_x < \rho_c$, the scheme's perturbation parameter is $\varepsilon = \sigma_f/\rho_x$ for one- dimensional flow, and $\varepsilon = \sigma_f/\rho_x^2$ for two-dimensional flow with mean flow in the x direction. For $\rho_x > \rho_c$, the parameter $\varepsilon = \sigma_f/\rho_x^3$ may be thought as the perturbation parameter for two-dimensional flow. The shape of the log conductivity fluctuation two point correlation function, and boundary conditions influence the convergence of the perturbation scheme.

Key words: Flow, stochastic, perturbation, velocity, head gradient

1. Introduction

In choosing a perturbation parameter to use when approximating the solution to a PDE, it is advantageous to nondimensionalize the equation. But even after nondimensionalizing a problem, it still may remain unclear what is the allowable magnitude of the perturbation parameter. In some cases the perturbation process may be valid even for unexpectedly large values of the coefficient in question (Carrier and Pearson, 1988). A case in point is the steady flow in nondeformable porous media where σ_f (the fluctuating log-conductivity standard deviation) is often used as the perturbation parameter, yet the schemes work for $\sigma_f > 1$ (e.g., Dagan, 1989;

Cushman, 1990; Serrano, 1992; Gelhar, 1993; Neuman and Orr, 1993; Hsu *et al.*, 1996; Cushman, 1997; Hassan *et al.*, 1998). In this case, disparate length scales (domain size, anisotropic integral scales, etc.) can explain the robust character of the perturbation scheme. In subsequent analysis rather than working with σ_f alone, we study σ_{∇_f} and $\rho = \sqrt{\lambda/L}$, with λ the integral scale and L the domain size.

There exists extensive research on perturbation solutions to the stochastic flow problem (e.g., Dagan, 1989; Gelhar, 1993; Cushman, 1997). But it is not known how large σ_f can be and yet have a convergent expansion. For example, Hassan *et al.* (1998) have reported that small integral scales generate very slow convergence and sometimes instability in the solution to stochastic flow via Monte Carlo simulations, even using small values for σ_f. Other researchers (e.g. Hsu *et al.*, 1996; Zhang and Winter, 1999) have shown that values of $\sigma_f > 1$ produce agreement between perturbation and the Monte Carlo simulations. Zhang and Winter (1999) found this to hold for $\sigma_f^2 = 4.38$. The work presented here addresses convergence, and clarifies a commonly misunderstood issue. Further, the parameter bounds provided here offer a quick way to assess the magnitude of terms of higher order, an issue associated directly with the accuracy of truncated perturbation solutions.

We first examine a one-dimensional (1D) bounded domain, and the solution to the Dirichlet problem. This solution is used to provide a hint as to what are the appropriate scalings. The dependence on length scale parameters is then studied and two distinct yet similar scalings are applied. Constraints on the validity of the perturbation scheme are developed and then expounded upon. A second set of Neuman–Dirichlet boundary conditions is considered and analyzed. Two-dimensional (2D) flow is subsequently analyzed. Based on the consistency of the expansion for the 1D problem, attention is placed on the head variance to develop a convergence criterion and the adequate expansion parameter. The 1D case can be viewed as a limit of the 2D case, and the results are consistent.

2. Head Moments for a Heterogeneous Bounded Domain

Let $\mathbf{x}^* = (x_1^*, \ldots, x_m^*)$ represent an m-dimensional Cartesian coordinate system ($m = 1, 2, 3$). Define the dimensionless coordinate system $\mathbf{x} = (x_1^*/\lambda_s, \ldots, x_m^*/\lambda_s)$ where λ_s is a representative length scale involved in the problem. Consider the hydraulic head h^* and define the dimensionless head as $h = h^*/L^*$ with L^* corresponding to the domain's length in the main direction of flow. The steady flow of incompressible fluid through a locally isotropic medium is assumed to satisfy the continuity equation

$$\frac{\partial}{\partial x_i}\left(K(\mathbf{x})\frac{\partial h}{\partial x_i}\right) = 0, \quad \mathbf{x} \in \Omega, \tag{1}$$

where $K(\mathbf{x})$ is the hydraulic conductivity and Ω represents the domain. Suitable boundary conditions need to be specified to fully define the problem and they will be elucidated upon below. At larger scales the medium may possess anisotropic structure which is consistent with the above equation, provided one indexes K. By differentiating by parts and simplifying (1) reduces to

$$\frac{\partial \ln(K(\mathbf{x}))}{\partial x_i}\frac{\partial h}{\partial x_i} + \frac{\partial^2 h}{\partial x_i \partial x_i} = 0, \quad \mathbf{x} \in \Omega, \tag{2}$$

which is a Poisson equation; the solution of which may be approximated as $h = h_0 + h_1 + h_2 + \cdots$ with $h_n = O(\varepsilon^n)$ where ε is some, as yet to be determined, small parameter. This small parameter is related to the 'magnitude' of $\partial \ln(K(\mathbf{x}))/\partial x_i$ which is a random function of space. We need to define better what we mean by magnitude or norm of this random function, and the concept of norm is also instrumental in defining a dominating series for $h = h_0 + h_1 + h_2 + \cdots$. The following subsection is dedicated to set out this concept.

2.1. L^q NORMS

To study convergence of h we rely on a dominating series given in terms of max L^q norms. The head h is a random function and so we simplify the analysis by studying its moments. For the analysis of convergence, we will concentrate on the head mean, $\langle h \rangle$, and the head variance, σ_h^2. Associated with these moments are the L^q norms (q is an integer),

$$\|g(\mathbf{x})\|_q = \left(\int_{\omega \in S} | g(\mathbf{x}, \omega) |^q \; P(\mathrm{d}\omega) \right)^{1/q} \tag{3}$$

where q is an integer, S is the sample space and P is the probability measure. The head mean, $\langle h \rangle = h_0 + \langle h_2 \rangle + \langle h_4 \rangle + \cdots$ is the L^1 norm of h; $\langle h \rangle = \|h\|_1$ according to Equation (3). The head standard deviation is a centered moment, that is, the L^2 norm of $h - \langle h \rangle$; $\sigma_h = \|h - \langle h \rangle\|_2 = \|h - \|h\|_1\|_2$. An L^1 convergent perturbation expansion of h, $h = h_0 + h_1 + h_2 + \cdots$ is defined as one producing finite mean head, and an L^2 convergent perturbation expansion is defined as one producing finite head variance.

No general proof of L^1 or L^2 convergence for h exist or can be presented here. However, the perturbation methods literature suggests, as a test for convergence, to compute as many terms in the expansion as possible and to verify they are decreasing faster than those of a convergent series (Hinch, 1990). Convergence is studied here following this approach.

An additional refinement is needed to avoid the complexities posed by non-stationarity of the head field. In bounded domains, the head mean and variance are functions of space (non-stationary), that is, $\langle h \rangle = \langle h(\mathbf{x}) \rangle$ and $\langle \sigma_h^2 \rangle = \langle \sigma_h^2(\mathbf{x}) \rangle$. It is therefore, cumbersome to seek for a dominating convergent expansion for h at every location $\mathbf{x}$. This complexity can be circumvented by redefining the norm presented in Equation (3) as (Eckhaus, 1973)

$$\|g\|_q = \max_{\mathbf{x} \in \Omega} \left(\int_{\omega \in S} | g(\mathbf{x}, \omega) |^q \; P(\mathrm{d}\omega) \right)^{1/q} . \tag{4}$$

This definition ensures that L^q convergence will be attained at every point $\mathbf{x}$ if a dominating L^q convergent series h^M is obtained such that $\|h\|_q \leqslant \|h^M\|_q$. We use the following inequality for the mean head,

$$\begin{aligned} \|h\|_1 &= \|h_0 + h_1 + h_2 + \cdots \|_1 \\ &\leqslant \|h_0\|_1 + \|h_1\|_1 + \|h_2\|_1 + \|h_3\|_1 + \|h_4\|_1 + \cdots . \end{aligned} \tag{5}$$

If a convergent series h^M in the expansion parameter ε is found (h^M is a deterministic series) such that $\|h_n\|_1 \leqslant \|h_n^M\|_1$ for all $n = 1, 2, 3, \ldots$, then $\|h\|$ is L^1 convergent or equivalently, it has finite mean. Now turn the attention to L^2 convergence, which is a more restrictive condition than L^1 convergence. In general, L^k convergence implies convergence of inferior orders than k (result from Hölder's inequality $\|h\|_1 \leqslant \|h\|_q$ with $q \geqslant 1$), but the contrary is not true (Ash (1972), Lemma 5.10.5). If the mean is finite, the head variance σ_h^2 will be finite if $\|h\|_2$ is finite. The condition for L^2 convergence can be written as

$$\|h\|_2 \leqslant \|h_0\|_2 + \|h_1\|_2 + \|h_2\|_2 + \|h_3\|_2 + \|h_4\|_2 + \cdots \leqslant \sum_{n=0,\ldots} \|h_n^M\|_2 . \tag{6}$$

The above equation defines the convergence criterion that will be used through the analysis presented here. Note that h_0 is a deterministic function and therefore $\|h_0\|_1 = \|h_0\|_2 = \cdots \|h_0\|_q$. If we are therefore, satisfied looking at the first two terms of the expansion, h_0 and h_1, then the condition from the above equation provides an L_2 convergence criterion. The consistency in the higher orders we will observe when computing the mean head gives confidence the expansion for the head variance is consistent. Higher order computations are necessary to develop more confidence in the $\mathrm{O}(\varepsilon^2)$ result presented here.

2.2. MEAN HEAD AND VARIANCE

Having defined the norm of non stationary random variable moments, we look for an appropriate parameter, ϵ, for the head expansion. Consider again Equation (2). A first reasonable choice of expansion parameter is $\epsilon' = \|\partial \ln(K(\mathbf{x}))/\partial x_i\|_2$ (the prime, $'$, does not denote spatial differentiation, but it is used because we redefine the parameter below).

It is convenient at this point to characterize the field $K(\mathbf{x})$ as random and fluctuating around its mean: $Y = \ln K(\mathbf{x}) = \langle Y \rangle + f(\mathbf{x})$. With this decomposition, and assuming $\langle Y \rangle$ is constant, (2) becomes

$$\frac{\partial f}{\partial x_i}\frac{\partial h}{\partial x_i} + \frac{\partial^2 h}{\partial x_i \partial x_i} = 0, \quad \mathbf{x} \in \Omega . \tag{7}$$

In this case we measure the smallness of ∇f by the standard deviation $\epsilon = \sigma_{\nabla f}$ (redefining ϵ), which is assumed independent of $\mathbf{x}$ as indicated by Equation (4). In the most general case, the random fields $f(\mathbf{x})$ and $\partial f(\mathbf{x})/\partial x_i$ are independent. In order to develop a logical perturbation scheme for (7), we must guarantee ϵ is small. In most practical computations, the statistical model assigned to $f(\mathbf{x})$ from data is robust enough to admit differentiation (at least in a mean square sense), and therefore, an estimate of $\epsilon = \sigma_{\nabla f}$ can be had. The log conductivity fluctuation $f(\mathbf{x})$ is often assumed to be characterized by its first two moments; $\langle f(\mathbf{x})\rangle = 0$, and $\sigma_f^2 = \langle f(\mathbf{x}) f(\mathbf{x})\rangle$. A higher order characterization is sometimes sought for $f(\mathbf{x})$ through its multipoint correlation functions. There is evidence (e.g., Dagan, 1989; Gelhar, 1993) that at least for 1D conductivity data, the two point correlation function can be modeled as Gaussian. However, we should expect such hypothesis to be modified in the future as several authors have recently suggested alternative models (Neuman, 1995; SanchezVila *et al.*, 1996; Mukhopadhyay and Cushman, 1998; Gómez-Hernández and Wen, 1998; Benson *et al.*, 1999). Here, we analyze only the most popular weakly stationary, multidimensional, exponential and Gaussian models. The Gaussian model is represented by

$$\langle f(\mathbf{x})\rangle = 0, \tag{8a}$$

$$R_{ff}(\mathbf{x},\mathbf{y}) \equiv \langle f(\mathbf{x}) f(\mathbf{y})\rangle = R_{ff}(\mathbf{x}-\mathbf{y}), \tag{8b}$$

and

$$\langle f(\mathbf{x}) f(\mathbf{y}) f(\mathbf{z})\rangle = 0, \tag{9a}$$

$$\begin{aligned}\langle f(\mathbf{x}) f(\mathbf{y}) f(\mathbf{z}) f(\mathbf{w})\rangle &= R_{ff}(\mathbf{x},\mathbf{y})R_{ff}(\mathbf{z},\mathbf{w}) + \\ &\quad + R_{ff}(\mathbf{x},\mathbf{z})R_{ff}(\mathbf{y},\mathbf{w}) + R_{ff}(\mathbf{x},\mathbf{w})R_{ff}(\mathbf{y},\mathbf{z})\end{aligned} \tag{9b}$$

with

$$R_{ff}(\mathbf{x}-\mathbf{y}) = R_{ff}(\mathbf{r}) = \sigma_f^2 \exp\left(-\lambda_s^2\left[\sum_{i=1}^{m}\frac{r_i^2}{\lambda_i^2}\right]\right), \tag{10}$$

where λ_i is the integral scale in the ith direction, m is the dimension of the problem ($m = 1, 2, 3$), $\mathbf{x}$ and $\mathbf{y}$ are two dimensionless spatial coordinates. The separation vector $\mathbf{r}$ is defined $r_i = x_i - y_i$, and λ_s is a representative length scale for the problem. In Appendix A we provide an analysis for the R_{ff} exponential, defined by Equations (8a), (9a), and (73) (1D flow) or (67) (2D flow). Since we are pursuing an estimate for $\sigma_{\nabla f}$, from (10) we find

$$\frac{\partial^2 R_{ff}(\mathbf{r})}{\partial r_i \partial r_i} = \sigma_f^2 \exp\left(-\lambda_s^2\left[\sum_{i=1}^{m}\frac{r_i^2}{\lambda_i^2}\right]\right)\left[\frac{4r_i^2\lambda_s^4}{\lambda_i^4} - \frac{2\lambda_s^2}{\lambda_i^2}\right] \tag{11}$$

with no implied summation on i. The log conductivity fluctuation gradient, $\epsilon_i = \|\partial f/\partial x_i/\|_2 = \sigma_{\nabla f_i}$, is of vectorial character. To find a meaningful perturbation parameter we look at its maximum norm as defined by Equation (4) for each component of the vector $\sigma_{\nabla f_i}$. With this objective, consider the maximum of (11). Differentiating (11) with respect to r_i, $i = 1 \cdots m$, setting these derivatives to zero, and solving we find the extreme values are at the origin $r_i = 0 \wedge r_j = 0$ (maximum) and $r_i = \pm\sqrt{\frac{3}{2}\frac{\lambda_i}{\lambda_s}} \wedge r_j = 0 \,\forall\, j \neq i$ (minimum). The value of the maximum at the origin is the largest absolute value of the function (11) and it is an estimate of the magnitude of components $\epsilon_i = \sigma_{\nabla f_i}$;

$$\epsilon_i = \left\| \frac{\partial f}{\partial x_i} \right\|_2 = \sigma_{\nabla f_i} = \frac{\sqrt{2}\sigma_f \lambda_s}{\lambda_i}. \tag{12}$$

Furthermore, consider the maximum component of such vector

$$\epsilon = \sigma_{\nabla f} = \max_{i=1,\dots,m} \sigma_{\nabla f_i}, \tag{13}$$

where m represents the dimension of the problem, $m = 1, 2, 3$. Unless noted otherwise, ϵ is defined according to (13). This simple result suggests that if no length scaling is performed ($\lambda_s = 1$), the perturbation scheme for a Gaussian log conductivity fluctuation two point correlation function should be based on the parameter

$$\epsilon = \max_{i=1,\dots,m} \frac{\sqrt{2}\sigma_f}{\lambda_i}. \tag{14}$$

We reiterate that since $\sigma_{\nabla f_i}$ is a vector, we are selecting its maximum component to represent the parameter ϵ. Anisotropy however, does have an effect and an additional parameter to account for it will be discussed later. Length scales also influence the magnitude of the stochastic gradient, $\partial h/\partial x_i$, in the flow equation

$$\frac{\partial^2 h}{\partial x_i \partial x_i} = -\frac{\partial f}{\partial x_i}\frac{\partial h}{\partial x_i}, \quad \mathbf{x} \in \Omega, \tag{15}$$

and therefore, the order of magnitude of the rhs (15) is unknown.

If we develop a perturbation around σ_f, we arrive at expressions containing implicitly the parameter $\sigma_{\nabla f}$. To illustrate this, consider Equation (52) in Dagan (1982) which rewritten by Bonilla and Cushman (2000) for the head covariance is

$$R_{hh}^{(2)}(\mathbf{x}, \mathbf{y}) = \int_{S_{\mathbf{x}''}} \int_{S_{\mathbf{x}'}} G_1(\mathbf{x}, \mathbf{x}'') G_1(\mathbf{y}, \mathbf{x}') \frac{\partial^2 R_{ff}(\mathbf{x}', \mathbf{x}'')}{\partial x_i' \partial x_j''} \times \\ \times \frac{\partial h_0(\mathbf{x}')}{\partial x_i'} \frac{\partial h_0(\mathbf{x}'')}{\partial x_j''} \,\mathrm{d}\mathbf{x}' \,\mathrm{d}\mathbf{x}''. \tag{16}$$

This result is supposedly $O(\sigma_f^2)$ (e.g., Dagan, 1982; Rubin and Dagan, 1992; Deng and Cushman, 1995; Bonilla and Cushman, 2000). However, by inspecting the norm of $\| \frac{\partial^2 R_{ff}(\mathbf{x}',\mathbf{x}'')}{\partial x_i' \partial x_j''} \frac{\partial h_0(\mathbf{x}')}{\partial x_i'} \frac{\partial h_0(\mathbf{x}'')}{\partial x_j''} \|_2$, we find

$$\left\| \frac{\partial^2 R_{ff}(\mathbf{x}', \mathbf{x}'')}{\partial x_i' \partial x_j''} \frac{\partial h_0(\mathbf{x}')}{\partial x_i'} \frac{\partial h_0(\mathbf{x}'')}{\partial x_j''} \right\|_2 = \mathrm{O}\left(\frac{\sqrt{2}\sigma_f \lambda_s}{\lambda_i} J_i J_j \right), \tag{17}$$

where J_i represents the deterministic zero order gradient, or equivalently, the order $O(\epsilon^2)$ mean gradient. This indicates that unless the Green's function $G_1(\mathbf{x}, \mathbf{y})$ and subsequent integration eliminate the length scales λ_i and λ_s, they will play a role in the magnitude of $R_{hh}^{(2)}(\mathbf{x}, \mathbf{y})$, and in general on each h_n in the decomposition $h = h_0 + h_1 + h_2 + \cdots$. In Equation (17), J_i is the order of magnitude of the dimensionless mean head gradient in the ith direction. In the above discussion, $R_{hh}^{(2)}(\mathbf{x}, \mathbf{y})$ has been used as an example. Equations of higher order (as reported in Appendix B of Bonilla and Cushman, 2000) indicate length scales also affect the magnitude of h_n, for $n > 1$.

2.3. MATCHING ORDER OF MAGNITUDE ESTIMATES FOR VELOCITY MOMENTS

So far we have concentrated on obtaining a reasonable expansion for the dimensionless head h. Next we turn our attention to the velocity field which controls the dispersive process. After introducing the log conductivity fluctuation gradient in the perturbation solutions to flow, we explore the complexity that this introduces into the velocity expansion, i.e. according to Darcy's law the ith component of the velocity vector, V_i, is the product of random functions $\mathbf{K}$, expanded in parameter σ_f, and the random gradient function, $\partial h/\partial x_i$, expanded in parameter $\sigma_{\nabla f_i}$. This results in an expansion of 'mixed' orders in the two parameters. This issue is addressed in a preliminary manner in this section, and, later on, in Section 3.4, a more satisfactory treatment is provided.

As we have shown, this is not a trivial issue when $\sigma_{\nabla f}$ is included in the perturbation scheme. Suppose we had expanded the head as $h = h_0 + h_1 + h_2 + \cdots$, with $h_n = O(\sigma_{\nabla f}^n)$, then by using Darcy's law

$$\begin{aligned} V_i &= -\frac{K}{n}\frac{\partial h}{\partial x_i} \\ &= -\frac{K_G}{n}\left(1 + f + \frac{1}{2}f^2 + \frac{1}{6}f^3 + \cdots\right) \times \\ &\quad \times \left(\frac{\partial h_0}{\partial x_i} + \frac{\partial h_1}{\partial x_i} + \frac{\partial h_2}{\partial x_i} + \frac{\partial h_3}{\partial x_i} \cdots\right) \\ &= V_i^{[0]} + V_i^{[1]} + V_i^{[2]} + \cdots, \end{aligned} \tag{18}$$

where the velocity in the ith direction, V_i, has been expanded as $V_i = V_i^{[0]} + V_i^{[1]} + V_i^{[2]} + \cdots$. The term - $\frac{K_G}{n}\frac{\partial h_0}{\partial x_i}$ is O(1), $\frac{K_G}{n}\frac{\partial h_1}{\partial x_i}$ is $O(\sigma_{\nabla f})$, $\frac{K_G}{n} f \frac{\partial h_0}{\partial x_i}$ is $O(\sigma_f)$, and the next three terms are of orders $O(\sigma_{\nabla f}^2)$, $O(\sigma_{\nabla f}\sigma_f)$, and $O(\sigma_f^2)$. We are therefore, left with a somewhat impractical scheme for the velocity and its covariance (though we still can compute it). It is not possible to tell the order of magnitude of the resulting velocity fluctuations or the velocity covariance as a function of $\sigma_{\nabla f}$ alone. The order of the resulting velocity depends on both σ_f^2 and $\sigma_{\nabla f}^2$. So it is neither correct to say $V_i^{[n]} = O(\sigma_f^n)$ or $V_i^{[n]} = O(\sigma_{\nabla f}^n)$. The separation of orders in the expansion for V_i would be consistent in the sense given to it in most of the literature only when $\sigma_f \approx \sigma_{\nabla f}$. If we could choose a scaling factor λ_s such that (using Equation 12) $\sigma_{\nabla f} \approx \sigma_f$ we would have a consistent scheme for V_i and R_{ij}, the velocity covariance function. Choosing λ_s to be the largest or smallest integral scale involved in the problem does not help. The relation between the velocity terms and the head gradient and conductivity terms will be revisited in Section 3.4.

3. One Dimensional Flow

To understand the role of length scales on the small parameter ϵ, we start by considering the simpler case of 1D flow. We follow the outline laid out by Bonilla and Cushman (2000).

Rewrite Equation (7) for 1D flow in the dimensionless coordinate system

$$\frac{\mathrm{d}^2 h}{\mathrm{d}x^2} = -\frac{\mathrm{d}f}{\mathrm{d}x}\frac{\mathrm{d}h}{\mathrm{d}x}, \qquad x \in (0, L), \tag{19a}$$

$$h = h^D, \qquad x = 0 \ (or\ x = L), \tag{19b}$$

$$-\frac{K}{\lambda_s}\frac{\mathrm{d}h}{\mathrm{d}x} = q, \qquad x = L \ (or\ x = 0), \tag{19c}$$

where $L = L^*/\lambda_s$ is the dimensionless domain size, h^D and q are random functions defining Dirichlet and Neuman boundary conditions, respectively. Again, let $h = h_0 + h_1 + h_2 + \cdots$ with $h_n = O(\sigma_{\nabla f}^n = (\sqrt{2}\sigma_f \lambda_s/\lambda_x)^n)$. Substituting this decomposition in Equations (19a) and separating terms of the same order, the following hierarchy of equations is obtained

$$n = 0, \qquad \frac{\mathrm{d}^2 h_0}{\mathrm{d}x^2} = 0, \quad x \in (0, L), \tag{20a}$$

$$h_0 = h_0^D, \quad x = 0 \ (or\ x = L), \tag{20b}$$

$$\frac{\mathrm{d}h_0}{\mathrm{d}x} = -\lambda_s \frac{q_0}{K_G}, \quad x = L \ (or\ x = 0), \tag{20c}$$

and

$$n \geqslant 1 \quad \frac{\mathrm{d}^2 h_n}{\mathrm{d}x^2} = -\frac{\mathrm{d}f}{\mathrm{d}x}\frac{\mathrm{d}h_{n-1}}{\mathrm{d}x} \quad x \in (0, L), \tag{21a}$$

$$h_n = h_n^D, \quad x = 0 \ (or\ x = L) \tag{21b}$$

$$-\frac{K_G}{\lambda_s}\frac{\mathrm{d}h_n}{\mathrm{d}x} = s_n = q_n + \frac{K_G}{\lambda_s}\left(\sum_{j=1}^{n}\frac{1}{j!}f^j\frac{\mathrm{d}h_{n-j}}{\mathrm{d}x}\right), \quad x = L \ (or\ x = 0), \tag{21c}$$

where the random boundary conditions h^D and q must have a representation in powers of $\sigma_{\nabla f}$ (Bonilla and Cushman, 2000), $h^D = h_0^D + h_1^D + h_2^D + \cdots$, and $q = q_0 + q_1 + q_2 + \cdots$. The solution for the zero order problem ($n = 0$) using Green's functions is

$$h_0(v) = \left(h_0^D(x)\frac{\mathrm{d}G_0(v, x)}{\mathrm{d}x}\right)\Bigg|_{x=0}^{x=L} + \left(\frac{\lambda_s}{K_G}q_0(x)G_0(v, x)\right)\Bigg|_{x=0}^{x=L}, \tag{22}$$

where $G_0(v, x)$ is the 1D Green's function for the problem

$$\begin{aligned}
&\frac{\mathrm{d}^2 G_0(v, x)}{\mathrm{d}x^2} = \delta(v - x) \quad v,\ x \in (0, L),\\
&G_0(v, x) = 0 \ \text{ for } \ x = 0 \ (or\ x = L),\\
&\frac{\mathrm{d}G_0(v, x)}{\mathrm{d}x} = 0 \ \text{ for } \ x = L \ (or\ x = 0),
\end{aligned} \tag{23}$$

with $\delta(v - x)$ the Dirac delta function. The solutions to the higher order problems ($n \geqslant 1$) are given by

$$\begin{aligned}
h_n(v) &= -\int_0^L G_n(v, x)\frac{\mathrm{d}f(x)}{\mathrm{d}x}\frac{\mathrm{d}h_{n-1}(x)}{\mathrm{d}x}\,\mathrm{d}x + \left(h_n^D(x)\frac{\mathrm{d}G_n(v, x)}{\mathrm{d}x}\right)\Bigg|_{x=0}^{x=L} + \\
&\quad + \left(\frac{\lambda_s}{K_G}s_n(x)G_n(v, x)\right)\Bigg|_{x=0}^{x=L},
\end{aligned} \tag{24}$$

where the Green's function $G_n(v, x)$ is stochastic in the most general case, but can be approximated to first order in $\sigma_{\nabla f}$ by a deterministic Green's function defined as in Equation (23).

3.1. DIRICHLET BOUNDARY CONDITIONS

We next solve the problem using several different nondimensionalizations. For simplicity we use Dirichlet boundary conditions

$$n = 0, \qquad \frac{d^2 h_0}{dx^2} = 0, \quad x \in (0, L), \tag{25a}$$

$$h_0(x = 0) = J \quad \text{and} \quad h_0(x = L) = 0, \tag{25b}$$

with dimensionless head $h_0 = J = H_0/L^*$ at boundary $x = 0$, where H_0 is the hydraulic head prescribed at that boundary. The zero order Green's function is defined by

$$\frac{d^2 G_0(v, x)}{dx^2} = \delta(v - x) \quad v, \ x \in (0, L),$$

$$G_0(v, x = 0) = 0 \quad \text{and} \quad G_0(v, x = L) = 0, \tag{26}$$

with solution

$$G_0(v, x) = \frac{2}{L} \sum_{n=1}^{\infty} \frac{\sin(\alpha_n v) \sin(\alpha_n x)}{\alpha_n^2}, \tag{27}$$

with $\alpha_n = n\pi/L$, or alternatively

$$\begin{aligned} G_0(v, x) &= v\left(1 - \frac{x}{L}\right) \quad if \ \ v < x, \\ G_0(v, x) &= x\left(1 - \frac{v}{L}\right) \quad if \ \ v > x. \end{aligned} \tag{28}$$

For higher order problems ($n \geqslant 1$), homogenous Dirichlet boundary conditions will be used, $h_n^D = 0$ in Equation (21b),

$$n \geqslant 1, \qquad \frac{d^2 h_n}{dx^2} = -\frac{df}{dx}\frac{dh_{n-1}}{dx}, \qquad x \in (0, L) \tag{29a}$$

$$h_n(x = 0) = 0; \quad \text{and} \quad h_n(x = L) = 0 \tag{29b}$$

Bonilla and Cushman (2000) have studied this decomposition and its physical implications. Conditions in Equation (29b) constitute a deterministic limit case for which there is no head fluctuation at the boundary. With these boundary conditions, the Green's functions for the higher order problems are identical to those of the zero order problem $G_n(v, x) = G_0(v, x)$. In each of the following subsections, a different length scaling is studied. The variables v, x, y, and z denote points within the domain $\Omega = (0, L)$, and dimensionless L is defined according to the length scale used, λ_s, in $L = L^*/\lambda_s$.

3.2. UNSCALED ONE DIMENSIONAL FLOW

We solve Equations (20)–(29) with $\lambda_s = 1$ (dimensionless) and $L = L^*$. To be consistent, the hydraulic head is not scaled, $h = h^*$, and in Equation (25) $h_0^D(x = 0) = H_0$. Under these conditions, using Equations (25), (22) and (28), the solution to the zero order problem is;

$$h_0(v) = H_0\left(1 - \frac{v}{L^*}\right) \quad 0 < v < L^*. \tag{30}$$

Next we quantify the order of h_1 by examining $\sigma_h(v) = \sqrt{\langle h_1(v)h_1(v)\rangle}$ (the algebra was worked with Maple and will be provided upon request to the authors)

$$\begin{aligned}\|h_1\|_2 &= \sigma_f^2 H_0^2\Big[(3/2 + 1/2\, e^{-L^*/\lambda_x} - 2\, e^{-1/2\, L^*/\lambda_x}) \times \\ &\quad \times \frac{{\lambda_x}^2}{L^{*2}} - 1/2\,\frac{\lambda_x}{L^*} + 1/24\,\frac{L^*}{\lambda_x},\Big]\end{aligned} \tag{31}$$

where $\lambda_1 = \lambda_x$ is the integral scale and the norm is evaluated at maximum $v = L^*/2$ according to Equation (4). An exponential two point correlation function, as developed in Appendix A, has been used because it allows all integrals to be evaluated analytically. The Gaussian model will be studied in the 2D case. The leading term within the parenthesis in Equation (31) is $\frac{1}{24}\frac{L^*}{\lambda_x}$ (see Figure 2) and therefore

$$\sigma_h^{[2]} = \|h_1\|_2 \approx \frac{H_0}{2\sqrt{6}}\sigma_f\sqrt{\frac{L^*}{\lambda_x}}, \tag{32}$$

which suggests $\varepsilon = \sigma_f\sqrt{L^*/\lambda_x}$ as a small parameter for the perturbation scheme. The superscript $^{[n]}$ denotes $O(\varepsilon^n)$. Instead of continuing this development here, the number of parameters is reduced and the analysis simplified by introducing $\rho_x^2 = \lambda_x/L^*$ in the following section.

3.3. SCALED ONE DIMENSIONAL FLOW

Next consider the 1D flow problem defined in Equations (25) and (29) in dimensionless coordinate system $x = x^*/\sqrt{\lambda_x L^*}$. Define the parameter $\rho_x = \sqrt{\lambda_x/L^*}$. Equations (19)–(29) are valid with $\lambda_s = \sqrt{\lambda_x L^*}$ (λ_s has length units) and $L = \sqrt{L^*/\lambda_x} = 1/\rho_x$. The hydraulic head h^* is scaled, with dimensionless head $h = h^*/L^*$, so in Equation (25) $h_0^D(x = 0) = H_0/L^* = J$. In this system, Equations (25), (22) and (28) render the following solution for h_0

$$h_0(v) = J(1 - \rho_x v) \quad 0 < v < \sqrt{\frac{L^*}{\lambda_x}} = \frac{1}{\rho_x} = L, \tag{33}$$

with zero order gradient $\mathrm{d}h_0/\mathrm{d}v = -J\rho_x$. The Green's function in Equation (28) becomes

$$\begin{aligned} G_0(v,x) &= v(1-\rho_x x) \quad \text{if} \quad v < x, \\ G_0(v,x) &= x(1-\rho_x v) \quad \text{if} \quad v > x, \end{aligned} \tag{34}$$

and the higher order Green's functions, $G_n(v,x)$ will be identical to $G_0(v,x)$. As in the previous unscaled case, a two point exponential correlation is used to render the moments analytical. After this scaling it can be written as

$$R_{ff}(x,z) = \sigma_f^2 \exp\left(-\frac{|x-z|}{\rho_x}\right) \tag{35}$$

with first and second spatial derivatives given by

$$\frac{\mathrm{d}R_{ff}(x,z)}{\mathrm{d}x} = -\frac{\sigma_f^2}{\rho_x}\,\mathrm{sign}\,(x-z)\exp\left(-\frac{|x-z|}{\rho_x}\right) \tag{36a}$$

$$\frac{\mathrm{d}^2 R_{ff}(x,z)}{\mathrm{d}x\,\mathrm{d}z} = -\frac{\sigma_f^2}{\rho_x^2}\exp\left(-\frac{|x-z|}{\rho_x}\right) \tag{36b}$$

and we have defined all the functions required to compute the mean head norms $\|h_1\|_1 = 0$, $\|h_2\|_1$, $\|h_3\|_1 = 0$, $\|h_4\|_1, \ldots,$ and L^2 head norms $\|h_1\|_2$ necessary to compute mean head and covariance. Using the expression for the head variance;

$$\begin{aligned} \sigma_h^2(v) = \; & \sigma_f^2 J^2[(-2\,e^{-\rho_x{}^{-2}}+2)v^2\rho_x{}^6 + \\ & +(-2\,e^{-1+v\rho_x/\rho_x{}^2} - 2 + 2\,e^{-v/\rho_x} + 2\,e^{-\rho_x{}^{-2}})v\rho_x{}^5 + \\ & +(2\,v^2 - 2\,e^{-v/\rho_x}+2)\rho_x{}^4 - 2\,v\rho_x{}^3 + \\ & +2/3\,v^4\rho_x{}^2 - 4/3\,v^3\rho_x + 2/3\,v^2] \end{aligned} \tag{37}$$

with maximum $\|h_1\|_2$ (norm) at $v = 1/2\rho_x = L/2$ (see Figure 1)

$$\begin{aligned} \|h_1\|_2 = \; & (\sigma_f^2 J^2[(3/2 + 1/2\,e^{-\rho_x{}^{-2}} - 2\,e^{-1/2\,\rho_x{}^{-2}})\times \\ & \times \rho_x{}^4 - 1/2\,\rho_x{}^2 + 1/24\,\rho_x{}^{-2}])^{1/2} \end{aligned} \tag{38}$$

with leading term $\sigma_f^2 J^2 1/24\rho_x^2$ (see Figure 2) so the magnitude of σ_h is

$$\|h_1\|_2 \approx \frac{J}{2\sqrt{6}}\frac{\sigma_f}{\rho_x}. \tag{39}$$

Likewise we evaluate $\langle h_2(v)\rangle$ (Figure 3) as

$$\begin{aligned} \langle h_2(v)\rangle = \; & \sigma_f^2 J\bigg[(-2\,e^{-\rho_x{}^{-2}}+2)v\rho_x{}^5 + \\ & +(e^{-\rho_x{}^{-2}} - 1 - e^{-1+v\rho_x/\rho_x{}^2} + e^{-v/\rho_x})\rho_x{}^4 + \\ & +1/3\,v^3\rho_x - 1/2\,v^2 + 1/6\,\frac{v}{\rho_x}\bigg]. \end{aligned} \tag{40}$$

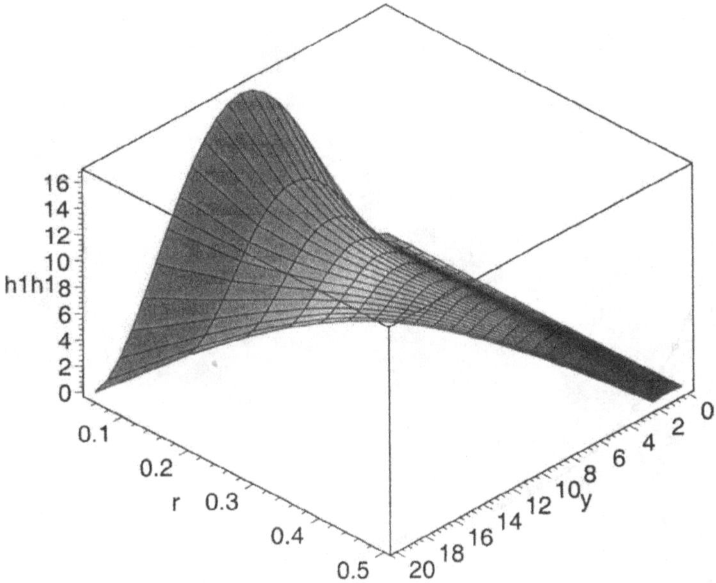

Figure 1. Head variance, $\sigma_h^2(r = \rho_x, y)/\sigma_f J$, (normalized by $\sigma_f J$) as a function of dimensionless parameter $r = \rho_x$ and coordinate $y \in (0, \frac{1}{\rho_x})$ for 1D flow. The maximum, $\|h_1\|_2^2$, occurs at $y = 1/2\rho_x$.

And the norm of the above expression can be found by evaluating it at the maximum, $v = 1/4\rho_x = L/4$, to obtain (Figure 3),

$$\|h_2\|_1 = \sigma_f^2 J\Big[(1/2\, e^{-\rho_x{}^{-2}} - 1/2 - e^{-3/4\,\rho_x{}^{-2}} + \\ + e^{-1/4\,\rho_x{}^{-2}})\rho_x{}^4 + \frac{1}{64}\,\rho_x{}^{-2}\Big], \tag{41}$$

with leading term $\sigma_f^2 J \frac{1}{64\rho_x^2}$ (Figure 4) and thus

$$\|h_2\|_1 \approx \frac{J}{64}\frac{\sigma_f^2}{\rho_x^2}. \tag{42}$$

Going through analogous steps (performed using Maple; procedure available upon request to the authors), it is possible to quantify the magnitude of the L^1 norm for h_4 (Figures 5 and 6) and the leading term for the region $\rho_x < \rho_c \approx 0.16$,

$$\|h_4\|_1 \approx J\left(\frac{1}{4.0}\right)^4 \frac{\sigma_f^4}{\rho_x^4}. \tag{43}$$

For the region $\rho_x > \rho_c \approx 0.16$, $\|h_4\|$ is dominated by a leading term proportional to $1/\rho_x^2$ and further analysis is necessary to determine which may be the most

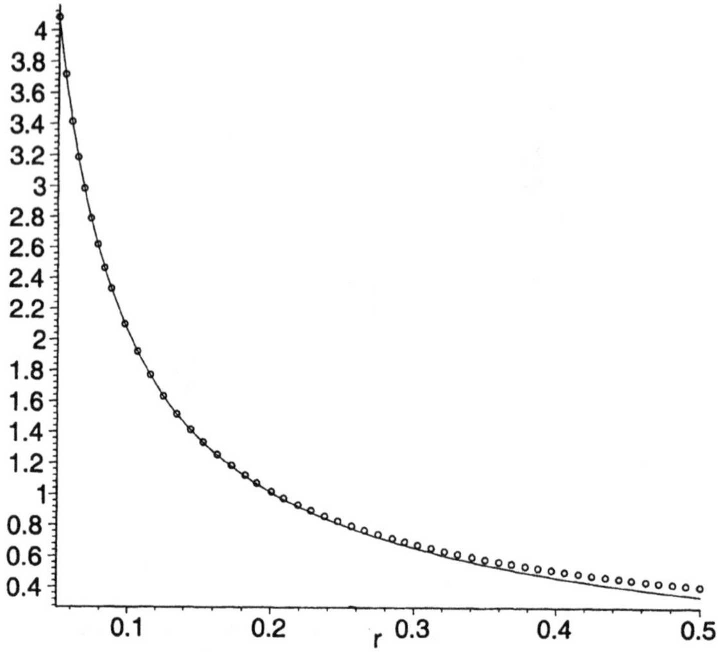

Figure 2. Head standard deviation (first order) norm as a function of dimensionless parameter $r = \rho_x$, $\frac{\|h_1\|_2}{\sigma_f J}$ (normalized by $\sigma_f J$) for 1D flow. Exact solution (solid line) and leading term, $1/2\sqrt{6}r$ (circles) shown.

judicious choice for the perturbation parameter. In our 1D analysis we concentrate in the region $\rho_x < \rho_c$ and reserve the analysis for the region $\rho_x > \rho_c$ for the more interesting case of 2D flow, where the result for 1D can be thought of as a limit case.

3.3.1. *Convergence*

Now since the norms $\|h\|_1 \leqslant \|h_0\|_1 + \|h_1\|_1 + \|h_2\|_1 + \|h_3\|_1 + \cdots$ and $\|h\|_2 \leqslant \|h_0\|_2 + \|h_1\|_2 + \cdots$ can be used as convergence criteria for the mean head and variance expansions, respectively, we are able to develop bounds for the parameter $\varepsilon = \sigma_f/\rho_x$. Comparing the $\|h\|_1$ series with the convergent geometric series $h^M = 1 + w + w^2 + w^3 + \cdots$ with $|w| < 1$, and restricting ε so that

$$\begin{aligned}\|h\|_1 &\leqslant \|h_0\|_1 + \|h_1\|_1 + \|h_2\|_1 + \|h_3\|_1 + \|h_4\|_1 + \\ &+ \cdots < J(1 + w + w^2 + w^3 + w^4 + \cdots)\end{aligned} \tag{44}$$

with

$$J\left(1 + \left(\frac{1}{8}\right)^2 \frac{\sigma_f^2}{\rho_x^2} + \left(\frac{1}{4}\right)^4 \frac{\sigma_f^4}{\rho_x^4} + \cdots\right) < J(1 + w^2 + w^4 + \cdots), \tag{45}$$

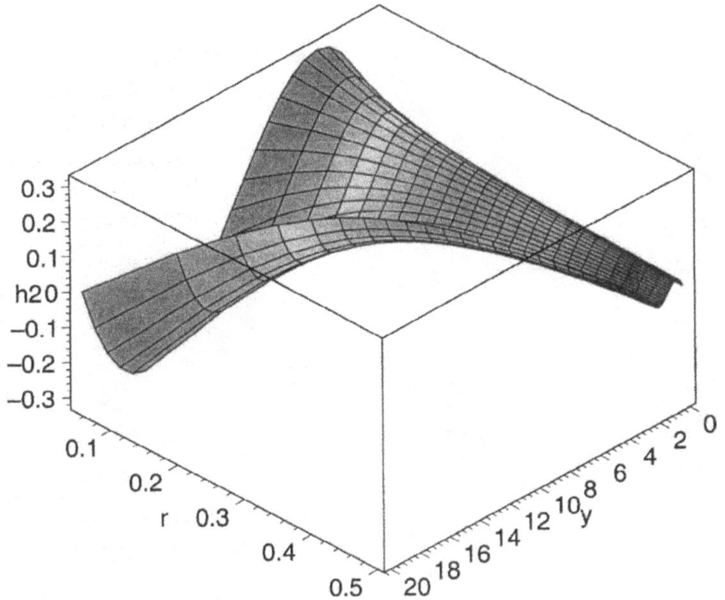

Figure 3. Second order head fluctuation average, $\langle h_2(y,r)\rangle/\sigma_f^2 J$ (normalized) as a function of dimensionless parameter $r = \rho_x$ and coordinate $y \in (0, \frac{1}{\rho_x})$ for 1D flow. The maximum $\|h_2\|_1/\sigma_f^2 J$ occurs at $y = 1/4\rho_x$.

where $\|h_1\|_1 = \|h_3\|_1 = 0$. The coefficients multiplying the powers of ε govern convergence for the perturbation scheme. We conjecture convergence for the mean head is governed by $\varepsilon = \sigma_f/\rho_x < 2\sqrt{2}$ (derived by bounding it with the more restrictive term, $O(\varepsilon)$). Next, comparing the $\|h\|_2$ series with the convergent geometric series $h^M = 1 + w + w^2 + w^3 + \cdots$ with $|w| < 1$, and restricting ε so that

$$\|h\|_2 \leqslant \|h_0\|_2 + \|h_1\|_2 + \cdots < J(1 + w + \cdots) \tag{46}$$

with

$$J\left(1 + \frac{1}{2\sqrt{6}}\frac{\sigma_f}{\rho_x} + \cdots\right) < J(1 + w + \cdots). \tag{47}$$

Here convergence is conjectured for the head variance by $\varepsilon = \sigma_f/\rho_x < 2\sqrt{6}$. A shortcoming of adopting the norm (4) should be pointed out. We expected the resultant convergence condition to be more restrictive for the head variance than for the mean head. However, the result is contrary to this. It should be noted that the mean head norm, Equation (45), does not occur as a mean head anywhere in the domain. The norm (maximum value) of each one of the expansion terms h_n occurs at different points in the domain, that serves our purpose of finding a dominating series to all points in the domain, but leaves us defeated, as the bound

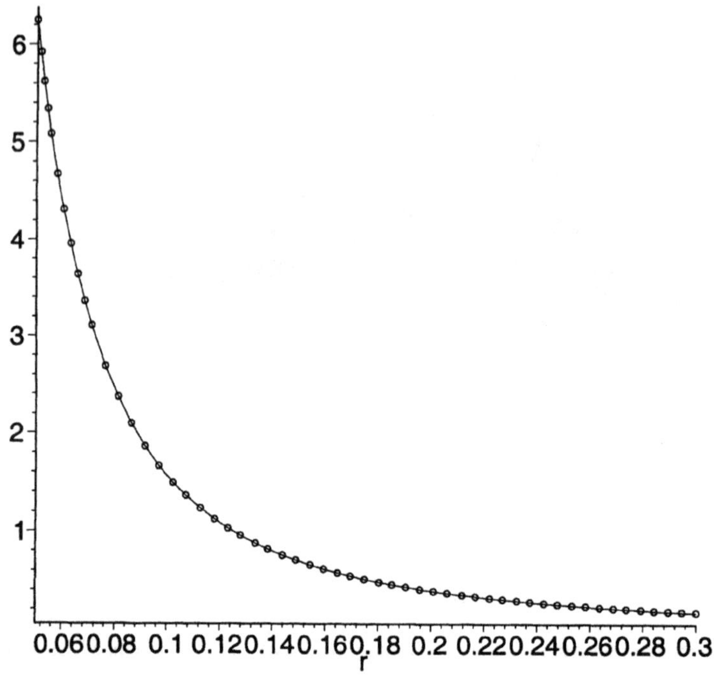

Figure 4. Second order head fluctuation norm a function of dimensionless parameter $r = \rho_x$, $\|h_1\|_2^2/\sigma_f^2 J$ (normalized) for 1D flow. Exact solution (solid line) and leading term, $1/64r^2$ (circles) shown.

obtained is most probably not the best bound. The bound still provides insight into the parameter ε dependence on integral scale λ_x and size of the domain L^*. It can be argued on this basis that the variance convergence condition $\varepsilon = \sigma_f/\rho_x < 2\sqrt{6}$ may be an appropriate choice of the convergence bound. It suggests that the analysis for the mean velocity and mean variance may produce different bounds. Full proof of the hypothesized convergence condition, $\varepsilon = \sigma_f/\rho_x < 2\sqrt{6}$ requires computations (or determination of bounds) of the leading term in the coefficient of h_n for arbitrary n. While such proof is not available, one should be satisfied that the expansions in (45), and (45) are consistent to orders $O(\sigma_{\nabla f}^4)$ and $O(\sigma_{\nabla f}^2)$, respectively (Hinch, 1990). The bound is very restrictive for 1D flow, but it will be shown in Section 4 that the convergence condition is more relaxed for 2D flow with exponential log conductivity, and even more relaxed for the Gaussian case. From now on we will look exclusively at the variance convergence condition, that is, L^2 convergence.

3.4. VELOCITY EXPANSION

The relatively smaller range of the parameter ρ_x after scaling makes the velocity-expansion order-matching possible. Consider Equation (18) in one dimension. After

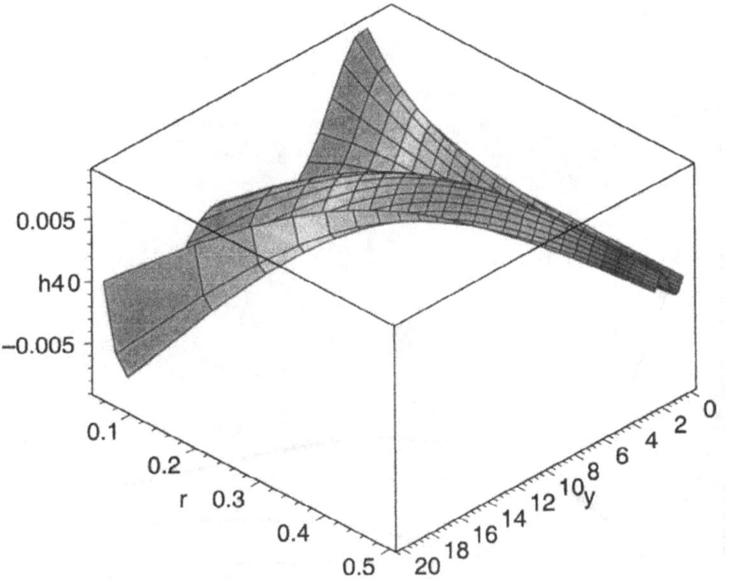

Figure 5. Fourth order head fluctuation average, $\langle h_4(y)\rangle/\sigma_f^4 J$, (normalized by $\sigma_f^4 J$) as a function of dimensionless parameter $r = \rho_x$ and coordinate $y \in (0, \frac{1}{\rho_x})$, for 1D flow. The maximum of the function occurs at $y = 1/4\rho_x$ for $\rho_x < \rho_c \approx 0.16$ and at $4/10\rho_x$ for $\rho_x > \rho_c \approx 0.16$.

scaling the coordinates and head as in Section 3.3, we find

$$V(x) = -\frac{K_G}{n}\left(\frac{1}{\rho_x}\frac{dh_0}{dx} + \frac{1}{\rho_x}f\frac{dh_0}{dx} + \frac{1}{\rho_x}\frac{dh_1}{dx} + \frac{1}{\rho_x}\frac{dh_2}{dx} + \right.$$
$$\left. + \frac{1}{\rho_x}f\frac{dh_1}{dx} + \frac{1}{\rho_x}\frac{1}{2}f^2\frac{dh_0}{dx} + \cdots\right), \tag{48}$$

where we have not factored out $1/\rho_x$ intentionally so as to include it in the analysis of each term's magnitude. We proceed to estimate the L^1 norm (mean velocity) of each term involved in Equation (48) using the corresponding integral equations for their solution. For $\rho_x < \rho_c = 0.16$

$$\left\|\frac{1}{\rho_x}\frac{dh_0}{dx}\right\| = J, \tag{49a}$$

$$\left\|\frac{1}{\rho_x}\frac{dh_2}{dx}\right\| \approx \frac{1}{6}J\frac{\sigma_f^2}{\rho_x^2}, \tag{49b}$$

$$\left\|\frac{1}{\rho_x}f\frac{dh_1}{dx}\right\| \approx \frac{9}{10}J\sigma_f^2\rho_x, \tag{49c}$$

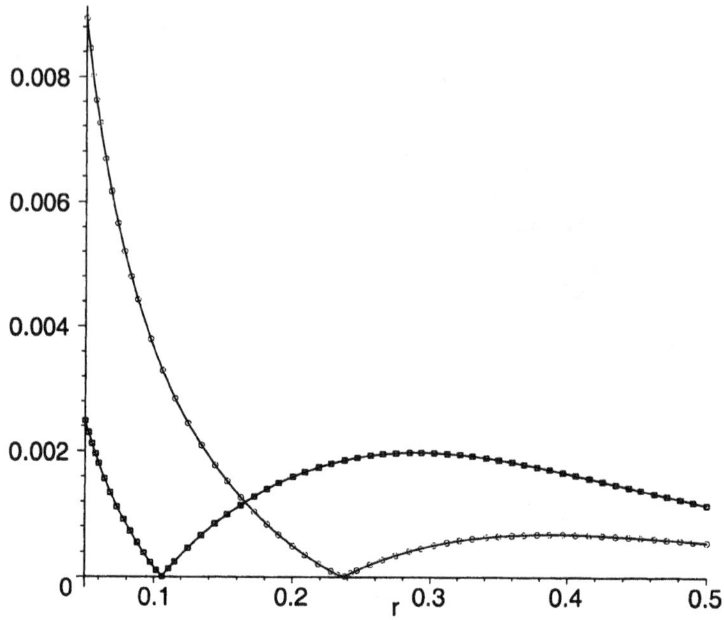

Figure 6. Norm of fourth order head fluctuation average, $\|\langle h_4(y)\rangle\|/\sigma_f^4 J$, (normalized by $\sigma_f^4 J$) as a function of dimensionless parameter $r = \rho_x$ for 1D flow. The maximum, $\|h_4\|_1$, occurs at $y = 1/4\rho_x$ for $\rho_x < \rho_c \approx 0.16$ (circles) and at $4/10\rho_x$ for $\rho_x > \rho_c \approx 0.16$ (squares).

$$\left\| \frac{1}{\rho_x}\frac{1}{2}f^2\frac{\mathrm{d}h_0}{\mathrm{d}x} \right\| = \frac{1}{2}J\sigma_f^2, \tag{49d}$$

$$\left\| \frac{1}{\rho_x}\frac{\mathrm{d}h_4}{\mathrm{d}x} \right\| \approx \frac{1}{60}J\frac{\sigma_f^4}{\rho_x^4}, \tag{49e}$$

$$\left\| \frac{1}{\rho_x}\frac{1}{2}f^2\frac{\mathrm{d}h_2}{\mathrm{d}x} \right\| \approx \frac{1}{12}J\frac{\sigma_f^4}{\rho_x^2}, \tag{49f}$$

$$\left\| \frac{1}{\rho_x}\frac{1}{6}f^3\frac{\mathrm{d}h_1}{\mathrm{d}x} \right\| \approx \frac{9}{60}J\sigma_f^4\rho_x, \tag{49g}$$

$$\left\| \frac{1}{\rho_x}\frac{1}{24}f^4\frac{\mathrm{d}h_0}{\mathrm{d}x} \right\| \approx \frac{1}{24}J\sigma_f^4, \tag{49h}$$

where all the norms considered are L^1, that is $\|\ \| = \|\ \|_1$ in the above equation. From these results we conclude that the velocity is expressed as a sum of two series in the parameters σ_f and σ_f/ρ_x. The first parameter appears in terms of the form $f^a(\mathrm{d}h_b/\mathrm{d}x)$ with $a > 0$, $b \geqslant 0$ (e.g. $\|f^2(\mathrm{d}h_2/\mathrm{d}x)\|$). As seen in Equations (49c, and g) the parameter with power $a + b$ is σ_f while ρ_x may be multiplying or dividing the term, but with a constant power that does not increase as $a + b$. The second series on the small parameter σ_f/ρ_x is formed by terms of the form $\|dh_c/dx\|$ with $c \geqslant 0$ as seen in Equations (49a, b, and e). The parameter with power c for these terms is $\varepsilon = \sigma_f/\rho_x$. Two conditions are required therefore to render a convergent scheme for the velocity; $\sigma_f < \nu_1$, and $\varepsilon < \nu_2$. If the same convergence test, i.e. bounding via a geometric series, is used for the two series, we can hypothesize

$$\nu_1 = \left(\frac{1}{\frac{1}{24} + \frac{9\rho_x}{60} + \frac{1}{12\rho_x^2}} \right)^{1/4} \quad \text{and} \quad \nu_2 = 6.0.$$

These conditions establish that both the head gradient expansion, and the expansion of the correlation between head gradient and conductivity should be convergent to render a scheme that is convergent for the velocity. Note that the second condition is novel and it is in addition to the condition for a convergent expansion of the head. This novel result is very important as it alerts researchers of two important features of perturbation approaches to flow: (i) the range of convergence will be affected by both the magnitude of log-conductivity fluctuations (σ_f) and the length scales involved in the problem (integral scales λ_i and domain size L_i^*); and (ii) the bounds determined herein provide an estimate of the magnitude of higher order corrections; the closer the parameters are to the convergence condition, the greater the magnitude of higher order corrections. The insufficient restriction imposed on σ_f by itself in the literature has left previous results uninformed of both features of perturbative solutions.

3.5. DIRICHLET–NEUMANN BOUNDARY CONDITIONS

Using the same scaling as in Section 3.3, we now consider a different case with Dirichlet boundary condition at $x = 0$ and Neumann boundary condition at $x = L$. Again, let $\lambda_s = \sqrt{\lambda_x L^*}$ so that

$$\frac{d^2 h}{dx^2} = -\frac{df}{dx}\frac{dh}{dx} \quad x \in (0, L) \ , \tag{50a}$$

$$h = h^D, \quad x = 0 \tag{50b}$$

$$-\frac{K}{\rho_x}\frac{dh}{dx} = q, \quad x = L, \tag{50c}$$

where $L = \sqrt{L^*/\lambda_s}$ is the dimensionless domain size, and $h = h^*/L^*$ is dimensionless head. Further, define the boundary functions h^D and q within the resulting hierarchy of equations;

$$n = 0 \quad \frac{d^2 h_0}{dx^2} = 0, \quad x \in (0, L) \tag{51a}$$

$$h_0(x=0)=J \quad \text{and} \quad -\frac{K_G}{\rho_x}\frac{\mathrm{d}h_0(x=L)}{\mathrm{d}x}=JK_G \tag{51b}$$

and

$$n\geqslant 1 \quad \frac{\mathrm{d}^2h_n}{\mathrm{d}x^2}=-\frac{\mathrm{d}f}{\mathrm{d}x}\frac{\mathrm{d}h_{n-1}}{\mathrm{d}x} \quad x\in(0,L) \tag{52a}$$

$$h_n(x=0)=h_n^D=0, \tag{52b}$$

$$-\frac{K_G}{\rho_x}\frac{\mathrm{d}h_n(x=L)}{\mathrm{d}x}=q_n+\frac{K_G}{\rho_x}\left(\sum_{j=1}^{n}\frac{1}{j!}f^j\frac{\mathrm{d}h_{n-j}(x=L)}{\mathrm{d}x}\right)=0 \tag{52c}$$

by which we mean there is no head fluctuation at $x = 0$ and there is no head gradient fluctuation at $x = L$. Arguing as before, one obtains the corresponding bound for $\|h\|_1$ and $\|h\|_2$ and as in Section 3.3.1 for the Dirichlet case, we find for the head mean $\|h\|_1$

$$J\left(1+\frac{1}{2}\frac{\sigma_f^2}{\rho_x^2}+\frac{1}{6}\frac{\sigma_f^4}{\rho_x^4}+\cdots\right)<J(1+w^2+w^4+\cdots) \tag{53}$$

and for the head variance

$$J\left(1+\sqrt{\tfrac{2}{3}}\frac{\sigma_f}{\rho_x}+\cdots\right)<J(1+w+\cdots) \tag{54}$$

with bounds for ε; (i) mean head convergence $\varepsilon < 6^{1/4} = 1.57$, and (ii) head variance convergence $\varepsilon < \sqrt{\frac{3}{2}} = 1.22$. It is observed that the coefficients in the parameter ϵ are greater for this case than for the Dirichlet case of Section 3.3. It is therefore, hypothesized that boundary conditions influence the convergence of the perturbation expansion. In particular, Neumann boundary conditions induce slower convergence, in this case a convergence condition would be $\sigma_f < 1.22\rho_x$ (head variance).

4. Two Dimensional Flow

The 1D analysis of the previous section sets the stage for the 2D problem. Henceforth, we concentrate on bounding the head variance using L^2 convergence. Therefore, all the norms below are L^2, that is read $\| \;\|_2$ where $\| \;\|$ is found.

Rewrite Equation (7) in an unscaled coordinate system, this time denoted by $\mathbf{x}' = x_i'\ i = 1,2$

$$\frac{\partial f}{\partial x_i'}\frac{\partial h'}{\partial x_i'}+\frac{\partial^2 h'}{\partial x_i'\partial x_i'}=0 \quad \mathbf{x}'\in\Omega'. \tag{55}$$

The primed notation $\mathbf{x}' = (x_1', x_2')$ (the x_i' have length units) has been introduced for convenience, so that after scaling, a non-primed system, $\mathbf{x}$, is obtained. Introduce the dimensionless system $\mathbf{x} = (x, y)$ with $x = x_1'/\sqrt{\lambda_x L_x}$ and $y = x_2'/\sqrt{\lambda_y L_y}$ where λ_x and λ_y denote the integral scales in the two directions of the unscaled system $((x_1', x_2') = (x', y'))$, and L_x and L_y are the side lengths of a rectangular domain $\Omega' = [0..L_x, 0..L_y]$. As in the 1D case, the goal is to obtain a dimensionless head that is independent of the length scales of the problem. To make a judicious scaling, it is important to consider gradients in both directions, x, and y. However, for environmental applications it is common to study the case for which mean flow occurs in one direction (say x direction). General mean head gradients with non zero components in both directions and space dependence have been analyzed by Bonilla and Cushman (2000) for steady saturated flow and by Zhang (1999) for transient unsaturated flow. From their analysis the most convenient scaling for the hydraulic head is $h = h'/L_x$ where x is the dominant direction of flow. With this scaling, (55) becomes

$$\xi^2 \frac{\partial^2 h}{\partial x^2} + \frac{\partial^2 h}{\partial y^2} = -\xi^2 \frac{\partial f}{\partial x}\frac{\partial h}{\partial x} - \frac{\partial f}{\partial y}\frac{\partial h}{\partial y}, \tag{56}$$

where $\xi^2 = \lambda_y L_y/\lambda_x L_x$ is the square of a modified anisotropy ratio, ξ. As in the 1D problem, define $\rho_x^2 = \lambda_x/L_x$, and $\rho_y^2 = \lambda_y/L_y$. Then ξ can be rewritten as $\xi = \rho_y L_y/\rho_x L_x$. Equation (56) must be supplemented with boundary conditions (BC's), and they may be stochastic or deterministic. These may be Dirichlet, Neumann or mixed. We consider stochastic Dirichlet and Neumann:

$$h = h^D \quad \mathbf{x} \in \Gamma_{\mathrm{D}} \tag{57a}$$

$$-K\frac{\partial h}{\partial \eta} = q \quad \mathbf{x} \in \Gamma_{\mathrm{N}} \tag{57b}$$

where h^{D} is random dimensionless head on Dirichlet boundary Γ_{D}, $h^{\mathrm{D}} = h'^{\mathrm{D}}/L_x$, and q is random scaled recharge through Neumann boundary Γ_N with outward unit normal η, $\eta = x$ or $\eta = y$ for the rectangular domain being considered. $q = q'\rho_x$ if $\eta = x$, and $q = q'\rho_x\xi$ if $\eta = y$.

After this scaling, the two point conductivity fluctuation (multi-Gaussian) from Equation (10) is

$$R_{ff}(\mathbf{x} - \mathbf{u}) = \sigma_f^2 \exp\left(-\frac{(x-u)^2}{\rho_x^2} - \frac{(y-w)^2}{\rho_y^2}\right) \tag{58}$$

with points $\mathbf{x} = (x, y)$, and $\mathbf{u} = (u, w)$. The above Poisson equation (56) involves an 'anisotropic' Laplace operator in the left hand side and a non-zero fluctuation in the right hand side. From Equation (12) (multi-Gaussian model), it follows that two parameters may be used for the perturbation expansion, $\sigma_{\nabla f_x}$ and $\sigma_{\nabla f_y}$

$$\left\| \frac{\partial f}{\partial x} \right\| = \sigma_{\nabla f_x} = \frac{\sqrt{2}\sigma_f}{\rho_x}, \tag{59a}$$

$$\left\| \frac{\partial f}{\partial y} \right\| = \sigma_{\nabla f_y} = \frac{\sqrt{2}\sigma_f}{\rho_y}. \tag{59b}$$

Consider the case in which the mean flow is in the x direction, and consider $\sigma_{\nabla f} \equiv \sigma_{\nabla f_x}$. By substituting the expansion $h = h_0 + h_1 + h_2 + \cdots$, with $h_n = \mathrm{O}(\sigma^n_{\nabla f})$ into Equations (56) and (57a), and separating terms of order $n = 0, 1, 2, \ldots$ we obtain the hierarchy of equations

$$\mathrm{O}(1): \quad \xi^2 \frac{\partial^2 h_0}{\partial x^2} + \frac{\partial^2 h_0}{\partial y^2} = 0, \tag{60a}$$

$$n \geqslant 1\, \mathrm{O}(\sigma^n_{\nabla f}): \quad \xi^2 \frac{\partial^2 h_n}{\partial x^2} + \frac{\partial^2 h_n}{\partial y^2} = -\xi^2 \frac{\partial f}{\partial x}\frac{\partial h_{n-1}}{\partial x} - \frac{\partial f}{\partial y}\frac{\partial h_{n-1}}{\partial y} \tag{60b}$$

with associated hierarchical boundary conditions

$$n \geqslant 0 \quad h_n = h_n^D \quad \mathbf{x} \in \Gamma_D \tag{61a}$$

$$\mathrm{O}(1): \quad -K_G \frac{\partial h_0}{\partial \eta} = q_0, \quad \mathbf{x} \in \Gamma_N \tag{61b}$$

$$n \geqslant 1 \quad \mathrm{O}(\sigma^n_{\nabla f}): -K_G \frac{\partial h_n}{\partial \eta} = q_n + K_G \left(\sum_{j=1}^{n} \frac{1}{j!} f^j \frac{\partial h_{n-j}}{\partial \eta} \right) \quad \mathbf{x} \in \Gamma_N \tag{61c}$$

where the random boundary functions h^{D} and q have the expansions $h^{\mathrm{D}} = h_0^{\mathrm{D}} + h_1^{\mathrm{D}} + h_2^{\mathrm{D}} + \cdots$ and $q = q_0 + q_1 + q_2 + \cdots$ as a result of the expansion for h. The zeroth order, deterministic problem, described by equations (60a), (61a), and (61b) has solution

$$\begin{aligned} h_0(x, y) = & \int_{\mathbf{x}'' \in \Gamma_D} \left[h_0^D(x'', y'') \frac{\partial G(x, y | x'', y'')}{\partial x''} \right] \Bigg|_{\mathbf{x}'' \in \Gamma_D} \mathrm{d}\Gamma_D + \\ & + \frac{1}{K_G} \int_{\mathbf{x}'' \in \Gamma_N} [q_0(x'', y'') G(x, y | x'', y'')] \Bigg|_{\mathbf{x}'' \in \Gamma_N} \mathrm{d}\Gamma_N, \end{aligned} \tag{62}$$

where $\mathbf{x}''$ denotes an integration variable in the scaled system (not to confuse it with unscaled system previously denoted $\mathbf{x}'$), and the Greens function $G(\mathbf{x}, \mathbf{x}'')$ is obtained by solving the problem

$$\begin{aligned} & \Delta_\xi G(\mathbf{x}, \mathbf{x}'') = \xi^2 \frac{\partial^2 G}{\partial x^2} + \frac{\partial^2 G}{\partial y^2} = \delta(\mathbf{x} - \mathbf{x}'') \quad \mathbf{x} \in \Omega, \\ & G(\mathbf{x}, \mathbf{x}'') = 0 \ \text{ for } \mathbf{x} \in \Gamma_D, \\ & \frac{\partial G(\mathbf{x}, \mathbf{x}'')}{\partial \eta} = 0 \ \text{ for } \mathbf{x} \in \Gamma_N, \end{aligned} \tag{63}$$

where $\delta(\mathbf{x} - \mathbf{x}'')$ is the 2D delta function.

Problems of order $n \geqslant 1$ are prescribed by Equations (60b), (61a), and (61c). The right hand side of Poisson equation (60b) is a random forcing term. Using Greens functions the solution is

$$h_n = -\int\int_D G_n(x, y|x'', y'')\left(\xi^2 \frac{\partial f}{\partial x''}\frac{\partial h_{n-1}}{\partial x''} + \frac{\partial f}{\partial y''}\frac{\partial h_{n-1}}{\partial y''}\right) \mathrm{d}x''\,\mathrm{d}y'' + h_{n_{bc}}, \tag{64}$$

where $h_{n_{bc}}$ represents the boundary integrals incorporating the conditions given by Equations (61a) with $n \geqslant 1$, and (61c). These are completely analogous to the boundary integrals presented in Equation (62). When higher order ($n \geqslant 1$) problems have homogeneous boundary conditions then $h_{n_{bc}} = 0$. The Greens function $G_n(x, y|x'', y'')$ may be different for each n depending on the model for processes occurring at the boundaries.

4.1. 2D FLOW WITH DIRICHLET BOUNDARY CONDITIONS

Consider the case of flow in a rectangular scaled domain $\mathbf{x} \in \Omega = [0..\ell_x, 0..\ell_y]$, with all boundaires Dirichlet. Here $\ell_x = 1/\rho_x$ and $\ell_y = 1/\rho_y$. To simplify the analysis, consider deterministic boundary conditions, that is, $h_0^D \neq 0$, and $h_n^D = 0$ for $n \geqslant 1$. Equations (60a) and (61) become

$$\xi^2 \frac{\partial^2 h_0}{\partial x^2} + \frac{\partial^2 h_0}{\partial y^2} = 0 \quad \mathbf{x} \in \Omega \tag{65a}$$

$$h_0(x = 0, y) = J, \tag{65b}$$

$$h_0(x = \ell_x, y) = 0, \tag{65c}$$

$$h_0(x, y = 0) = J(1 - \rho_x x), \tag{65d}$$

$$h_0(x, y = \ell_y) = J(1 - \rho_x x), \tag{65e}$$

so that the flow is in the x direction, with constant gradient $\partial h_0/\partial x = -J\rho_x$. Equations (60b) with homogeneous boundary conditions $h_n^D = 0$ for $n \geqslant 1$ complete the statement of the problem. The Greens function for Dirichlet BC's may be represented as

$$\begin{aligned} G(x, y|x'', y'') &= \frac{2}{\pi}\sum_{n=1}^{\infty} \frac{\sinh(\gamma_n x)\sinh(\gamma_n[\ell_x - x''])\sin(\alpha_n y)\sin(\alpha_n y'')}{n\sinh(\gamma_n \ell_x)} \\ &\quad \text{if } 0 \leqslant y < y'' \\ G(x, y|x'', y'') &= \frac{2}{\pi}\sum_{n=1}^{\infty} \frac{\sinh(\gamma_n x'')\sinh(\gamma_n[\ell_x - x])\sin(\alpha_n y)\sin(\alpha_n y'')}{n\sinh(\gamma_n \ell_x)} \\ &\quad \text{if } y'' < y \leqslant \ell_x \end{aligned} \tag{66}$$

where $\alpha_n = n\pi/\ell_y$ and $\gamma_n = \alpha_n/\xi$. The above Greens function includes the non-local anisotropy, that without scaling is manifest only in the two point correlation functions for log conductivity fluctuations, R_{ff}. After the scaling, both the Greens function and R_{ff} are anisotropic with a different anisotropy ratio, as shown by Equation (58).

4.2. HEAD VARIANCE

Using Equations (16) and (64), a solution is obtained for the $O(\sigma^2_{\nabla f})$ head covariance. First use an exponential two point correlation function (also named separated exponential in the literature)

$$R_{ff}(\mathbf{x}-\mathbf{u}) = \sigma_f^2 \exp\left(-\frac{|x-u|}{\rho_x} - \frac{|y-w|}{\rho_y}\right). \tag{67}$$

The resulting expression is computed using numerical integration techniques including Gaussian quadrature, Romberg integration, and importance sampling. Importance sampling is used to decrease the number of sample points. For instance, Equations (67) and (58) suggest that if one integrates first over u (w), for a fixed x (y), one can restrict the sampling points to a region (interval in u) where $\exp(-\frac{|x-u|}{\rho_x})$ ($\exp(-\frac{|y-w|}{\rho_y})$) is not negligible (exponential case). Likewise, Greens function (66) has a singularity when $y \to y''$, and $x \to x''$. To address this singularity, when integrating over x'' (y'') for x (y) fixed , it is useful to divide the integration interval $[0..\ell_x]$ ($[0..\ell_y]$) in two subintervals, $[0..x]$ and $[x..\ell_x]$ ($[0..y]$ and $[y..\ell_y]$). The subintervals are best handled using Romberg integration excluding the right and left points respectively, where the singularity is found. Once it has been integrated over x'' and y'', a smoother function of x and y is obtained, and it is expensive and unnecessary to use Romberg integration anymore, so a Gaussian quadrature rule is used. A further refinement can be achieved by approximating the Greens function (66)

$$\begin{aligned} G(x, y|x'', y'') = \frac{1}{4\pi}\Bigg[&\ln\left(\frac{\cosh(\gamma(x-x'')) - \cos(\alpha[y+y''])}{\cosh(\gamma[x-x'']) - \cos(\alpha[y-y''])}\right) + \\ &+ \ln\left(\frac{\cosh(\gamma[x+x'']) - \cos(\alpha[y-y''])}{\cosh(\gamma[x+x'']) - \cos(\alpha[y+y''])}\right) + \\ &+ \ln\left(\frac{\cosh(\gamma[\ell_x-(x-x'')]) - \cos(\alpha[y-y''])}{\cosh(\gamma[\ell_x-(x-x'')]) - \cos(\alpha(y+y''))}\right) + \\ &+ \ln\left(\frac{\cosh(\gamma[\ell_x+(x+x'')]) - \cos(\alpha(y+y''))}{\cosh(\gamma[\ell_x+(x+x'')]) - \cos(\alpha(y-y''))}\right)\Bigg], \end{aligned} \tag{68}$$

where $\alpha = \pi/\ell_y$ and $\gamma = \alpha/\xi$. This expression is very accurate (error under 0.05%) for $L_x \geqslant L_y$. This however is not a restriction as an equivalent expression may be found to approximate the Greens function for the case $L_y \geqslant L_x$.

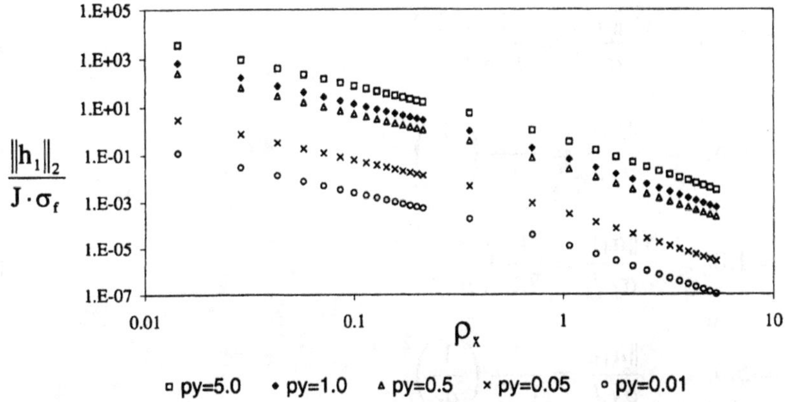

Figure 7. Dependence of $\|h_1\|_2$ on the parameter ρ_x for exponential log conductivity two point correlation function, two dimensional (2D) flow.

For this 2D case we have four independent parameters to analyze: the log-conductivity fluctuation standard deviation, σ_f, and three scale-related parameters, $\rho_x = \sqrt{\lambda_x/L_x}$, $\rho_y = \sqrt{\lambda_y/L_y}$, and $\xi = \rho_y L_y/\rho_x L_x$. The dependence of the perturbation expansion on σ_f is similar to the 1D case. For the length scale parameters, start the analysis for dependence on the single parameter ρ_x. Selecting $L_x = L_y$, a 'square domain' (in the unscaled system) is considered, so the effect of ρ_x and ρ_y can be studied. First, fixing ρ_y at different values, compute the head covariance norm as defined in Equation (4). The covariance (16) has a maximum at $\mathbf{x} = \mathbf{y}$, i.e. the variance. Under the boundary conditions given, the variance has a maximum at the domain's center, $\mathbf{x} = (x, y) = (\ell_x/2, \ell_y/2)$.

Figure 7 shows the head variance norm $\|h\|$ as a function of the parameter ρ_x for different values of the parameter ρ_y on a log-log scale. An exponential two point correlation function, Equation (67) was used. Two linear regions of different slopes are noted to the left and right of the point $\rho_x = \rho_c$ (the transition for the exponential case happens at a very well-defined region, unlike the Gaussian case analyzed below), with $0.35 < \rho_c < 0.7$ for different values of ρ_y. Let us first analyze the region $\rho_x < \rho_c$. The straight lines suggest a relationship of the type

$$\frac{\|h\|}{J\sigma_f} = C(\rho_y)\frac{1}{\rho_x^n} \tag{69}$$

holds as in the 1D analysis, but, C depends on the parameter ρ_y, $C = C(\rho_y)$, and possibly on the parameter ξ. Performing regressions to fit relationship (69) in the data plotted in Figure 7, renders the following relationships:

$$\rho_y = 0.01, \qquad \frac{\|h_1\|}{\sigma_f J} = \frac{1}{42000}\left(\frac{1}{\rho_x}\right)^2 \tag{70a}$$

$$\rho_y = 0.05, \qquad \frac{\|h_1\|}{\sigma_f J} = \frac{1}{1700}\left(\frac{1}{\rho_x}\right)^2 \tag{70b}$$

$$\rho_y = 0.15, \qquad \frac{\|h_1\|}{\sigma_f J} = \frac{1}{186}\left(\frac{1}{\rho_x}\right)^2 \tag{70c}$$

$$\rho_y = 0.5, \qquad \frac{\|h_1\|}{\sigma_f J} = \frac{1}{19.5}\left(\frac{1}{\rho_x}\right)^2 \tag{70d}$$

$$\rho_y = 1.0, \qquad \frac{\|h_1\|}{\sigma_f J} = \frac{1}{7.5}\left(\frac{1}{\rho_x}\right)^2 \tag{70e}$$

$$\rho_y = 5.0, \qquad \frac{\|h_1\|}{\sigma_f J} = \frac{1}{1.37}\left(\frac{1}{\rho_x}\right)^2. \tag{70f}$$

Notice that the dependence on $1/\rho_x$ is quadratic ($n = 2$ in Equation (69)), unlike the 1D case where $\|h_1\|$ depended linearly on $1/\rho_x$. If we are to use the criterion $\frac{\|h_1\|}{J} < 1$ as discussed for the 1D case, and rely purely on $\|h_1\|$ as the bounding term for setting a convergence criterion, then several conclusions follow. For the discussion, one can think of this 2D case as a vertical section of a confined aquifer. For the limiting case of a 'strongly layered' formation, Equation (70a) (considering y to be the vertical direction, $\rho_y \ll 1$, that is, very small y integral scale, and the mean flow to be aligned with the horizontal direction x), convergence will be achieved even with very large σ_f. On the other hand, if there exist vertical mean flows, or local gradients in the y direction, a conclusion may be drawn for this 'strongly layered' case from the previous analysis. Consider the case $\rho_y = 5.0$, Equation (70f). This corresponds to flow in a 'strongly layered' formation in the direction perpendicular to the layering. The convergence condition for this case is much more restrictive, $\sigma_f < 1.37\rho_x^2$. For environmental conditions, it is expected the flow will align in the direction of layering, but recharge and boundary conditions will induce vertical gradients, and the bound on σ_f for the perturbation expansion to be convergent will be reduced. Figure 9 shows the dependence of $\|h_1\|$ on ρ_y for fixed $L_x/L_y = 1$ and two different values of ρ_x. A fitted relationship for these two values of ρ_x is

$$\rho_x = 0.018 \qquad \frac{\|h_1\|}{\sigma_f J} = 626\rho_y^2, \tag{71a}$$

$$\rho_x = 0.28 \qquad \frac{\|h_1\|}{\sigma_f J} = 2.72\rho_y^2. \tag{71b}$$

For $\rho_x > \rho_c$, a different relationship is observed. If Equation (69) is fitted an exponent $n = 3$ is obtained, along with a set of coefficients that guarantee convergence for very large values of σ_f. This case corresponds to the condition in which the integral scale is of the same or greater order of magnitude than the domain size we are looking at, $\lambda_i \approx L_i$. It is intuitively correct, since the domain of interest

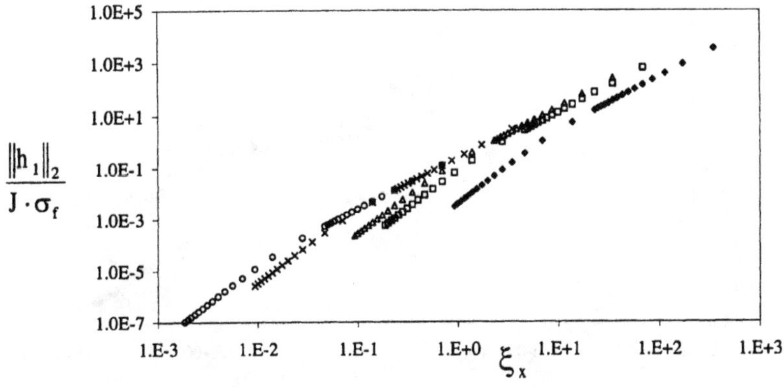

Figure 8. Dependence of $\|h_1\|_2$ on the parameter $\xi = \rho_y L_y / \rho_x L_x$ for exponential log conductivity two point correlation function, 2D flow.

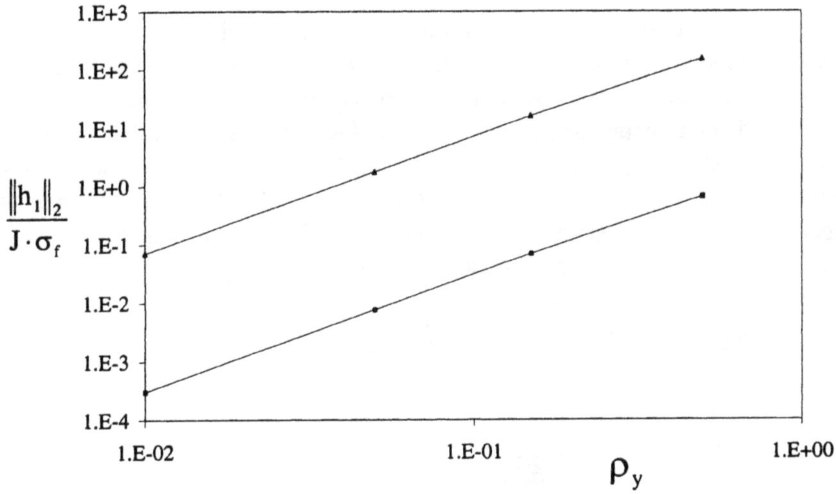

Figure 9. Dependence of $\|h_1\|_2$ on the parameter ρ_y for exponential log conductivity two point correlation function, 2D flow.

will exhibit very smooth change in hydraulic conductivity. As an example of such a bound, for $\rho_y = 0.5$, the convergence condition is $\frac{\|h_1\|}{J} = \frac{\sigma_f}{30}\left(\frac{1}{\rho_x}\right)^3 < 1$, with $\rho_x > 0.35$. For example, if $\rho_x = 0.5$, the above convergence condition establishes that the solution will be convergent if $\sigma_f < 30(0.5)^3 = 3.75$. Finally the relationship between the parameter ξ and $\|h_1\|$ is shown in Figure 8. The dependence is also quadratic for large values of ξ (small values of $\rho_x < \rho_c$) and cubic for small values of $\xi = \rho_y L_y / \rho_x L_x$, and the convergence condition changes for different values of ρ_y.

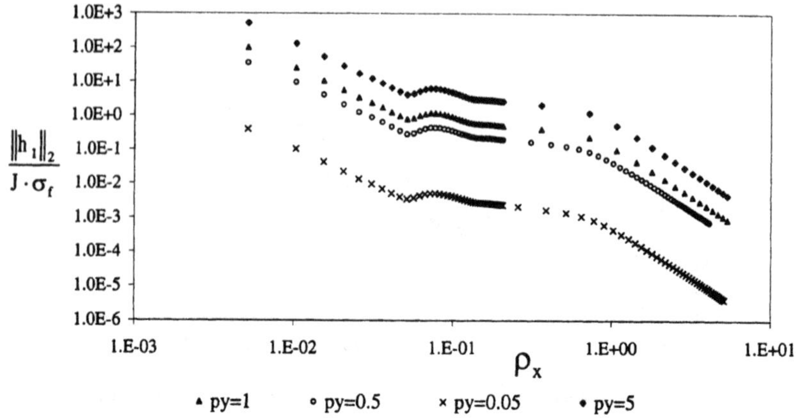

Figure 10. Dependence of $\|h_1\|_2$ on the parameter ρ_x for Gaussian log conductivity two point correlation function, 2D flow.

If we compute $\|h_1\|$ using a Gaussian two point correlation function, Equation (58), we can see the large impact of this correlation function's shape on the convergence of the perturbation scheme. Figure 10 shows the dependence of $\|h_1\|$ on ρ_x. A fitted relationship to powers of $1/\rho_x$ does not seem suitable except at the limits $\rho_x \to 0$ and $\rho_x > 1$ where such fitting follows a trend (69) with n integer. Figure 10 shows a complex dependence of $\|h_1\|$ on the length parameters in the transition region $\rho_{cl} = 0.05 < \rho_x < 1.0 = \rho_{cu}$. For the limiting case $\rho_x < 0.05$, the following convergence relationships are found;

$$\rho_y = 0.05, \qquad \frac{\|h_1\|}{\sigma_f J} = \frac{1}{145000}\left(\frac{1}{\rho_x}\right)^2, \tag{72a}$$

$$\rho_y = 1.0, \qquad \frac{\|h_1\|}{\sigma_f J} = \frac{1}{675}\left(\frac{1}{\rho_x}\right)^2, \tag{72b}$$

$$\rho_y = 5.0, \qquad \frac{\|h_1\|}{\sigma_f J} = \frac{1}{144}\left(\frac{1}{\rho_x}\right)^2. \tag{72c}$$

Notice that convergence occurs for much larger values of σ_f, than for the exponential case, while keeping all the other parameters constant. For a Gaussian two point correlation function, large values of the anisotropy ratio and/or vertical gradients will render a scheme with greater convergence bounds than for the exponential case. For the case $\rho_x > 1$, the dependence is again cubic, for example, for $\rho_y = 1.0$, $\|h_1\|/\sigma_f J = \frac{1}{20}\left(\frac{1}{\rho_x}\right)^3$, with $\rho_x > 1$. Figure 11 shows the dependence of $\|h_1\|$ on the parameter ξ for different values of ρ_y.

Further analysis for higher head corrections should be performed before the dependence of each one of the terms $h_1, h_2, h_3, \ldots$ on the parameters ρ_x, ρ_y, and ξ

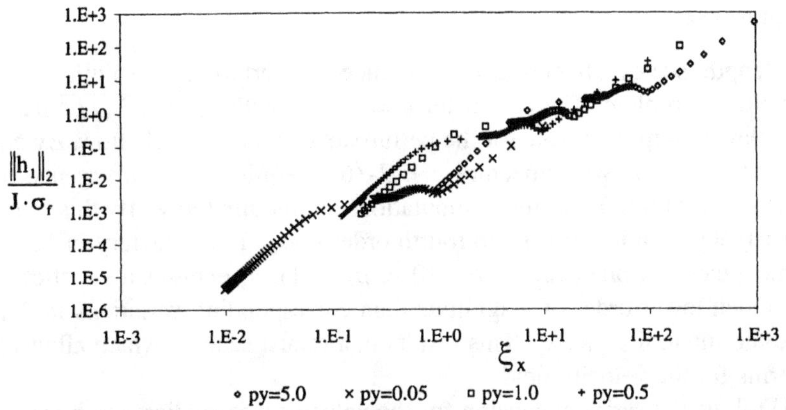

Figure 11. Dependence of $\|h_1\|_2$ on the parameter $\xi = \rho_y L_y / \rho_x L_x$ for Gaussian log conductivity two point correlation function, 2D flow.

can be fully assessed. However, the preliminary results for h_1 support the findings of numerical Monte Carlo techniques, where it has been found that Gaussian two point correlation functions converge faster than the exponential model, e.g., Hassan *et al.* (1998). It also reinforces the idea that perturbation schemes for stochastic flow should rely on the gradient of the log conductivity fluctuation for their expansion. $\sigma_{\nabla f}$ is the small parameter needed in the perturbation analysis. However, the spatial pattern of the two point correlation is not captured by its value at the maximum point, $\sigma_{\nabla f}$. Additional computations are required to improve knowledge of the 'shape' effect on convergence. Hassan *et al.* (1998) have also analyzed fractal conductivity fluctuation distributions and have found that convergence is difficult to obtain for that case. The effect of boundary conditions in 2D flow will be presented at a later date.

Since the perturbation expansion for the head and velocity is generally consistent (Deng and Cushman, 1995; Hsu *et al.*, 1996) (that is, terms of higher order $O(\sigma_f^2)$ have been computed and found to be of smaller magnitude as the order increases for parameter values ranging within values of practical interest), some of the conclusions drawn in this paper for the 2D case have been based on such consistency and computation of higher order will be performed in the future to further test the outcome of the present research. This does not constitute a proof of convergence but, as pointed out in the literature of perturbation methods, one should be satisfied if the expansion is consistent as far as we have proceeded (Hinch, 1990). For the 1D case this consistency is clear, and it is expected that for the 2D case terms of higher order will reinforce these findings. There are significant differences in convergence between one- and two- dimensional flow. Convergence limits for σ_f in two dimensions are greater than those for one dimension. The literature (e.g., Mizell *et al.*, 1982) suggest convergence for the 3D case should be close to the criterion for the 2D case. For instance, observe the slopes of Figure 2 of Mizell *et al.* (1982).

5. Conclusions

Various length scales affect the convergence of perturbation solutions to the stochastic flow problem. The parameter $\varepsilon = \sigma_f/\rho_x$ with $\rho_x = \sqrt{\lambda_x/L^*}$ has been found to play an important role for the perturbative solution of 1D flow. By choosing a maximum norm, we conjecture $\varepsilon < 2\sqrt{6}$ is required for convergence of the head potential (1D flow). Further computations are required to verify this scenario. The consistency seen in terms up to fourth order in $\sigma_{\nabla f}$ is satisfactory in 1D flow.

For most cases of practical interest ($0 < \rho_x < 1$), ρ_x exhibits a variation in its value of much lower order of magnitude than corresponding variations in domain size, L^*, and integral scale, λ_x. This results in a robust scheme which allows us to match terms for the velocity field.

For 1D flow, the series expansion for the velocity was studied and bounded by the sum of two series in the parameters σ_f and $\varepsilon = \sigma_f/\rho_x$. By analogy, the two conditions $\varepsilon < 6.0$, and $\sigma_f < \left(\frac{1}{\frac{1}{24}+\frac{9\rho_x}{60}+\frac{1}{12\rho_x^2}}\right)^{1/4}$ are hypothesized to result in a convergent velocity expansion.

For one dimension, in the limit case, $\rho_x \to 0$, the analysis indicates divergence. On the other hand, long range correlation of the hydraulic conductivity increases the robust character of the perturbation scheme, as it converges for $\sigma_f^2 > 1$. This result helps explain previous findings (Deng and Cushman, 1995; Hsu *et al.*, 1996; Serrano, 1992; Zhang and Winter, 1999; Hassan *et al.*, 1998).

For the 2D case, convergence limits are greater, showing dependence in the parameters $\rho_x^2 = \lambda_x/L_x$, $\rho_y^2 = \lambda_y/L_y$, and $\xi = \rho_y L_y/\rho_x L_x$. A criterion for selecting a value of σ_f that allows a convergent scheme is $\sigma_f < C(\rho_y)\rho_x^2$ where $C(\rho_y)$ is a function of ρ_y. Some values have been given in Equation (70).

For 2D flow, a Gaussian log conductivity two point correlation function produces greater convergence than that found for an exponential function, as can be seen by comparing Equations (70) against Equations (72). This result suggests that the log-conductivity gradient plays an important role in convergence of perturbation expansions for flow and velocity.

Boundary conditions have a strong effect on convergence, as well as the shape of the two point correlation function, i.e., the log- conductivity fluctuation gradient. Neumann BC's induce more restrictive convergence bounds for the parameter σ_f (keeping all other parameters constant). The results presented here suggest significant variations of conductivity over small distances may cause the perturbation expansion to diverge, linking the issue of convergence to both length scales involved in the problem and the shape of the derivatives of the log conductivity two point correlation function.

Current and future work will address the effect of other boundary conditions on convergence for the two and three dimensional cases.

Acknowledgements

This work has been supported through DOE/EM contract DE-FG07-97ER62354. The authors wish to thank the very valuable comments by T. R. Ginn which greatly improved the manuscript and triggered important modifications, as well as those of two anonymous reviewers.

Appendix A. Two Point Correlation Functions and Computational Aspects

A.1. TWO POINT EXPONENTIAL CORRELATION

The two point exponential correlation

$$R_{ff}(x-y) = \sigma_f^2 \exp\left(-\lambda_s \frac{|x-y|}{\lambda_x}\right) \tag{73}$$

with spatial derivatives

$$\frac{\mathrm{d}R_{ff}(x,z)}{\mathrm{d}x} = -\frac{\sigma_f^2 \lambda_s}{\lambda_x} \mathrm{sign}(x-z) \exp\left(-\lambda_s \frac{|x-y|}{\lambda_x}\right), \tag{74a}$$

$$\frac{\mathrm{d}^2 R_{ff}(x,z)}{\mathrm{d}x\,\mathrm{d}z} = -\frac{\sigma_f^2 {\lambda_s}^2}{{\lambda_x}^2} \exp\left(-\lambda_s \frac{|x-y|}{\lambda_x}\right), \tag{74b}$$

and the norm for $\mathrm{d}^2 R_{ff}(x,z)/\mathrm{d}x\,\mathrm{d}z$ occurs at $x - z = 0$

$$\|R_{ff}\| = \frac{\sigma_f^2 {\lambda_s}^2}{{\lambda_x}^2}. \tag{75}$$

For a dimensionless coordinate $x = x^*/\lambda_x$, that is, $\lambda_s = \lambda_x$, the two point correlation becomes isotropic, while the hydraulic conductivity tensor at the local scale can be thought to be anisotropic. For such scaling the parameter $L = L^*/\lambda_x$ appears in the solution for $\|h_n\|$. Take for instance the solution to $\|h_1(y)h_1(y)\|$,

$$\begin{aligned} \sigma_h^2(y) = \sigma_f^2 J^2 \Bigg[& 2/3\,\frac{y^2}{L} + \frac{-4/3\,y^3 - 2\,y - 2\,e^{-y} + 2}{L^2} + \\ & + \frac{2/3\,y^4 + \left(-2\,e^{-L+y} - 2 + 2\,e^{-y} + 2\,e^{-L}\right) y + 2\,y^2}{L^3} + \\ & + \frac{\left(2 - 2\,e^{-L}\right) y^2}{L^4} \Bigg] \end{aligned} \tag{76}$$

with norm $\|h_1\|_2$ (occurs at $y = L/2$)

$$\|h_1\|_2 = 1/24\,L - 1/2\,L^{-1} + \frac{-2\,e^{-1/2\,L} + 3/2 + 1/2\,e^{-L}}{L^2}. \tag{77}$$

It is noted that the leading term has the same coefficient $1/24$ as in the scaling presented in the text, and since $L = 1/\rho_x^2$, the analysis of convergence presented is independent of the scaling performed.

Similarly for $\langle h_2(y, L)\rangle$,

$$\begin{aligned}\langle h_2(y, L)\rangle &= \sigma_f^2 J\Bigg[1/6\,vL - 1/2\,v^2 + \\ &\quad + \frac{-1 + 1/3\,v^3 + e^{-L} + e^{-v} - e^{-L+v}}{L} + \frac{\left(2 - 2\,e^{-L}\right)v}{L^2}\Bigg] \end{aligned} \tag{78}$$

with norm

$$\|h_2(L)\|_1 = \frac{1}{64}\,L + \frac{-e^{-3/4\,L} - 1/2 + 1/2\,e^{-L} + e^{-1/4\,L}}{L^2} \tag{79}$$

and the leading term has the same coefficient as before, $1/64$.

References

Ash, R. B.: 1972, *Real Analysis and Probability*. Academic Press, London, San Diego.

Benson, D. A., Wheatcraft, S. W. and Meerschaert, M. M.: 1999, Development and application of the fractional advection–dispersion equation. *Water Resour. Res.* (submitted).

Bonilla, F. A. and Cushman, J. H.: 2000, Role of boundary conditions on convergence and nonlocality of solutions to stochastic flow problems in bounded domains. *Water Resour. Res.* **36**(4), 981–997.

Carrier, G. F. and Pearson, C. E.: 1988 *Partial Differential Equations Theory and Technique*. Academic Press, Boston, San Diego.

Cushman, J. H. (ed.): 1990, *Dynamics of Fluids in Hierarchical Porous Media*. Academic Press, London, San Diego.

Cushman, J. H.: 1997, *The Physics of Fluids in Hierarchical Porous Media: Angstroms to Miles*. Kluwer Academic Publishers, NY, 1997.

Cushman, J. H., Hu, B. X. and Deng, F. W.: 1995, Nonlocal reactive transport with physical and chemical heterogeneity: Localization errors. *Water Resour. Res.* **31**(9), 2219–2237.

Dagan, G.: 1982, Stochastic modeling of groundwater flow by unconditional and conditional probabilities 1. Conditional simulation and the direct problem. *Water Resour. Res.* **18**(4), 813–833.

Dagan, G.: 1989, *Flow and Transport in Porous Formations*. Springer-Verlag, New York.

Dagan, G.: 1993, Higher-Order correction of effective permeability of heterogeneous isotropic formations of lognormal conductivity distribution. *Transport in Porous Media* **12**, 279–290.

Deng, F. W. and Cushman, J. H.: 1995, On higher-order corrections to the flow velocity covariance tensor. *Water Resour. Res.* **31**(7), 1659–1672.

Eckhaus, W.: 1973, *Matched Asymptotic Expansions and Singular Perturbations*. North-Holland Pub. Co., American Elsevier Pub. Co., Amsterdam, New York.

Gelhar, L. W.: 1993, *Stochastic Subsurface Hydrology*. Prentice-Hall, Englewood Cliffs, NJ.

Gómez-Hernández, J. J. and Wen, X. H.: 1998, To be or not to be multi-Gaussian? A reflection on stochastic hydrogeology. *Adv. In Water Resour.* **21**, 47–61.

Hassan, A., Cushman, J. H. and Delleur, J. W.: 1998, A Monte Carlo assessment of Eulerian flow and transport perturbation models. *Water Resour. Res.* **34**(5), 1143–1163.

Hinch, E. J.: 1990, *Perturbation Methods*. Cambridge University Press, Cambridge, New York.

Hsu, K. C., Zhang, D. X. and Neuman, S. P.: 1996, Higher-order effects on flow and transport in randomly heterogeneous porous media. *Water Resour. Res.* **32**(3), 571–582.

Magnus, W., Oberhettinger, F. and Soni, R. P.: 1966, *Formulas and Theorems for the Special Functions of Mathematical Physics.* Springer-Verlag, New York.

Mizell, S. A., Gutjahr, A. L. and Gelhar, L. W.: 1982, Stochastic analysis of spatial variability in two-dimensional steady groundwater flow assuming stationary and nonstationary heads. *Water Resour. Res.* **18**(4), 1053–1067.

Mukhopadhyay S. and Cushman, J. H.: 1998, Diffusive transport of volatile pollutants in nonaqueous-phase liquid contaminated soil: A fractal model. *Transport in Porous Media* **30**, 125–154.

Neuman, S. P.: 1995, On advective transport in fractal permeability and velocity fields. *Water Resour. Res.* **31**(6), 1455–1460.

Neuman, S. P. and Orr, S.: 1993, Prediction of steady state flow in nonuniform geologic media by conditional moments -exact nonlocal formalism, effective conductivities, and weak approximation. *Water Resour. Res.* **29**(2), 341–364.

Roach, G. F.: 1982, *Green's Functions.* Cambridge University Press, London.

Rubin, Y. and Dagan, G.: 1992, A note on head and velocity covariances in three-dimensional flow through heterogeneous anisotropic porous media. *Water Resour. Res.* **28**, 1463–1470.

SanchezVila, X., Carrera, J. and Girardi, J. P.: 1996, Scale effects in transmissivity. *J. Hydrology* **183**(1–2), 1–22.

Serrano, S. E.: 1992, Semianalytical methods in stochastic groundwater transport. *Applied Mathematical Modeling* **16**(4), 181–191.

Zhang, D. X.: 1999, Nonstationary stochastic analysis of transient unsaturated flow in randomly heterogeneous media. *Water Resour. Res.* **35**, 1127–1141.

Zhang, D. X. and Winter, C. L.: 1999, Moment-equation approach to single-phase fluid flow in heterogeneous reservoirs. *Soc. Petroleum Eng.* **3**(2).

Transport in Porous Media **42:** 37–67, 2001.

Recursive Conditional Moment Equations for Advective Transport in Randomly Heterogeneous Velocity Fields

ALBERTO GUADAGNINI[1] and SHLOMO P. NEUMAN[2]

[1]*Dipartimento di Ingegneria Idraulica Ambientale e del Rilevamento, Politecnico di Milano, Piazza L. Da Vinci, 32, 20133 Milano, Italy, e-mail: alberto.guadagnini@rolini.it*

[2]*Department of Hydrology and Water Resources, The University of Arizona, Tucson, Arizona 85721, U.S.A., e-mail: neuman@hwr.arizona.edu*

(Received: 10 April 1999; in final form: 23 September 1999)

Abstract. Flow and transport parameters such as hydraulic conductivity, seepage velocity, and dispersivity have been traditionally viewed as well-defined local quantities that can be assigned unique values at each point in space-time. Yet in practice these parameters can be deduced from measurements only at selected locations where their values depend on the scale (support volume) and mode (instruments and procedure) of measurement. Quite often, the support of the measurements is uncertain and the data are corrupted by experimental and interpretive errors. Estimating the parameters at points where measurements are not available entails an additional random error. These errors and uncertainties render the parameters random and the corresponding flow and transport equations stochastic. The stochastic flow and transport equations can be solved numerically by conditional Monte Carlo simulation. However, this procedure is computationally demanding and lacks well-established convergence criteria. An alternative to such simulation is provided by conditional moment equations, which yield corresponding predictions of flow and transport deterministically. These equations are typically integro-differential and include nonlocal parameters that depend on more than one point in space-time. The traditional concept of a REV (representative elementary volume) is neither necessary nor relevant for their validity or application. The parameters are nonunique in that they depend not only on local medium properties but also on the information one has about these properties (scale, location, quantity, and quality of data). Darcy's law and Fick's analogy are generally not obeyed by the flow and transport predictors except in special cases or as localized approximations. Such approximations yield familiar-looking differential equations which, however, acquire a non-traditional meaning in that their parameters (hydraulic conductivity, seepage velocity, dispersivity) and state variables (hydraulic head, concentration) are information-dependent and therefore, inherently nonunique. Nonlocal equations contain information about predictive uncertainty, localized equations do not. We have shown previously (Guadagnini and Neuman, 1997, 1998, 1999a, b) how to solve conditional moment equations of steady-state flow numerically on the basis of recursive approximations similar to those developed for transient flow by Tartakovsky and Neuman (1998, 1999). Our solution yields conditional moments of velocity, which are required for the numerical computation of conditional moments associated with transport. In this paper, we lay the theoretical groundwork for such computations by developing exact integro-differential expressions for second conditional moments, and recursive approximations for all conditional moments, of advective transport in a manner that complements earlier work along these lines by Neuman (1993).

Key words: solute transport, dispersion heterogeneity, random media, stochastic equations, moment equations, conditioning.

1. Introduction

Flow and transport parameters such as hydraulic conductivity, seepage velocity, and dispersivity have been traditionally viewed as well-defined local quantities that can be assigned unique values at each point in space-time. Yet in practice these parameters can be deduced from measurements only at selected well locations and depth intervals where their values depend on the scale (support volume) and mode (instruments and procedure) of measurement. Quite often, the support of the measurements is uncertain and the data are corrupted by experimental and interpretive errors. Estimating the parameters at points where measurements are not available entails an additional random error. These errors and uncertainties render the parameters random and the corresponding flow and transport equations stochastic. The stochastic flow and transport equations can be solved numerically by conditional Monte Carlo simulation. This approach is conceptually straightforward and has the advantage of applying to a very broad range of both linear and nonlinear flow and transport problems. It is however computationally intensive, and lacks well-established convergence criteria. An alternative to such simulation is provided by conditional moment equations, which yield corresponding predictions of flow and transport deterministically. Exact conditional moment equations for steady state flow in saturated media have been developed by Neuman and Orr (1993) and Neuman *et al.* (1996); for nonsteady state flow in such media by Tartakovsky and Neuman (1998, 1999); for advective transport by Neuman (1993); and for advective–dispersive transport by Zhang and Neuman (1996). Similar equations have been developed by Tartakovsky *et al.* (1999) for a class of steady-state unsaturated flow problems. For a more thorough review of related developments by others, the reader is referred to Neuman (1997).

All these conditional moment equations are integro-differential and include nonlocal parameters that depend on more than one point in space-time. The traditional concept of an REV (representative elementary volume) is neither necessary nor relevant for their validity or application. The corresponding parameters are nonunique in that they depend not only on medium properties but also on the information one has about these properties (scale, location, quantity, and quality of data). Darcy's law and Fick's analogy are generally not obeyed by the flow and transport predictors except in special cases or as localized approximations. Such approximations yield familiar-looking differential equations which, however, acquire a non-traditional meaning in that their parameters (hydraulic conductivity, seepage velocity, dispersivity) and state variables (hydraulic head, concentration) are information-dependent and therefore, inherently nonunique. Whereas nonlocal equations contain information about predictive uncertainty, localized equations do not.

We have shown previously (Guadagnini and Neuman, 1997, 1998, 1999a, b) how to solve conditional moment equations of steady state saturated flow numerically on the basis of recursive approximations similar to those developed for transient flow by Tartakovsky and Neuman (1998, 1999). In particular, we considered the effect of measuring randomly varying local hydraulic conductivities $K(\mathbf{x})$ on one's ability to predict deterministically, without upscaling, steady-state flow in bounded domains driven by random source and boundary terms. Our aim was to allow optimum unbiased prediction of hydraulic heads $h(\mathbf{x})$ and Darcy fluxes $\mathbf{q}(\mathbf{x})$ by means of their ensemble moments, $\langle h(\mathbf{x})\rangle_c$ and $\langle \mathbf{q}(\mathbf{x})\rangle_c$, conditioned on measurements of $K(\mathbf{x})$. According to Neuman and Orr (1993) and Neuman *et al.* (1996), these predictors satisfy a deterministic flow equation that contain an integro-differential *residual flux* term. This term renders $\langle \mathbf{q}(\mathbf{x})\rangle_c$ nonlocal and non-Darcian so that the concept of effective hydraulic conductivity looses meaning in all but a few special cases. Instead, the residual flux contains kernels which constitute nonlocal parameters that are conditional on hydraulic conductivity data, and therefore, nonunique. The kernels include symmetric and nonsymmetric second-rank tensors as well as vectors. We started by developing exact nonlocal equations for second conditional moments of head and flux which constitute measures of predictive uncertainty. Next, we developed recursive closure approximations for the moment equations through expansion in powers of a small parameter σ which represents the standard estimation error of $\ln K(\mathbf{x})$. We then solved these equations to first order in σ^2 by finite elements on a rectangular grid in two dimensions. Upon comparing our finite element solution to conditional Monte Carlo results we found that the former yields highly accurate results, under either mean uniform or convergent flows, for both mildly and strongly heterogeneous media with σ^2 as large as 4–5. Since conditional mean quantities are smooth relative to their random counterparts our method allows, in principle, resolving them on relatively coarse grids without upscaling. We have not yet explored this option.

We (Guadagnini and Neuman, 1997, 1998, 1999a, b; Neuman and Guadagnini, 1999) also examined situations under which the residual flux can be localized so as to render it approximately Darcian. One way to achieve such localization is to treat hydraulic conductivity as if it was locally homogeneous and mean flow as if it was locally uniform. This renders the flux predictor Darcian according to $\langle \mathbf{q}(\mathbf{x})\rangle_c = -\mathbf{K}_c(\mathbf{x})\nabla\langle h(\mathbf{x})\rangle_c$ where $\mathbf{K}_c(\mathbf{x})$ is a conditional hydraulic conductivity tensor which depends on measurements of $K(\mathbf{x})$ and is therefore a nonunique deterministic function of space. This function can be estimated by means of either stochastically-derived analytical formulae (for bounded or unbounded domains) or standard inverse methods (in which case localization coincides with common groundwater modeling practice). We used the first approach and solved the corresponding localized conditional mean equation by finite elements on a rectangular grid in two dimensions. Here the conditional hydraulic conductivity is given by the local geometric mean $K_G(\mathbf{x})$. Upon comparing our localized finite element solution with the aforementioned nonlocal finite element solution and conditional Monte Carlo results, we found that the first is generally less accurate than the second. The accuracy of the localized solution deteriorates relative to that of the nonlocal solution as one approaches points of conditioning and singularity, or as the variance and correlation scale of the log hydraulic conductivity increase. Contrary to the nonlocal solution, localization does not (as already stated) yield information about predictive uncertainty.

In this paper, we lay the theoretical groundwork for the numerical solution of conditional moment equations associated with transport. In particular, we develop exact integro-differential expressions for second conditional moments, and recursive approximations for all conditional moments, of advective transport in a manner that complements the earlier work of Neuman (1993). Our next step will be to develop analogous expressions that include local dispersion in a way that complements the more recent work of Zhang and Neuman (1996). We will then solve the corresponding recursive transport equations numerically.

2. Exact Conditional Mean Transport Equations

Following Neuman (1993), we start by defining an advective velocity vector $\mathbf{v}(\mathbf{x}, t)$ on a local volume support ω, centered about point $\mathbf{x}$ in a Cartesian coordinate system, where t represents time. In this paper we disregard local dispersion, hence the solute mass flux $\mathbf{J}(\mathbf{x}, t)$ is given by

$$\mathbf{J}(\mathbf{x}, t) = \mathbf{v}(\mathbf{x}, t)c(\mathbf{x}, t), \tag{1}$$

where $c(\mathbf{x}, t)$ is concentration. We postulate that all quantities in (1) are representative of ω and can be evaluated locally by means of appropriate measurement and interpretive methods. Our definition of ω is pragmatic and operational. All we require is that it correspond to a volume at which quantities of interest are amenable in principle to direct measurement, or indirect evaluation, at each point in the flow domain, not that it correspond to some abstract representative elementary volume (REV). The question whether ω is or is not a valid support scale is determined by testing and measurement, not by abstract reasoning. This renders the traditional concept of an REV irrelevant to our discussion.

Nonreactive solute mass conservation implies that

$$\frac{\partial c(\mathbf{x}, t)}{\partial t} + \nabla \cdot \mathbf{J}(\mathbf{x}, t) = g(\mathbf{x}, t), \tag{2}$$

which we wish to solve subject to the initial condition

$$c(\mathbf{x}, 0) = C_0(\mathbf{x}). \tag{3}$$

Here $C_0(\mathbf{x}, 0)$ is a random initial solute concentration in an infinite domain Ω_∞, and $g(\mathbf{x}, t)$ is a random source of solute, both defined on the scale ω. Though our theory does not require it, we assume for simplicity that the statistics of C_0 and g are prescribed independently of each other. The fluid velocity is a space-time non-stationary function satisfying a stochastic flow equation;

$$\nabla \cdot \mathbf{v}(\mathbf{x}, t) = f(\mathbf{x}, t), \tag{4}$$

in which $f(\mathbf{x}, t)$ is an ω-scale random fluid source (and/or accumulation term involving the time derivative of head) whose statistics, including joint moments with $\mathbf{v}$, are known. These moments are determined by solving the stochastic flow equation (4), subject to appropriate (generally random) initial and boundary conditions, conditioned on measurements of hydraulic conductivity and kinematic porosity on support, ω. Methods to compute them include conditional Monte Carlo simulation or the solution of corresponding conditional moments equations in the manner of Guadagnini and Neuman (1997, 1998, 1999a, b).

We write

$$\mathbf{v}(\mathbf{x}, t) = \langle \mathbf{v}(\mathbf{x}, t)\rangle_c + \mathbf{v}'(\mathbf{x}, t) \quad \langle \mathbf{v}'(\mathbf{x}, t)\rangle_c \equiv 0 \tag{5}$$

$$c(\mathbf{x}, t) = \langle c(\mathbf{x}, t)\rangle_c + c'(\mathbf{x}, t) \quad \langle c'(\mathbf{x}, t)\rangle_c \equiv 0 \tag{6}$$

$$\mathbf{J}(\mathbf{x}, t) = \langle \mathbf{J}(\mathbf{x}, t)\rangle_c + \mathbf{J}'(\mathbf{x}, t) \quad \langle \mathbf{J}'(\mathbf{x}, t)\rangle_c \equiv 0 \tag{7}$$

where $\langle\rangle_c$ denotes ensemble mean conditioned on measurements, and primed quantities are random fluctuations about the corresponding conditional means. The former can be viewed as unbiased predictors of their random counterparts, and the latter as the associated prediction errors. Our aim is to compute the first two conditional moments of $c(\mathbf{x}, t)$ and $\mathbf{J}(\mathbf{x}, t)$. Taking the conditional mean of (1) and (2) yields

$$\langle \mathbf{J}(\mathbf{x}, t)\rangle_c = \langle \mathbf{v}(\mathbf{x}, t)\rangle_c \langle c(\mathbf{x}, t)\rangle_c + \mathbf{Q}_c(\mathbf{x}, t), \tag{8}$$

$$\frac{\partial \langle c(\mathbf{x}, t)\rangle_c}{\partial t} + \nabla \cdot \langle \mathbf{J}(\mathbf{x}, t)\rangle_c = \langle g(\mathbf{x}, t)\rangle, \tag{9}$$

subject to

$$\langle c(\mathbf{x}, 0)\rangle_c \equiv \langle C_0(\mathbf{x})\rangle, \tag{10}$$

where $\mathbf{Q}_c(\mathbf{x}, t)$ is a conditional dispersive flux defined as

$$\mathbf{Q}_c(\mathbf{x}, t) = \langle \mathbf{v}'(\mathbf{x}, t) c'(\mathbf{x}, t)\rangle_c. \tag{11}$$

Consider a random auxiliary function $G(\chi, \tau|\mathbf{x}, t)$ that satisfies the stochastic equation (though χ and $\tilde{\chi}$ are not bold, they nevertheless are all vectors),

$$\frac{\partial G(\chi, \tau|\mathbf{x}, t)}{\partial \tau} + \mathbf{v}(\chi, \tau) \cdot \nabla_\chi G(\chi, \tau|\mathbf{x}, t) = \delta(\chi - \mathbf{x})\delta(t - \tau), \tag{12}$$

subject to the terminal condition

$$G(\chi, \tau|\mathbf{x}, t) = 0 \quad \text{for } \tau \geqslant t. \tag{13}$$

This function is given by

$$G(\chi, \tau|\mathbf{x}, t) = [H(\tau - t) - 1]\delta(\chi - \tilde{\chi}), \tag{14}$$

where H is the Heaviside function; $\tilde{\chi}(t-\tau|\mathbf{x})$ is the random upstream position at time τ ($\tau \leqslant t$) of an indivisible, inert solute particle, of unit mass normalized with respect to porosity, that reaches the downstream point $\mathbf{x}$ at time t; and $\delta(\chi - \tilde{\chi})$ represents the concentration of this particle at (χ, τ). With the aid of this auxiliary function, Neuman (1993) was able to demonstrate in a rigorous manner that (Neuman did so for zero initial concentration; we show in Appendix A that the same holds true for arbitrary initial concentration).

$$\begin{aligned}\mathbf{Q}_c(\mathbf{x}, t) &= \iint_0^t \alpha_c(\mathbf{x}, t, \chi, \tau)\nabla_\chi \cdot \mathbf{Q}_c(\chi, \tau)\, \mathrm{d}\tau\, \mathrm{d}\chi - \\ &\quad - \iint_0^t \beta_c(\mathbf{x}, t, \chi, \tau)\nabla_\chi \langle c(\chi, \tau)\rangle_c\, \mathrm{d}\tau\, \mathrm{d}\chi - \\ &\quad - \iint_0^t \gamma_c(\mathbf{x}, t, \chi, \tau)\langle c(\chi, \tau)\rangle_c\, \mathrm{d}\tau\, \mathrm{d}\chi, \end{aligned} \tag{15}$$

where $\alpha_c(\mathbf{x}, t, \chi, \tau)$, $\beta_c(\mathbf{x}, t, \chi, \tau)$, $\gamma_c(\mathbf{x}, t, \chi, \tau)$ are kernels independent of $g(\mathbf{x}, t)$ and defined by the formal expressions

$$\alpha_c(\mathbf{x}, t, \chi, \tau) = \langle \tilde{c}(\mathbf{x}, t|\chi, \tau)\tilde{\mathbf{v}}'(\mathbf{x}, t)\rangle_c, \tag{16}$$

$$\beta_c(\mathbf{x}, t, \chi, \tau) = \langle \tilde{c}(\mathbf{x}, t|\chi, \tau)\tilde{\mathbf{v}}'(\mathbf{x}, t)\mathbf{v}'^{\mathrm{T}}(\chi, \tau)\rangle_c, \tag{17}$$

$$\gamma_c(\mathbf{x}, t, \chi, \tau) = \langle \tilde{c}(\mathbf{x}, t|\chi, \tau)\tilde{\mathbf{v}}'(\mathbf{x}, t) f'(\chi, \tau)\rangle_c. \tag{18}$$

Here $\tilde{c}(\mathbf{x}, t, |\chi, \tau)$, the adjoint of $\delta(\chi - \tilde{\chi})$, is a fundamental advective concentration satisfying (1)–(3) for $f \equiv 0$, $C_0 \equiv 0$ (zero initial concentration), and an instantaneous point source of unit solute mass at (χ, τ), $g(\mathbf{x}, t) = \delta(\mathbf{x} - \chi)\, \delta(t - \tau)$, $0 \leqslant \tau \leqslant t$. The tilde over $\mathbf{v}'$ implies that $\mathbf{v}'$ is immune to operations with respect to $\mathbf{x}$ and t, and f' is zero mean random fluctuation in f about its conditional mean $\langle f\rangle$. The vectors α_c and γ_c, and the tensor β_c, constitute nonlocal parameters which depend on the flow field but not on the solute source or its concentration. They are however conditioned on measurements, as is the local parameter $\langle \mathbf{v}\rangle_c$, which renders all of these parameters inherently nonunique.

The parameter α_c represents dispersive flux due to an instantaneous point source of unit normalized solute mass at (χ, τ). We outline in Appendix B how its components can be expressed in integro-differential form as

$$\begin{aligned}\alpha_{ic}(\mathbf{x}, t, \chi, \tau) &= \langle \tilde{v}_i'(\mathbf{x}, t)\tilde{c}(\mathbf{x}, t|\chi, \tau)\rangle_c \\ &= \iint_0^t \langle \psi(\mathbf{y}, s|\mathbf{x}, t)\tilde{v}_i'(\mathbf{x}, t)\rangle_c \frac{\partial}{\partial y_j}\alpha_{jc}(\mathbf{x}, t, \mathbf{y}, s)\, \mathrm{d}s\, \mathrm{d}\mathbf{y} - \\ &\quad - \iint_0^t \langle \psi(\mathbf{y}, s|\mathbf{x}, t)\tilde{v}_i'(\mathbf{x}, t)v_j'(\mathbf{y}, s)\rangle_c \frac{\partial}{\partial y_j}\langle \tilde{c}(\mathbf{x}, t|\mathbf{y}, s)\rangle_c\, \mathrm{d}s\, \mathrm{d}\mathbf{y} - \\ &\quad - \iint_0^t \langle \psi(\mathbf{y}, s|\mathbf{x}, t)\tilde{v}_i'(\mathbf{x}, t) f'(\mathbf{y}, s)\rangle_c \langle \tilde{c}(\mathbf{x}, t|\mathbf{y}, s)\rangle_c\, \mathrm{d}s\, \mathrm{d}\mathbf{y}, \end{aligned} \tag{19}$$

where $i = 1, 2, 3$, $\psi(\mathbf{y}, s|\mathbf{x}, t) = \tilde{c}(\mathbf{x}, t|\mathbf{y}, s)$, and repeated indices imply summation. Alternatively, α_c satisfies the differential equation (Appendix B)

$$\begin{aligned}&\frac{\partial \alpha_c^T(\mathbf{x}, t, \chi, \tau)_c}{\partial t} + \nabla_x^T[\langle \mathbf{v}(\mathbf{x}, t)\rangle_c \alpha_c^T(\mathbf{x}, t, \chi, \tau)]\\ &= -\nabla_x^T \langle \mathbf{v}'(\mathbf{x}, t)\tilde{\mathbf{v}}'^T(\mathbf{x}, t)\tilde{c}'(\mathbf{x}, t|\chi, \tau)\rangle_c - \\ &\quad - \nabla_x^T[\langle \mathbf{v}'(\mathbf{x}, t)\tilde{\mathbf{v}}'^T(\mathbf{x}, t)\rangle_c \langle \tilde{c}(\mathbf{x}, t|\chi, \tau)\rangle_c],\end{aligned} \tag{20}$$

subject to zero initial conditions.

3. Recursive Conditional Mean Transport Approximations

Though the above conditional mean transport equations are exact, they cannot be evaluated numerically in their exact form. To render them workable, we derive below a set of recursive approximations for these equations by expanding all random quantities in terms of a small parameter σ, representing a measure of the standard deviation of $Y'(\mathbf{x}) = Y(\mathbf{x}) - \langle Y(\mathbf{x})\rangle_c$, where $Y(\mathbf{x}) = \ln K(\mathbf{x})$. This nominally limits our approximations to either mildly heterogeneous media, or well conditioned strongly heterogeneous media, with $\sigma \ll 1$. Corresponding recursive approximations for joint conditional moments of $\mathbf{v}(\mathbf{x}, t)$ and $f(\mathbf{x}, t)$ are assumed to be given. The latter can be computed on the basis of recursive approximations similar to those developed for transient flow by Tartakovsky and Neuman (1998, 1999) and implemented computationally for steady state flow by Guadagnini and Neuman (1997, 1998, 1999a, b). Though these authors dealt only with moments of the flux $\mathbf{q}(\mathbf{x}, t)$, it is relatively simple to include in their analyses the effect of a random porosity $\phi(\mathbf{x})$ on moments of the velocity $\mathbf{v}(\mathbf{x}, t)$.

A perturbation expansion of (15) leads to the following ith order approximation for the dispersive solute flux,

$$\begin{aligned}\mathbf{Q}_c^{(i)}(\mathbf{x}, t) &= \iint_0^t \sum_{n=0}^{i} \left[\alpha_c^{(n)}(\mathbf{x}, t, \chi, \tau)\nabla_\chi \cdot \mathbf{Q}_c^{(i-n)}(\chi, \tau)\right] \mathrm{d}\tau\, \mathrm{d}\chi - \\ &\quad - \iint_0^t \sum_{n=0}^{i} \left[\beta_c^{(n)}(\mathbf{x}, t, \chi, \tau)\nabla_\chi \langle c^{(i-n)}(\chi, \tau)\rangle_c\right] \mathrm{d}\tau\, \mathrm{d}\chi - \\ &\quad - \iint_0^t \sum_{n=0}^{i} \left[\gamma_c^{(n)}(\mathbf{x}, t, \chi, \tau)\langle c^{(i-n)}(\chi, \tau)\rangle_c\right] \mathrm{d}\tau\, \mathrm{d}\chi,\end{aligned} \tag{21}$$

where the superscript (i) indicates terms of ith order in σ, and $\langle c^{(i)}(\chi, \tau)\rangle_c$ is the ith order predictor of concentration. To order $i = 0$, the dispersive flux predictor $\mathbf{Q}_c^{(0)}(\mathbf{x}, t)$ is identically equal to zero, and therefore the mean transport equations become

$$\langle \mathbf{J}^{(0)}(\mathbf{x}, t)\rangle_c = \langle \mathbf{v}^{(0)}(\mathbf{x}, t)\rangle_c \langle c^{(0)}(\mathbf{x}, t)\rangle_c, \tag{22}$$

$$\frac{\partial \langle c^{(0)}(\mathbf{x},t)\rangle_c}{\partial t} + \nabla\cdot\langle \mathbf{J}^{(0)}(\mathbf{x},t)\rangle_c = \langle g(\mathbf{x},t)\rangle \tag{23}$$

subject to the initial condition

$$\langle c^{(0)}(\mathbf{x},0)\rangle_c = \langle C_0(\mathbf{x})\rangle. \tag{24}$$

To all higher orders $i \geqslant 1$, the dispersive flux predictor satisfies

$$\langle \mathbf{J}^{(i)}(\mathbf{x},t)\rangle_c = \sum_{n=0}^{i} \langle \mathbf{v}^{(n)}(\mathbf{x},t)\rangle_c \langle c^{(i-n)}(\mathbf{x},t)\rangle_c + \mathbf{Q}_c^{(i)}(\mathbf{x},t), \tag{25}$$

$$\frac{\partial \langle c^{(i)}(\mathbf{x},t)\rangle_c}{\partial t} + \nabla\cdot\langle \mathbf{J}^{(i)}(\mathbf{x},t)\rangle_c = 0, \tag{26}$$

subject to the initial condition

$$\langle c^{(i)}(\mathbf{x},t)\rangle_c \equiv 0. \tag{27}$$

We outline in Appendix B how

$$\begin{aligned}
&\alpha_c^{(0)} = \alpha_c^{(1)} = 0,\\
&\alpha_c^{(2)}(\mathbf{x},t,\chi,\tau) = [\langle \tilde{v}_i'(\mathbf{x},t)\tilde{c}(\mathbf{x},t|\chi,\tau)\rangle_c]^{(2)}\\
&\quad = -\iint_0^t \langle \psi^{(0)}(\mathbf{y},s|\mathbf{x},t)\rangle_c [\langle \tilde{v}_i'(\mathbf{x},t)v_j'(\mathbf{y},s)\rangle_c]^{(2)} \times\\
&\qquad \times \frac{\partial}{\partial y_j}\langle \tilde{c}^{(0)}(\mathbf{x},t|\mathbf{y},s)\rangle_c \,\mathrm{d}s\,\mathrm{d}\mathbf{y} - \iint_0^t \langle \psi^{(0)}(\mathbf{y},s|\mathbf{x},t)\rangle_c \times\\
&\qquad \times [\langle \tilde{v}_i'(\mathbf{x},t)f'(\mathbf{y},s)\rangle_c]^{(2)} \langle \tilde{c}^{(0)}(\mathbf{x},t|\mathbf{y},s)\rangle_c \,\mathrm{d}s\,\mathrm{d}\mathbf{y},
\end{aligned} \tag{28}$$

$$\begin{aligned}
&\beta_c^{(0)} = \beta_c^{(1)} = 0,\\
&\beta_c^{(2)}(\mathbf{x},t,\chi,\tau) = \langle \tilde{c}^{(0)}(\mathbf{x},t|\chi,\tau)\rangle_c [\langle \tilde{\mathbf{v}}'(\mathbf{x},t)\mathbf{v}'^{\mathrm{T}}(\chi,\tau)\rangle_c]^{(2)},
\end{aligned} \tag{29}$$

$$\begin{aligned}
&\gamma_c^{(0)} = \gamma_c^{(1)} = 0,\\
&\gamma_c^{(2)}(\mathbf{x},t,\chi,\tau) = \langle \tilde{c}^{(0)}(\mathbf{x},t|\chi,\tau)\rangle_c [< \tilde{\mathbf{v}}'(\mathbf{x},t)f'(\chi,\tau)\rangle_c]^{(2)},
\end{aligned} \tag{30}$$

where $\tilde{\mathbf{v}}'(\mathbf{x},t)$ is the function $\mathbf{v}'(\boldsymbol{\xi},\sigma)$ evaluated at $(\mathbf{x},t)$, which renders it immune to operations with respect to $\mathbf{x}$ and t, and $\langle \tilde{c}^{(0)}(\mathbf{x},t|\chi,\tau)\rangle_c$ is the solution of the zero-order equation

$$\begin{aligned}
&\frac{\partial \langle \tilde{c}^{(0)}(\mathbf{x},t|\chi,\tau)\rangle_c}{\partial t} + \nabla\cdot(\langle \mathbf{v}^{(0)}(\mathbf{x},t)\rangle_c \langle \tilde{c}^{(0)}(\mathbf{x},t|\chi,\tau)\rangle_c)\\
&\quad = \delta(\mathbf{x}-\chi)\delta(t-\tau)
\end{aligned} \tag{31}$$

$$\langle \tilde{c}^{(0)}(\mathbf{x},0|\chi,\tau)\rangle_c = 0. \tag{32}$$

The vector $\alpha_c^{(2)}$ forms the solution of a differential equation (Appendix C),

$$\frac{\partial \alpha_c^{(2)T}(\mathbf{x}, t|\chi, \tau))_c}{\partial t} + \nabla^T\{\langle \mathbf{v}(\mathbf{x}, t)^{(0)}\rangle_c \alpha_c^{(2)T}(\mathbf{x}, t|\chi, \tau) + \\ + \langle \tilde{c}^{(0)}(\mathbf{x}, t|\chi, \tau)\rangle_c [\langle \mathbf{v}'(\mathbf{x}, t)\tilde{\mathbf{v}}'^T(\mathbf{x}, t)\rangle_c]^{(2)}\} = 0 \tag{33}$$

subject to the zero initial condition

$$\alpha_c^{(2)}(\mathbf{x}, 0|\chi, \tau) = 0. \tag{34}$$

This equation is expected to be more conducive to the evaluation of $\alpha_c^{(2)}$ than (28) due to its relative simplicity and the fact that it does not contain a mixed moment of $\mathbf{v}$ and f. Note that the moments $[\langle \tilde{\mathbf{v}}'(\mathbf{x}, t)\mathbf{v}'^T(\chi, \tau)\rangle_c]^{(2)}$ and $[\langle \tilde{\mathbf{v}}'(\mathbf{x}, t) f'(\chi, \tau)\rangle_c]^{(2)}$ in (28)–(30) are Eulerian and therefore, much easier to compute than their Lagrangian counterparts in (16)–(18), which would require information about particle trajectories.

It follows from the above that the dispersive flux is given to second order explicitly by

$$\mathbf{Q}_c^{(0)} = \mathbf{Q}_c^{(1)} = 0$$

$$\mathbf{Q}_c^{(2)}(\mathbf{x}, t) = -\iint_0^t \langle \tilde{c}^{(0)}(\mathbf{x}, t|\chi, \tau)\rangle_c [\langle \tilde{\mathbf{v}}'(\mathbf{x}, t)\mathbf{v}'^T(\chi, \tau)\rangle_c]^{(2)} \times \\ \times \nabla_\chi \langle c^{(0)}(\chi, \tau)\rangle_c \, d\tau \, d\chi - \\ - \iint_0^t \langle \tilde{c}^{(0)}(\mathbf{x}, t|\chi, \tau)\rangle_c [\langle \tilde{\mathbf{v}}'(\mathbf{x}, \mathrm{t}) f'(\chi, \tau)\rangle_c]^{(2)} \langle c^{(0)}(\chi, \tau)\rangle_c \, d\tau \, d\chi. \tag{35}$$

In fact, approximating the dispersive flux in (21) to any order i renders it explicit, because α_c is at least of order 2, so $\mathbf{Q}_c$ on the right-hand side cannot be of order higher than $(i-1)$. It means that recursive transport equations (23)–(27) of order i are local (differential, as opposed to nonlocal or integro-differential) in $\langle c^{(i)}\rangle_c$, and can therefore be solved by standard numerical methods for advection-dominated transport, such as the method of characteristics. Since all quantities which enter into these deterministic equations are smooth relative to their random counterparts, one should be able to resolve them at coarser levels of space-time discretization than would be required for Monte Carlo solution of the corresponding stochastic equations (1)–(3).

4. Exact Conditional Second Moment Equations

Substituting (5)–(7) into (1)–(3) and subtracting (8)–(10) shows that the concentration perturbation $c'(\mathbf{x}, t)$ satisfies

$$\frac{\partial c'(\mathbf{x}, t)}{\partial t} + \frac{\partial}{\partial x_i}[v_i(\mathbf{x}, t)c'(\mathbf{x}, t)] \\ = \frac{\partial Q_{ic}(\mathbf{x}, t)}{\partial x_i} - v_i'(\mathbf{x}, t)\frac{\partial \langle c(\mathbf{x}, t)\rangle_c}{\partial x_i} - f'(\mathbf{x}, t)\langle c(\mathbf{x}, t)\rangle_c + g'(\mathbf{x}, t) \tag{36}$$

subject to the initial condition

$$c'(\mathbf{x}, 0) = C_0'(\mathbf{x}) \tag{37}$$

where $C_0'(\mathbf{x})$ is a zero mean random fluctuation of the initial concentration. Here, as elsewhere in the paper, repeated indices imply summation. Let $C_{cc}(\mathbf{x}, \mathbf{y}, t, s) = \langle c'(\mathbf{x}, t)c'(\mathbf{y}, s)\rangle_c$ be the conditional covariance of the concentration prediction errors. Neuman (1993) developed an explicit integral expression for C_{cc}. We show in Appendix D that C_{cc} satisfies exactly the differential equation

$$\begin{aligned}\frac{\partial C_{cc}(\mathbf{x}, \mathbf{y}, t, s)}{\partial t} &+ \frac{\partial}{\partial x_i}[\langle v_i(\mathbf{x}, t)\rangle_c C_{cc}(\mathbf{x}, \mathbf{y}, t, s)] \\ &= -\frac{\partial R_{ic}(\mathbf{x}, \mathbf{y}, t, s)}{\partial x_i} - Q_{ic}(\mathbf{x}, \mathbf{y}, t, s)\frac{\partial \langle c(\mathbf{x}, t)\rangle_c}{\partial x_i} - \\ &\quad - C_{fc}(\mathbf{x}, \mathbf{y}, t, s)\langle c(\mathbf{x}, t)\rangle_c + C_{gc}(\mathbf{x}, \mathbf{y}, t, s),\end{aligned} \tag{38}$$

subject to the initial condition

$$C_{cc}(\mathbf{x}, \mathbf{y}, t, s) = B_c(\mathbf{y}, \mathbf{x}, s), \quad t = 0, \ \mathbf{x} \in \Omega_\infty, \tag{39}$$

where R_{ic} is given explicitly by the integral expression

$$\begin{aligned}R_{ic}(\mathbf{x}, \mathbf{y}, t, s) &= \langle v_i'(\mathbf{x}, t)c'(\mathbf{x}, t)c'(\mathbf{y}, s)\rangle_c \\ &= \iint_0^t \langle \tilde{c}(\mathbf{y}, s|\chi, \tau)v_i'(\mathbf{x}, t)c'(\mathbf{x}, t)\rangle_c \frac{\partial}{\partial \chi_j} Q_{jc}(\chi, \tau)\, \mathrm{d}\tau\, \mathrm{d}\chi - \\ &\quad - \iint_0^t \langle \tilde{c}(\mathbf{y}, s|\chi, \tau)v_i'(\mathbf{x}, t)c'(\mathbf{x}, t)v_j'(\chi, \tau)\rangle_c \frac{\partial}{\partial \chi_j}\langle c(\chi, \tau)\rangle_c\, \mathrm{d}\tau\, \mathrm{d}\chi - \\ &\quad - \iint_0^t \langle \tilde{c}(\mathbf{y}, s|\chi, \tau)v_i'(\mathbf{x}, t)c'(\mathbf{x}, t)f'(\chi, \tau)\rangle_c \langle c(\chi, \tau)\rangle_c\, \mathrm{d}\tau\, \mathrm{d}\chi + \\ &\quad + \int \langle \tilde{c}(\mathbf{y}, s|\chi, \tau)v_i'(\mathbf{x}, t)c'(\mathbf{x}, t)C_0'(\chi)\rangle_c\, \mathrm{d}\chi,\end{aligned} \tag{40}$$

$$Q_{ic}(\mathbf{x}, \mathbf{y}, t, s) = \langle v_i'(\mathbf{x}, t)c'(\mathbf{y}, s)\rangle_c, \tag{41}$$

$$C_{fc}(\mathbf{x}, \mathbf{y}, t, s) = \langle f'(\mathbf{x}, t)c'(\mathbf{y}, s)\rangle_c, \tag{42}$$

$$C_{gc}(\mathbf{x}, \mathbf{y}, t, s) = \langle g'(\mathbf{x}, t)c'(\mathbf{y}, s)\rangle_c. \tag{43}$$

Explicit integral expressions for $\mathbf{Q}_c$, C_{fc} and C_{gc} are presented in Appendix D of Neuman (1993). Though Neuman considered only zero initial concentration, these expressions remain valid for nonzero initial concentration as we take the latter to be prescribed in a statistically independent manner (Appendix A). We develop instead a system of differential equations that these four quantities must satisfy. An explicit equation for $B_c(\mathbf{y}, \mathbf{x}, s) = \langle c'(\mathbf{y}, s)C_0'(\mathbf{x})\rangle_c$ is given in Appendix A.

Taking the initial concentration to be statistically independent of the velocity field, as represented by $\mathbf{v}$ and f, is justified when the solute acts as a tracer so that it does not impact fluid density, and when it is sampled at some reference time $t = 0$ without materially perturbing the velocity field (by remote sensing as in the case of radioactive tracers, or by extracting minute volumes of fluid by means of solution samplers). In such cases, sampling yields an estimate of concentration that is virtually independent of $\mathbf{v}$ and f, and can be regarded as initial concentration for purposes of predicting subsequent plume evolution. We also treat the solute source term g as being statistically independent of the velocity field. This is justified when tracer is introduced into groundwater without materially perturbing $\mathbf{v}$ and f, as by spreading contaminants on the soil surface without impacting infiltration rates, or introducing a tracer into an active well without affecting its flow rate.

As shown in Appendix D, $R_{ic}(\mathbf{x}, \mathbf{y}, t, s)$ is the solution of

$$\frac{\partial R_{ic}(\mathbf{x}, \mathbf{y}, t, s)}{\partial s} + \frac{\partial}{\partial y_j}[\langle v_j(\mathbf{y}, s)\rangle_c R_{ic}(\mathbf{x}, \mathbf{y}, t, s)]$$
$$= -\frac{\partial Q_{jc}(\mathbf{y}, s)}{\partial y_j} Q_{ic}(\mathbf{x}, t) - \frac{\partial \eta_{ijc}(\mathbf{x}, \mathbf{y}, t, s)}{\partial y_j} - \Delta_{ijc}(\mathbf{x}, \mathbf{y}, t, s)\frac{\partial \langle c(\mathbf{y}, s)\rangle_c}{\partial y_j} -$$
$$- \Gamma_{ic}(\mathbf{x}, \mathbf{y}, t, s)\langle c(\mathbf{y}, s)\rangle_c + \Theta_{ic}(\mathbf{x}, \mathbf{y}, t, s) \tag{44}$$

subject to the initial condition

$$R_{ic}(\mathbf{x}, \mathbf{y}, t, s) = \langle v_i'(\mathbf{x}, t)c'(\mathbf{x}, t)C_0'(\mathbf{y})\rangle_c, \quad s = 0, \quad \mathbf{y} \in \Omega_\infty, \tag{45}$$

where

$$\eta_{ijc}(\mathbf{x}, \mathbf{y}, t, s) = \langle v_i'(\mathbf{x}, t)v_j'(\mathbf{y}, s)c'(\mathbf{y}, s)c'(\mathbf{x}, t)\rangle_c, \tag{46}$$

$$\Delta_{ijc}(\mathbf{x}, \mathbf{y}, t, s) = \langle v_i'(\mathbf{x}, t)v_j'(\mathbf{y}, s)c'(\mathbf{x}, t)\rangle_c, \tag{47}$$

$$\Gamma_{ic}(\mathbf{x}, \mathbf{y}, t, s) = \langle v_i'(\mathbf{x}, t)c'(\mathbf{x}, t)f'(\mathbf{y}, s)\rangle_c, \tag{48}$$

$$\Theta_{ic}(\mathbf{x}, \mathbf{y}, t, s) = \langle v_i'(\mathbf{x}, t)c'(\mathbf{x}, t)g'(\mathbf{y}, s)\rangle_c. \tag{49}$$

The moment $\Delta_{ijc}(\mathbf{x}, \mathbf{y}, t, s)$ satisfies

$$\frac{\partial \Delta_{ijc}(\mathbf{x}, \mathbf{y}, t, s)}{\partial t} + \frac{\partial}{\partial x_k}[\langle v_k(\mathbf{x}, t)\rangle_c \Delta_{ijc}(\mathbf{x}, \mathbf{y}, t, s)]$$
$$= -\frac{\partial \varepsilon_{ijkc}(\mathbf{x}, \mathbf{y}, t, s)}{\partial x_k} - \frac{\partial Q_{kc}(\mathbf{x}, t)}{\partial x_k}\langle \tilde{v}_i'(\mathbf{x}, t)v_j'(\mathbf{y}, s)\rangle_c -$$
$$- \langle v_k'(\mathbf{x}, t)\tilde{v}_i'(\mathbf{x}, t)v_j'(\mathbf{y}, s)\rangle_c \frac{\partial \langle c(\mathbf{x}, t)\rangle_c}{\partial x_k} -$$
$$- \langle \tilde{v}_i'(\mathbf{x}, t)v_j'(\mathbf{y}, s)f'(\mathbf{x}, t)\rangle_c \langle c(\mathbf{x}, t)\rangle_c, \tag{50}$$

subject to the initial condition

$$\Delta_{ijc}(\mathbf{x}, \mathbf{y}, t, s) = \langle \tilde{v}_i'(\mathbf{x}, t)v_j'(\mathbf{y}, s)C_0'(\mathbf{x})\rangle_c, \quad t = 0, \quad \mathbf{x} \in \Omega_\infty \tag{51}$$

which vanishes because we consider the initial concentration to be prescribed independently of velocity. Here

$$\varepsilon_{ijkc}(\mathbf{x}, \mathbf{y}, t, s) = \langle \tilde{v}_i'(\mathbf{x}, t) v_j'(\mathbf{y}, s) v_k'(\mathbf{x}, t) c'(\mathbf{x}, t)\rangle_c. \tag{52}$$

The moment $\Gamma_{ic}(\mathbf{x}, \mathbf{y}, t, s)$ satisfies

$$\begin{aligned}
&\frac{\partial \Gamma_{ic}(\mathbf{x}, \mathbf{y}, t, s)}{\partial t} + \frac{\partial}{\partial x_k}[\langle v_k(\mathbf{x}, t)\rangle_c \Gamma_{ic}(\mathbf{x}, \mathbf{y}, t, s)] \\
&\quad = -\frac{\partial \kappa_{ikc}(\mathbf{x}, \mathbf{y}, t, s)}{\partial x_k} - \frac{\partial Q_{kc}(\mathbf{x}, t)}{\partial x_k}\langle \tilde{v}_i'(\mathbf{x}, t) f'(\mathbf{y}, s)\rangle_c - \\
&\quad - \langle v_k'(\mathbf{x}, t)\tilde{v}_i'(\mathbf{x}, t) f'(\mathbf{y}, s)\rangle_c \frac{\partial \langle c(\mathbf{x}, t)\rangle_c}{\partial x_k} - \\
&\quad - \langle \tilde{v}_i'(\mathbf{x}, t) f'(\mathbf{y}, s) f'(\mathbf{x}, t)\rangle_c \langle c(\mathbf{x}, t)\rangle_c - \langle \tilde{v}_i'(\mathbf{x}, t) f'(\mathbf{y}, s) g'(\mathbf{x}, t)\rangle_c,
\end{aligned} \tag{53}$$

subject to the initial condition

$$\Gamma_{ic}(\mathbf{x}, \mathbf{y}, t, s) = \langle \tilde{v}_i'(\mathbf{x}, t) f'(\mathbf{y}, s) C_0'(\mathbf{x})\rangle_c, \quad t = 0, \quad \mathbf{x} \in \Omega_\infty, \tag{54}$$

which vanishes because we consider the initial concentration to be statistically independent of $\mathbf{v}$ and f. The last term in (53) likewise vanishes because we consider g to be statistically independent of $\mathbf{v}$ and f, and

$$\kappa_{ikc}(\mathbf{x}, \mathbf{y}, t, s) = \langle \tilde{v}_i'(\mathbf{x}, t) v_k'(\mathbf{x}, t) f'(\mathbf{y}, s) c'(\mathbf{x}, t)\rangle_c. \tag{55}$$

The moment $\Theta_{ic}(\mathbf{x}, \mathbf{y}, t, s)$ is the solution of

$$\begin{aligned}
&\frac{\partial \Theta_{ic}(\mathbf{x}, \mathbf{y}, t, s)}{\partial t} + \frac{\partial}{\partial x_k}[\langle v_k(\mathbf{x}, t)\rangle_c \Theta_{ic}(\mathbf{x}, \mathbf{y}, t, s)] \\
&\quad = -\frac{\partial \vartheta_{ikc}(\mathbf{x}, \mathbf{y}, t, s)}{\partial x_k} - \frac{\partial Q_{kc}(\mathbf{x}, t)}{\partial x_k}\langle \tilde{v}_i'(\mathbf{x}, t) g'(\mathbf{y}, s)\rangle_c - \\
&\quad - \langle v_k'(\mathbf{x}, t)\tilde{v}_i'(\mathbf{x}, t) g'(\mathbf{y}, s)\rangle_c \frac{\partial \langle c(\mathbf{x}, t)\rangle_c}{\partial x_k} - \\
&\quad - \langle \tilde{v}_i'(\mathbf{x}, t) g'(\mathbf{y}, s) f'(\mathbf{x}, t)\rangle_c \langle c(\mathbf{x}, t)\rangle_c - \langle \tilde{v}_i'(\mathbf{x}, t) g'(\mathbf{y}, s) g'(\mathbf{x}, t)\rangle_c
\end{aligned} \tag{56}$$

subject to the initial condition

$$\Theta_{ic}(\mathbf{x}, \mathbf{y}, t, s) = \langle \tilde{v}_i'(\mathbf{x}, t) C_0'(\mathbf{x}) g'(\mathbf{y}, s)\rangle_c \quad t = 0, \quad \mathbf{x} \in \Omega_\infty \tag{57}$$

which vanishes because we consider the initial concentration to be statistically independent of $\mathbf{v}$ and g. The last four terms in (56) vanish because we take g to be statistically independent of $\mathbf{v}$ and f, and

$$\vartheta_{ikc}(\mathbf{x}, \mathbf{y}, t, s) = \langle \tilde{v}_i'(\mathbf{x}, t) v_k'(\mathbf{x}, t) g'(\mathbf{y}, s) c'(\mathbf{x}, t)\rangle_c. \tag{58}$$

The moment $Q_{ic}(\mathbf{x}, \mathbf{y}, t, s) = \langle v_i'(\mathbf{x}, t)c'(\mathbf{y}, s)\rangle_c$ satisfies

$$\begin{aligned}&\frac{\partial Q_{ic}(\mathbf{x}, \mathbf{y}, t, s)}{\partial s} + \frac{\partial}{\partial y_k}[\langle v_k(\mathbf{y}, s)\rangle_c Q_{ic}(\mathbf{x}, \mathbf{y}, t, s)] \\ &\quad = -\frac{\partial \Delta_{ikc}(\mathbf{y}, \mathbf{x}, s, t)}{\partial y_k} - \langle v_i'(\mathbf{x}, t)v_k'(\mathbf{y}, s)\rangle_c \frac{\partial \langle c(\mathbf{y}, s)\rangle_c}{\partial y_k} - \\ &\qquad - \langle v_i'(\mathbf{x}, t) f_j'(\mathbf{y}, s)\rangle_c \langle c(\mathbf{y}, s)\rangle_c,\end{aligned} \tag{59}$$

subject to the initial condition

$$Q_{ic}(\mathbf{x}, \mathbf{y}, t, s) = \langle v_i'(\mathbf{x}, t)C_0'(\mathbf{y})\rangle_c, \quad s = 0, \quad \mathbf{y} \in \Omega_\infty \tag{60}$$

which vanishes because we take the initial condition to be independent of velocity.

The moment $C_{fc}(\mathbf{x}, \mathbf{y}, t, s) = \langle f'(\mathbf{x}, t)c'(\mathbf{y}, s)\rangle_c$ is the solution of

$$\begin{aligned}&\frac{\partial C_{fc}(\mathbf{x}, \mathbf{y}, t, s)}{\partial s} + \frac{\partial}{\partial y_k}[\langle v_k(\mathbf{y}, s)\rangle_c C_{fc}(\mathbf{x}, \mathbf{y}, t, s)] \\ &\quad = -\frac{\partial \Gamma_{kc}(\mathbf{y}, \mathbf{x}, s, t)}{\partial y_k} - \langle v_k'(\mathbf{y}, s) f'(\mathbf{y}, s)\rangle_c \frac{\partial \langle c(\mathbf{y}, s)\rangle_c}{\partial y_k} - \\ &\qquad - \langle f'(\mathbf{x}, t) f'(\mathbf{y}, s)\rangle_c \langle c(\mathbf{y}, s)\rangle_c\end{aligned} \tag{61}$$

subject to the initial condition

$$C_{fc}(\mathbf{x}, \mathbf{y}, t, s) = \langle f'(\mathbf{x}, t)C_0'(\mathbf{y})\rangle_c \quad s = 0, \quad \mathbf{y} \in \Omega_\infty \tag{62}$$

which vanishes because we take f and C_0 to be independent.

The moment $C_{gc}(\mathbf{x}, \mathbf{y}, t, s) = \langle g'(\mathbf{x}, t)c'(\mathbf{y}, s)\rangle_c$ is the solution of

$$\begin{aligned}&\frac{\partial C_{gc}(\mathbf{x}, \mathbf{y}, t, s)}{\partial s} + \frac{\partial}{\partial y_k}[\langle v_k(\mathbf{y}, s)\rangle_c C_{gc}(\mathbf{x}, \mathbf{y}, t, s) \\ &\quad = -\frac{\partial \Theta_{kc}(\mathbf{y}, \mathbf{x}, s, t)}{\partial y_k} - \langle g'(\mathbf{x}, t) g'(\mathbf{y}, s)\rangle\end{aligned} \tag{63}$$

subject to the initial condition

$$C_{gc}(\mathbf{x}, \mathbf{y}, t, s) = \langle g'(\mathbf{x}, t)C_0'(\mathbf{y})\rangle \quad s = 0, \quad \mathbf{y} \in \Omega_\infty \tag{64}$$

which vanishes because we take g and C_0 to be independent.

In Appendix E, we derive the following explicit integral expressions for the fourth order mixed moments (46), (52), (55), and (58):

$$\begin{aligned}
\eta_{ijc}(\mathbf{x}, \mathbf{y}, t, s) &= \langle v_i'(\mathbf{x}, t) v_j'(\mathbf{y}, s) c'(\mathbf{y}, s) c'(\mathbf{x}, t) \rangle_c \\
&= \iint_0^t \langle \tilde{c}(\mathbf{x}, t | \chi, \tau) \tilde{v}_i'(\mathbf{x}, t) v_j'(\mathbf{y}, s) c'(\mathbf{y}, s) \rangle_c \frac{\partial}{\partial \chi_m} Q_{mc}(\chi, \tau) \, \mathrm{d}\tau \, \mathrm{d}\chi - \\
&\quad - \iint_0^t \langle \tilde{c}(\mathbf{x}, t | \chi, \tau) \tilde{v}_i'(\mathbf{x}, t) v_j'(\mathbf{y}, s) v_m'(\chi, \tau) c'(\mathbf{y}, s) \rangle_c \frac{\partial}{\partial \chi_m} \langle c(\chi, \tau) \rangle_c \, \mathrm{d}\tau \, \mathrm{d}\chi - \\
&\quad - \iint_0^t \langle \tilde{c}(\mathbf{x}, t | \chi, \tau) \tilde{v}_i'(\mathbf{x}, t) v_j'(\mathbf{y}, s) c'(\mathbf{y}, s) f'(\chi, \tau) \rangle_c \langle c(\chi, \tau) \rangle_c \, \mathrm{d}\tau \, \mathrm{d}\chi + \\
&\quad + \iint_0^t \langle \tilde{c}(\mathbf{x}, t | \chi, \tau) \tilde{v}_i'(\mathbf{x}, t) v_j'(\mathbf{y}, s) c'(\mathbf{y}, s) g'(\chi, \tau) \rangle_c \, \mathrm{d}\tau \, \mathrm{d}\chi + \\
&\quad + \int \langle \tilde{c}(\mathbf{x}, t | \chi, \tau) \tilde{v}_i'(\mathbf{x}, t) v_j'(\mathbf{y}, s) c'(\mathbf{y}, s) C_0'(\chi) \rangle_c \, \mathrm{d}\chi .
\end{aligned} \tag{65}$$

$$\begin{aligned}
\varepsilon_{ijkc}(\mathbf{x}, \mathbf{y}, t, s) &= \langle v_i'(\mathbf{x}, t) v_j'(\mathbf{y}, s) v_k'(\mathbf{x}, t) c'(\mathbf{x}, t) \rangle \\
&= \iint_0^t \langle \tilde{c}(\mathbf{x}, t | \chi, \tau) \tilde{v}_i'(\mathbf{x}, t) v_j'(\mathbf{y}, s) \tilde{v}_k'(\mathbf{x}, t) \rangle_c \frac{\partial}{\partial \chi_m} Q_{mc}(\chi, \tau) \, \mathrm{d}\tau \, \mathrm{d}\chi - \\
&\quad - \iint_0^t \langle \tilde{c}(\mathbf{x}, t | \chi, \tau) \tilde{v}_i'(\mathbf{x}, t) v_j'(\mathbf{y}, s) \tilde{v}_k'(\mathbf{x}, t) v_m'(\chi, \tau) \rangle_c \frac{\partial}{\partial \chi_m} \langle c(\chi, \tau) \rangle_c \, \mathrm{d}\tau \, \mathrm{d}\chi - \\
&\quad - \iint_0^t \langle \tilde{c}(\mathbf{x}, t | \chi, \tau) \tilde{v}_i'(\mathbf{x}, t) v_j'(\mathbf{y}, s) \tilde{v}_k'(\mathbf{x}, t) f'(\chi, \tau) \rangle_c \langle c(\chi, \tau) \rangle_c \, \mathrm{d}\tau \, \mathrm{d}\chi + \\
&\quad + \iint_0^t \langle \tilde{c}(\mathbf{x}, t | \chi, \tau) \tilde{v}_i'(\mathbf{x}, t) v_j'(\mathbf{y}, s) \tilde{v}_k'(\mathbf{x}, t) g'(\chi, \tau) \rangle_c \, \mathrm{d}\tau \, \mathrm{d}\chi .
\end{aligned} \tag{66}$$

$$\begin{aligned}
\kappa_{ikc}(\mathbf{x}, \mathbf{y}, t, s) &= \langle \tilde{v}_i'(\mathbf{x}, t) v_k'(\mathbf{x}, t) f'(\mathbf{y}, s) c'(\mathbf{x}, t) \rangle_c \\
&= \iint_0^t \langle \tilde{c}(\mathbf{x}, t | \chi, \tau) \tilde{v}_i'(\mathbf{x}, t) \tilde{v}_k'(\mathbf{x}, t) f'(\mathbf{y}, s) \rangle_c \frac{\partial}{\partial \chi_m} Q_{mc}(\chi, \tau) \, \mathrm{d}\tau \, \mathrm{d}\chi - \\
&\quad - \iint_0^t \langle \tilde{c}(\mathbf{x}, t | \chi, \tau) \tilde{v}_i'(\mathbf{x}, t) \tilde{v}_k'(\mathbf{x}, t) f'(\mathbf{y}, s) v_m'(\chi, \tau) \rangle_c \frac{\partial}{\partial \chi_m} \langle c(\chi, \tau) \rangle_c \, \mathrm{d}\tau \, \mathrm{d}\chi - \\
&\quad - \iint_0^t \langle \tilde{c}(\mathbf{x}, t | \chi, \tau) \tilde{v}_i'(\mathbf{x}, t) \tilde{v}_k'(\mathbf{x}, t) f'(\mathbf{y}, s) f'(\chi, \tau) \rangle_c \langle c(\chi, \tau) \rangle_c \, \mathrm{d}\tau \, \mathrm{d}\chi + \\
&\quad + \iint_0^t \langle \tilde{c}(\mathbf{x}, t | \chi, \tau) \tilde{v}_i'(\mathbf{x}, t) \tilde{v}_k'(\mathbf{x}, t) f'(\mathbf{y}, s) g'(\chi, \tau) \rangle_c \, \mathrm{d}\tau \, \mathrm{d}\chi + \\
&\quad + \int \langle \tilde{c}(\mathbf{x}, t | \chi, \tau) \tilde{v}_i'(\mathbf{x}, t) \tilde{v}_k'(\mathbf{x}, t) f'(\mathbf{y}, s) C_0'(\chi) \rangle_c \, \mathrm{d}\chi .
\end{aligned} \tag{67}$$

$$\vartheta_{ikc}(\mathbf{x}, \mathbf{y}, t, s) = \langle \tilde{v}_i'(\mathbf{x}, t) v_k'(\mathbf{x}, t) g'(\mathbf{y}, s) c'(\mathbf{x}, t) \rangle_c$$
$$= \iint_0^t \langle \tilde{c}(\mathbf{x}, t | \chi, \tau) \tilde{v}_i'(\mathbf{x}, t) \tilde{v}_k'(\mathbf{x}, t) g'(\mathbf{y}, s) \rangle_c \frac{\partial}{\partial \chi_m} Q_{mc}(\chi, \tau) \, d\tau \, d\chi -$$
$$- \iint_0^t \langle \tilde{c}(\mathbf{x}, t | \chi, \tau) \tilde{v}_i'(\mathbf{x}, t) \tilde{v}_k'(\mathbf{x}, t) g'(\mathbf{y}, s) v_m'(\chi, \tau) \rangle_c \frac{\partial}{\partial \chi_m} \langle c(\chi, \tau) \rangle_c \, d\tau \, d\chi -$$
$$- \iint_0^t \langle \tilde{c}(\mathbf{x}, t | \chi, \tau) \tilde{v}_i'(\mathbf{x}, t) \tilde{v}_k'(\mathbf{x}, t) g'(\mathbf{y}, s) f'(\chi, \tau) \rangle_c \langle c(\chi, \tau) \rangle_c \, d\tau \, d\chi +$$
$$+ \iint_0^t \langle \tilde{c}(\mathbf{x}, t | \chi, \tau) \tilde{v}_i'(\mathbf{x}, t) \tilde{v}_k'(\mathbf{x}, t) g'(\mathbf{y}, s) g'(\chi, \tau) \rangle_c \, d\tau \, d\chi. \tag{68}$$

The last two integrals in (66) and (67), as well as the first, second, third and last integral in (68), vanish due to our stipulation that g and C_0 be independent of each other and of $\mathbf{v}$ and f.

5. Recursive Conditional Second Moment Transport Approximations

Appendices E and F show how to formulate recursive approximations for conditional moments of transport to arbitrary order in σ. In this section we focus on equations of conditional variance–covariance for concentration to 2nd order in σ.

As shown in Appendix F, a perturbation expansion of (38)–(39) in terms of σ yields the following 2nd order approximation for the equation that governs the variance–covariance of concentration,

$$\frac{\partial C_{cc}^{(2)}(\mathbf{x}, \mathbf{y}, t, s)}{\partial t} + \frac{\partial}{\partial x_i} \left[\langle v_i^{(0)}(\mathbf{x}, t) \rangle_c C_{cc}^{(2)}(\mathbf{x}, \mathbf{y}, t, s) \right]$$
$$= -Q_{ic}^{(2)}(\mathbf{x}, \mathbf{y}, t, s) \frac{\partial \langle c^{(0)}(\mathbf{x}, t) \rangle_c}{\partial x_i} -$$
$$- C_{fc}^{(2)}(\mathbf{x}, \mathbf{y}, t, s) \langle c^{(0)}(\mathbf{x}, t) \rangle_c + C_{gc}^{(2)}(\mathbf{x}, \mathbf{y}, t, s) \tag{69}$$

subject to the initial condition

$$C_{cc}^{(2)}(\mathbf{x}, \mathbf{y}, t, s) = B_c^{(2)}(\mathbf{y}, \mathbf{x}, s) \quad t = 0, \quad \mathbf{x} \in \Omega_\infty \tag{70}$$

An expression for $B_c^{(2)}(\mathbf{y}, \mathbf{x}, s)$ is given in Appendix A. Appendix F includes recursive equations satisfied by the third mixed moments $\Delta_{ijc}(\mathbf{x}, \mathbf{y}, t, s)$, $\Gamma_{ic}(\mathbf{x}, \mathbf{y}, t, s)$ and $\Theta_{ic}(\mathbf{x}, \mathbf{y}, t, s)$ to arbitrary order of approximation. Fourth-order integral approximations for the fourth mixed moments (65)–(68) are listed in Appendix G.

Expansion of (59)–(60) yields, to 2nd order, the following equation for Q_{ic},

$$\frac{\partial Q_{ic}^{(2)}(\mathbf{x},\mathbf{y},t,s)}{\partial s}+\frac{\partial}{\partial y_k}\left[\langle v_k^{(0)}(\mathbf{y},s)\rangle_c Q_{ic}^{(2)}(\mathbf{x},\mathbf{y},t,s)\right]$$
$$=-[\langle v_i'(\mathbf{x},t)v_k'(\mathbf{y},s)\rangle_c]^{(2)}\frac{\partial\langle c^{(0)}(\mathbf{y},s)\rangle_c}{\partial y_k}-$$
$$-[\langle v_i'(\mathbf{x},t)f'(\mathbf{y},s)\rangle_c]^{(2)}\langle c^{(0)}(\mathbf{y},s)\rangle_c \tag{71}$$

subject to the initial condition

$$Q_{ic}^{(2)}(\mathbf{x},\mathbf{y},t,s)=\langle v_i'(\mathbf{x},t)C_0'(\mathbf{y})\rangle_c \quad s=0,\ \ \mathbf{y}\in\Omega_\infty \tag{72}$$

where the right hand side of (73) is considered known; in our case, it vanishes because we take $\mathbf{v}$ and C_0 to be independent.

Expansion of (61)–(62) gives a 2nd order equation for C_{fc},

$$\frac{\partial C_{fc}^{(2)}(\mathbf{x},\mathbf{y},t,s)}{\partial s}+\frac{\partial}{\partial y_k}\left[\langle v_k^{(0)}(\mathbf{y},s)\rangle_c C_{fc}^{(2)}(\mathbf{x},\mathbf{y},t,s)\right]$$
$$=-[\langle v_k'(\mathbf{y},s)f'(\mathbf{y},s)\rangle_c]^{(2)}\frac{\partial\langle c^{(0)}(\mathbf{y},s)\rangle_c}{\partial y_k}-$$
$$-[\langle f'(\mathbf{x},t)f'(\mathbf{y},s)\rangle_c]\langle c^{(0)}(\mathbf{y},s)\rangle_c \tag{73}$$

subject to the initial condition

$$C_{fc}^{(2)}(\mathbf{x},\mathbf{y},t,s)=\langle f'(\mathbf{x},t)C_0'(\mathbf{y})\rangle_c \quad s=0,\ \ \mathbf{y}\in\Omega_\infty, \tag{74}$$

where the right hand side of (75) is considered known. In our case, it vanishes because we take f and C_0 to be independent.

Finally, a 2nd order expansion of (63)–(64) gives for C_{gc}

$$\frac{\partial C_{gc}^{(2)}(\mathbf{x},\mathbf{y},t,s)}{\partial s}+\frac{\partial}{\partial y_k}\left[\langle v_k^{(0)}(\mathbf{y},s)\rangle_c C_{gc}^{(2)}(\mathbf{x},\mathbf{y},t,s)\right]$$
$$=-\langle g'(\mathbf{x},t)g'(\mathbf{y},s)\rangle \tag{75}$$

subject to the initial condition

$$C_{gc}^{(2)}(\mathbf{x},\mathbf{y},t,s)=\langle g'(\mathbf{x},t)C_0'(\mathbf{y})\rangle_c, \quad s=0,\ \ \mathbf{y}\in\Omega_\infty, \tag{76}$$

where the right hand side of (77) is considered known. In our case, it vanishes because we take g and C_0 to be independent.

6. Summary and Implications

Our paper is founded on the premise that all quantities (parameters, state variables, forcing terms) entering into the flow and transport equations are measurable, or can be deduced from measurements, on some consistent support scale ω at any point $(\mathbf{x}, t)$ in space time. These quantities are thus defined over a continuum without being necessarily associated with a representative elementary volume (REV). In other words, ω need not be an REV. In fact, quantities defined over ω can fluctuate randomly in space-time instead of varying smoothly, as is the case with quantities defined over REVs.

In particular, ω-scale hydraulic conductivity and kinematic porosity are taken to constitute multivariate random fields with well-defined geostatistical structures. These fields need not be (statistically) homogeneous but could, for example, be represented by random fractals such as those discussed by Di Federico and Neuman (1997, 1998a, b, 1999) and Neuman and Di Federico (1998). They could additionally be conditional on measurements, which would render them nonhomogeneous even if the corresponding unconditional fields were homogeneous.

Our second premise is that standard partial differential equations for flow and transport in porous continua apply on the scale of ω. These equations are typically based on Darcy's law and analogy to Fick's law, or variants thereof. As hydraulic conductivity is a random field, the ω-scale flow equation is stochastic. This, and the random nature of kinematic porosity, renders the flow velocity random. Introducing a random flow velocity into an ω-scale transport equation renders the latter stochastic, causing the corresponding ω-scale solute concentration and mass flux to vary randomly in space-time. Hence the solution of ω-scale flow and transport equations is inherently uncertain.

One way to solve stochastic, ω-scale flow and transport equations is numerically by means of conditional Monte Carlo simulations. Ideally, such simulations should be conducted on a computational grid with fine, ω-scale space discretization, so as to allow resolving all spatial frequencies of the corresponding ω-scale random functions that may manifest themselves within the flow domain. If these frequencies are low enough to extend across much of the flow domain, as is the case with random fractals, the simulations should be conducted on grids that are considerably larger than the flow domain, otherwise the low frequencies are not properly captured. When these computationally demanding conditions are satisfied, each conditional Monte Carlo simulation produces a realistically looking space-time image of ω-scale hydraulic head, hydraulic flux, velocity, solute concentration, and mass flux at a high resolution of scale ω. Each of these images is conditional on, and so makes full use of, the available data. As the images are equally likely, there is no way to decide which among them, if any, represents the real world most closely. The images are therefore nonunique. The usual approach is to generate a large number of images by running a large number of simulations, then computing corresponding sample statistics such as the average and variance of the results. For these statistics to approach their theoretical ensemble counterparts, the number of Monte Carlo simulations must be large, which adds to the computational burden. In addition to being computationally intensive, the Monte Carlo approach lacks well-established convergence criteria to tell when the number of simulations is large enough for the sample to become statistically meaningful.

When space-time images of ω-scale hydraulic head, hydraulic flux, velocity, solute concentration, and mass flux are averaged in the conditional Monte Carlo process, their random ω-scale fluctuations are smoothed out. The resulting conditional averages thus no longer form realistically looking images of possible reality, but smoothed versions of this reality. For example, the conditionally averaged concentration $\langle c(\mathbf{x}, t)\rangle_c$ no longer represents a likely realization of ω-scale concentration $c(\mathbf{x}, t)$ at the point $(\mathbf{x}, \mathrm{t})$ in space-time, but a conditional ensemble average (over all realizations) of $c(\mathbf{x}, t)$. In fact, $\langle c(\mathbf{x}, t)\rangle_c$ constitutes an optimum unbiased predictor of $c(\mathbf{x}, t)$ in the absence of knowledge that would allow one to discriminate between all equally likely but random, and therefore nonunique, realizations of $c(\mathbf{x}, t)$. Though the predictor $\langle c(\mathbf{x}, t)\rangle_c$ is smoother than is the realistic but random concentration $c(\mathbf{x}, t)$, it is nevertheless defined on the scale of ω. No spatial averaging, or upscaling, of any sort is used in the calculation of $\langle c(\mathbf{x}, t)\rangle_c$. The averaging is done solely in probability space, across realizations of quantities $c(\mathbf{x}, t)$ that are defined on the support scale ω. The conditional variance–covariance C_{cc} that one computes on the basis of Monte Carlo simulations represents a measure of prediction errors that one commits upon replacing the realistic but unknown random, ω-scale concentration $c(\mathbf{x}, t)$ by its smooth, deterministic predictor $\langle c(\mathbf{x}, t)\rangle_c$.

This paper was motivated by a desire to compute $\langle c(\mathbf{x}, t)\rangle_c$, and other conditional moment related to transport, deterministically rather than via a computationally intensive conditional Monte Carlo process. In particular, we developed integro-differential equations that are satisfied exactly by such moments, and presented recursive approximations that should allow one to compute these moments numerically. Our conditional moment equations include a local (dependent on only one point in space time) parameter $\langle \mathbf{v}(\mathbf{x}, t)\rangle_c$ that represents an optimum unbiased predictor of the velocity $\mathbf{v}(\mathbf{x}, t)$. As such, $\langle \mathbf{v}(\mathbf{x}, \mathrm{t})\rangle_c$ is not a spatially averaged or upscaled version of $\mathbf{v}(\mathbf{x}, t)$, but rather its probabilistic average that is defined on the scale ω. Whereas random realizations of the unknown velocity $\mathbf{v}(\mathbf{x}, t)$ provide realistically looking images of velocity with a spatial resolution of ω, $\langle \mathbf{v}(\mathbf{x}, t)\rangle_c$ provides a probabilistically smoothed average of these velocities with the same spatial resolution. Whereas the real but unknown concentration $c(\mathbf{x}, t)$ is advected at the unknown velocity $\mathbf{v}(\mathbf{x}, t)$, its predictor $\langle c(\mathbf{x}, t)\rangle_c$ is advected at the known velocity $\langle \mathbf{v}(\mathbf{x}, t)\rangle_c$. Since $\langle \mathbf{v}(\mathbf{x}, t)\rangle_c$ is conditional on data, it is information-dependent and thus nonunique. The same is true about nonlocal parameters in our conditional moment equations such as α_c, β_c, γ_c in (16)–(18), all of which are defined on the scale of ω.

Even though we took the real but unknown concentration $c(\mathbf{x}, \mathrm{t})$ to satisfy a purely advective transport equation, we found that its predictor $\langle c(\mathbf{x}, t)\rangle_c$ is controlled by an advective–dispersive process. The corresponding dispersive flux, $\mathbf{Q}_c(\mathbf{x}, t)$ defined in (15), depends on $\langle c(\mathbf{x}, t)\rangle_c$ and its gradient at points other than just $(\mathbf{x}, t)$, hence it is nonlocal in these quantities. As $\mathbf{Q}_c(\mathbf{x}, t)$ is not simply proportional to the gradient of $\langle c \rangle_c$ at $(\mathbf{x}, t)$, it does not obey Fick's law and causes the dispersion of $\langle c \rangle_c$ to be anomalous in this sense.

Only in a special case, where the first and third integrals in (15) are negligibly small in comparison to the second integral, and $\nabla_\chi \langle c(\chi, \tau) \rangle_c$ is more-or-less constant wherever the magnitude of $\beta_c(\mathbf{x}, t, \chi, \tau)$ is not vanishingly small, can the dispersive flux in (15) be approximated (localized) by the quasi-Fickian expression

$$\mathbf{Q}_c(\mathbf{x}, t) \approx -\mathbf{D}_c(\mathbf{x}, t) \nabla \langle c(\mathbf{x}, t) \rangle_c, \tag{77}$$

where

$$\mathbf{D}_c(\mathbf{x}, t) = \iint_0^t \beta_c(\mathbf{x}, t, \chi, \tau)\, \mathrm{d}\tau\, \mathrm{d}\chi, \tag{78}$$

is a dispersion tensor that is conditional on data, hence space-dependent and nonunique, which depends additionally on solute residence time (or, equivalently, mean travel distance or observation scale). Only in the limiting case of a statistically homogeneous (hence unconditional) velocity field is $\mathbf{D}$ independent of data and position in space. Even then, $\mathbf{D}$ remains a function of solute residence time for as long as the initial and current velocity remain autocorrelated. In a homogeneous velocity field, such correlation dies out asymptotically to render $\mathbf{D}$ constant and the dispersion of $\langle c \rangle_c$ Fickian. The localized dispersion process described by (78)–(79) has been implemented computationally by Zhang and Neuman (1995a, b, c, d).

We found that $\langle c \rangle_c$ disperses even if c does not. It is thus clear that this dispersion does not represent a physical spreading process but a loss of information. If $\mathbf{v}$ was perfectly known, $\mathbf{Q}_c$ and $\mathbf{D}_c$ would be identically equal to zero, and there would be no dispersion. The fewer data there are to define $\mathbf{v}$, or the poorer these data become, the smoother is the predictor $\langle \mathbf{v} \rangle_c$ and the larger is the corresponding random prediction error $\mathbf{v}'$. This causes the dispersion of $\langle c \rangle_c$ to increase, reflecting a corresponding loss of information about c that is contained in $\langle c \rangle_c$. A similar loss of information occurs with time, hence the time-dependence of $\mathbf{Q}_c$ and $\mathbf{D}_c$. What disperses here is not the solute (whose spread could be accounted for by adding a local dispersion coefficient to our transport equation, as was done by Zhang and Neuman, 1996) but information about the space-time distribution of its concentration.

Since (78)–(79) is associated with a quasi-Fickian advection–dispersion process, it is clear that force-fitting a traditional Fickian advection–dispersion model to measured ω-scale concentration data, c, implies treating these data as if they represented $\langle c \rangle_c$. Since this quasi-Fickian process is associated with an advective velocity $\langle \mathbf{v} \rangle_c$, such a fit further implies replacing the true ω-scale velocity $\mathbf{v}$ by its conditional smooth predictor. The fit returns a dispersion tensor that is closely related to $\mathbf{D}_c$, hence it should be of no surprise that its magnitude is large compared to laboratory-determined values, and that it appears to increase with residence time or travel distance, at least until the initial and current velocities become decorrelated. The latter may happen in statistically homogeneous media, but not in hierarchical media with multiple scales of heterogeneity (Neuman, 1990, 1995, 1997; Di Federico and Neuman, 1998b; Neuman and Di Federico, 1998), in which $\mathbf{D}_c$ may continue to grow indefinitely.

Since there is always uncertainty about local- or ω-scale medium heterogeneities, a deterministic analysis of transport is never warranted unless the advection–dispersion model is viewed and interpreted in the manner just described. This means that the computed concentration and mass flux are recognized to represent not actual but smooth predicted values with which there is associated a quantifiable error of prediction and smoothing, and that the computed concentrations are recognized to potentially spread, or disperse, to a much greater extent than do their real but unknown counterparts (recognizing that this enhanced spread, or dispersion, does not generally pertain to solute mass but to information about the space-time distribution of the same mass).

Only in quasi-ergodic situations can such enhanced dispersion be interpreted to imply physical dispersion of solute mass as measured on a scale of ω. Such situations arise as the mean travel distance becomes large enough for a plume to encounter (sample) heterogeneities of all relevant scales in the longitudinal direction of flow, and when the source of contamination is wide enough so that the plume can sample all such heterogeneities in the transverse direction. An extreme example of nonergodic transport is that of an imaginary solute particle of infinitesimal volume which never diffuses, disperses or dilutes. If the particle has unit mass normalized by porosity then its concentration, predicted deterministically with a relatively large 'field-scale' dispersion coefficient, merely represents the probability of finding this particle in the immediate vicinity of any given point in space-time.

We (Guadagnini and Neuman, 1997, 1998, 1999a, b) have shown previously how to solve conditional moment equations of steady state flow by finite elements. We have pointed out that since conditional moments are smooth in comparison with corresponding random functions defined on a scale ω, one should be able to resolve them on a relatively coarse computational grid with elements larger than ω. In this paper, we laid the theoretical groundwork for the numerical solution of conditional moment equations associated with transport, by developing exact integro-differential expressions for second conditional moments, and recursive approximations for all conditional moments, of advective transport in a manner that complements the earlier work of Neuman (1993). The obvious next step is to do the same for advective–dispersive transport in a way that complements the more recent work of Zhang and Neuman (1996), then implement our recursive transport equations numerically.

Appendix A

For an arbitrary initial concentration $c(\mathbf{x}, 0) = C_0(\mathbf{x})$ that is prescribed in a statistically independent manner, Equation (B7) of Neuman (1993) becomes

$$c(\mathbf{x}, t) = \iint_0^t g(\chi, \tau)\delta(\chi - \tilde{\chi})\,\mathrm{d}\tau\,\mathrm{d}\chi + \int C_0(\chi)\delta(\chi - \tilde{\chi})\,\mathrm{d}\chi. \tag{A.1}$$

This implies that

$$\begin{aligned} c'(\mathbf{x}, t) = & \iint_0^t \delta(\chi - \tilde{\chi})\nabla \cdot \mathbf{Q}_c(\chi, \tau)\,\mathrm{d}\tau\,\mathrm{d}\chi - \\ & - \iint_0^t \delta(\chi - \tilde{\chi})\mathbf{v}'^T(\chi, \tau)\nabla\langle c(\chi, \tau)\rangle_c\,\mathrm{d}\tau\,\mathrm{d}\chi - \\ & - \iint_0^t \delta(\chi - \tilde{\chi})f'(\chi, \tau)\langle c(\chi, \tau)\rangle_c\mathrm{d}\tau\,\mathrm{d}\chi + \\ & + \iint_0^t \delta(\chi - \tilde{\chi})g'(\chi, \tau)\,\mathrm{d}\tau\,\mathrm{d}\chi + \\ & + \int \delta(\chi - \tilde{\chi})C_0'(\chi)\,\mathrm{d}\chi, \end{aligned} \tag{A.2}$$

where the last integral, containing a zero mean random fluctuation in initial concentration, does not appear in Neuman's paper. Multiplying throughout by $\mathbf{v}'(\mathbf{y}, s)$, taking conditional expectation and evaluating the latter at $(\mathbf{y}, s) = (\mathbf{x}, t)$, while taking into account the statistical independence of $\mathbf{v}$ on C_0, yields (B15) of Neuman (1993). Expressions (C1), (C2), (C3) of Neuman (1993) remain unaltered, but his (C4) gains an additional term,

$$\int \langle \delta(\chi - \tilde{\chi})c'(\mathbf{y}, s)C_0'(\chi)\rangle\,\mathrm{d}\chi,$$

which is generally nonzero.

Multiplying (A.2) by $C_0'(\mathbf{s})$ and taking conditional expectation gives the following explicit equation for $B_c(\mathbf{x}, \mathbf{y}, t) = \langle c'(\mathbf{x}, t)C_0'(\mathbf{y})\rangle_c$,

$$\begin{aligned} B_c(\mathbf{x}, \mathbf{y}, t) &= \int \langle \delta(\chi - \tilde{\chi})C_0'(\mathbf{y})C_0'(\chi)\rangle_c\,\mathrm{d}\chi \\ &= \int \langle \tilde{c}(\mathbf{x}, t|\chi, \tau)C_0'(\mathbf{y})C_0'(\chi)\rangle_c\,\mathrm{d}\chi. \end{aligned} \tag{A.3}$$

Expanding in powers of σ yields for $l \geqslant 2$

$$B_c^{(l)}(\mathbf{x}, \mathbf{y}, t) = \int \langle \tilde{c}^{(l)}(\mathbf{x}, t|\chi, \tau)\rangle_c C_{C0}(\mathbf{y}, \chi)\,\mathrm{d}\chi \tag{A.4}$$

where

$$C_{C0}(\mathbf{y}, \chi) = \langle C_0'(\mathbf{y})C_0'(\chi)\rangle.$$

To second order,

$$B_c^{(2)}(\mathbf{x}, \mathbf{y}, t) = \int \langle \tilde{c}^{(0)}(\mathbf{x}, t|\chi, \tau)\rangle_c C_{C0}(\mathbf{y}, \chi)\,\mathrm{d}\chi. \tag{A.5}$$

Appendix B

The fundamental advective concentration $\tilde{c}(\mathbf{x}, t|\chi, \tau)$ is defined as the solution of

$$\frac{\partial \tilde{c}(\mathbf{x}, t|\chi, \tau)}{\partial t} + \nabla_x \cdot [\mathbf{v}(\mathbf{x}, t)\tilde{c}(\mathbf{x}, t|\chi, \tau)] = \delta(\mathbf{x} - \chi)\delta(t - \tau). \tag{B.1}$$

We express it as the sum of conditional mean and fluctuation,

$$\tilde{c}(\mathbf{x}, t|\chi, \tau) = \langle \tilde{c}(\mathbf{x}, t|\chi, \tau)\rangle_c + \tilde{c}'(\mathbf{x}, t|\chi, \tau) \quad \langle \tilde{c}'(\mathbf{x}, t|\chi, \tau)\rangle_c \equiv 0. \tag{B.2}$$

By means of the auxiliary function G in (12)–(14), one can follow the procedure of Neuman (1993; Appendix B) to obtain (19), the equivalent of his (B15). Rewriting (B.1) in terms of (5) and (B.2), post-multiplying throughout by $\mathbf{v}'^{\mathrm{T}}(\mathbf{y}, s)$ evaluating the latter at point $(\mathbf{x}, t)$, and taking conditional mean leads to (20). Substituting (B.2) into (16)–(18), and noting that the first term in (19) is at least of order 4, results immediately in (28)–(30).

Appendix C

Expansion of (20) in terms of σ yields the lth order approximation

$$\frac{\partial \alpha_c^{(l)T}(\mathbf{x}, t|\chi, \tau)\rangle_c}{\partial t} + \nabla^T \left\{ \sum_{n=0}^{l} \langle \mathbf{v}(\mathbf{x}, t)^{(n)}\rangle_c \alpha_c^{(l-n)T}(\mathbf{x}, t|\chi, \tau) + \right.$$
$$+ \langle \tilde{c}'(\mathbf{x}, t/\chi, \tau)\mathbf{v}'(\mathbf{x}, t)\tilde{\mathbf{v}}'^T(\mathbf{x}, t)\rangle_c^{(l)} +$$
$$\left. + \sum_{n=0}^{l} \langle \tilde{c}^{(n)}(\mathbf{x}, t|\chi, \tau)\rangle_c [\langle \mathbf{v}'(\mathbf{x}, t)\tilde{\mathbf{v}}'^T(\mathbf{x}, t)\rangle_c]^{(l-n)} \right\} = 0 \quad l \geqslant 2 \tag{C.1}$$

subject to initial condition

$$\alpha_c^{(l)}(\mathbf{x}, 0|\chi, \tau) = 0. \tag{C.2}$$

Setting $l = 2$ gives (33)–(34). The same applies to divergence-free $\mathbf{v}$.

Appendix D

Multiplying (36)–(37) by $c'(\mathbf{y}, s)$ and taking conditional mean gives (38)–(39).

Writing (36)–(37) in terms of $(\mathbf{y}, s)$, multiplying by $v_i'(\mathbf{x}, t)c'(\mathbf{x}, t)$ and taking conditional mean leads to (44)–(45). Multiplying instead by $v_i'(\boldsymbol{\xi}, \sigma)v_j'(\mathbf{y}, s)$, taking conditional mean, and evaluating $v_i'(\boldsymbol{\xi}, \sigma)$ at $(\mathbf{x}, t)$ yields (50)–(51). Doing the same with $v_i'(\boldsymbol{\xi}, \sigma) f'(\mathbf{y}, s)$ results in (53)–(54), while doing so with $v_i'(\boldsymbol{\xi}, \sigma) g'(\mathbf{y}, s)$ leads to (56)–(57).

Writing (36)–(37) in terms of $(\mathbf{y}, s)$, multiplying by $v_i'(\mathbf{x}, t)$, and taking conditional mean gives (59)–(60). Multiplying instead by $f'(\mathbf{x}, t)$ results in (61)–(62), while doing the same with $g'(\mathbf{x}, t)$ yields (63)–(64).

Appendix E

Multiplying (A.2) by $v_i'(\boldsymbol{\xi}, \sigma) v_j'(\mathbf{y}, s) c'(\mathbf{y}, s)$, taking conditional ensemble mean, and evaluating $v_i'(\boldsymbol{\xi}, \sigma)$ at $(\mathbf{x}, t)$ yields (65). Multiplying instead by $v_i'(\boldsymbol{\xi}, \sigma) v_j'(\mathbf{y}, s) v_k'(\mathbf{y}^*, s^*)$, taking conditional ensemble mean, and evaluating $v_i'(\boldsymbol{\xi}, \sigma)$ and $v_k'(\mathbf{y}^*, s^*)$ at $(\mathbf{x}, t)$ results in (66). Multiplying (A.2) by $v_i'(\boldsymbol{\xi}, \sigma) v_k'(\mathbf{y}, s) f'(\mathbf{y}, s)$, taking conditional ensemble mean, and evaluating $v_i'(\boldsymbol{\xi}, \sigma)$ and $v_k'(\mathbf{y}, s)$ at $(\mathbf{x}, t)$ gives (67). Doing the same with $v_i'(\boldsymbol{\xi}, \sigma) v_k'(\mathbf{y}, s) g'(\mathbf{y}, s)$, results in (68).

Appendix F

Expansion of (38)–(39) in terms of σ yields the following lth order approximations,

$$\begin{aligned} &\frac{\partial C_{cc}^{(l)}(\mathbf{x}, \mathbf{y}, t, s)}{\partial t} + \frac{\partial}{\partial x_i}\left[\sum_{n=0}^{l} \langle v_i^{(n)}(\mathbf{x}, t)\rangle_c C_{cc}^{(l-n)}(\mathbf{x}, \mathbf{y}, t, s)\right] \\ &\quad = -\frac{\partial R_{ic}^{(l)}(\mathbf{x}, \mathbf{y}, t, s)}{\partial x_i} - \\ &\quad - \sum_{n=0}^{l} Q_{ic}^{(l)}(\mathbf{x}, \mathbf{y}, t, s) \frac{\partial \langle c^{(l-n)}(\mathbf{x}, t)\rangle_c}{\partial x_i} - \\ &\quad - \sum_{n=0}^{l} C_{fc}^{(l)}(\mathbf{x}, \mathbf{y}, t, s) \langle c^{(l-n)}(\mathbf{x}, t)\rangle_c + \\ &\quad + C_{gc}^{(l)}(\mathbf{x}, \mathbf{y}, t, s) \quad l \geqslant 2 \end{aligned} \tag{F.1}$$

subject to the initial condition

$$C_{cc}^{(l)}(\mathbf{x}, \mathbf{y}, t, s) = B_c^{(l)}(\mathbf{y}, \mathbf{x}, s) \quad t = 0, \;\; \mathbf{x} \in \Omega_\infty. \tag{F.2}$$

Here, R_{ic} is of order 3 and higher, while Q_{ic}, C_{fc}, and C_{gc} are of order 2 or higher. Setting $l = 2$ gives (70)–(71).

Expansion of (44)–(45) yields the lth order approximations

$$\frac{\partial R_{ic}^{(l)}(\mathbf{x},\mathbf{y},t,s)}{\partial s}+\frac{\partial}{\partial y_j}\left[\sum_{n=0}^{l}\langle v_j^{(n)}(\mathbf{y},s)\rangle_c R_{ic}^{(l-n)}(\mathbf{x},\mathbf{y},t,s)\right]$$
$$=-\sum_{n=0}^{l}\frac{\partial Q_{jc}^{(n)}(\mathbf{y},s)}{\partial y_j}Q_{ic}^{(n-l)}(\mathbf{x},t)-\frac{\partial \eta_{ijc}^{(l)}(\mathbf{x},\mathbf{y},t,s)}{\partial y_j}-$$
$$-\sum_{n=0}^{l}\Delta_{ijc}^{(n)}(\mathbf{x},\mathbf{y},t,s)\frac{\partial\langle c^{(n-l)}(\mathbf{y},s)\rangle_c}{\partial y_j}-$$
$$-\sum_{n=0}^{l}\Gamma_{ic}^{(n)}(\mathbf{x},\mathbf{y},t,s)\langle c^{(n-l)}(\mathbf{y},s)\rangle_c+\Theta_{ic}^{(l)}(\mathbf{x},\mathbf{y},t,s)\quad l\geqslant 3 \tag{F.3}$$

subject to the initial condition

$$R_{ic}^{(l)}(\mathbf{x},\mathbf{y},t,s)=[\langle v_i'(\mathbf{x},t)c'(\mathbf{x},t)C_0'(\mathbf{y})\rangle_c]^{(l)}\quad s=0,\quad \mathbf{y}\in\Omega_\infty. \tag{F.4}$$

Here $\eta_{ijc}(\mathbf{x},\mathbf{y},t,s)$ is of order 4 or higher (an expansion of this term is given in Appendix G), while $\Delta_{ijc}(\mathbf{x},\mathbf{y},t,s)$, $\Gamma_{ic}(\mathbf{x},\mathbf{y},t,s)$, and $\Theta_{ic}(\mathbf{x},\mathbf{y},t,s)$ are of order 3 or higher.

Expansion of (50)–(51) results in

$$\frac{\partial \Delta_{ijc}^{(l)}(\mathbf{x},\mathbf{y},t,s)}{\partial t}+\frac{\partial}{\partial x_k}\left[\sum_{n=0}^{l}\langle v_k^{(n)}(\mathbf{x},t)\rangle_c\Delta_{ijc}^{(l-n)}(\mathbf{x},\mathbf{y},t,s)\right]$$
$$=-\frac{\partial \varepsilon_{ijkc}^{(l)}(\mathbf{x},\mathbf{y},t,s)}{\partial x_k}-\sum_{n=0}^{l}\frac{\partial Q_{kc}^{(n)}(\mathbf{x},t)}{\partial x_k}[\langle\tilde{v}_i'(\mathbf{x},t)v_j'(\mathbf{y},s)\rangle_c]^{(n-l)}-$$
$$-\sum_{n=0}^{l}[\langle v_k'(\mathbf{x},t)\rangle\tilde{v}_i'(\mathbf{x},t)v_j'(\mathbf{y},s)\rangle_c]^{(n)}\frac{\partial\langle c^{(l-n)}(\mathbf{x},t)\rangle_c}{\partial x_k}-$$
$$-\sum_{n=0}^{l}[\langle\tilde{v}_i'(\mathbf{x},t)v_j'(\mathbf{y},s)f'(\mathbf{x},t)\rangle_c]^{(n)}\langle c^{(l-n)}(\mathbf{x},t)\rangle_c\quad l\geqslant 3 \tag{F.5}$$

subject to the initial condition

$$\Delta_{ijc}^{(l)}(\mathbf{x},\mathbf{y},t,s)=[\langle\tilde{v}_i'(\mathbf{x},t)v_j'(\mathbf{y},s)C_0'(\mathbf{x})\rangle_c]^{(l)}\quad t=0,\ \ \mathbf{x}\in\Omega_\infty \tag{F.6}$$

which vanishes because we consider the initial concentration to be prescribed independently of velocity. Here $\varepsilon_{ijc}(\mathbf{x},\mathbf{y},t,s)$ is of order 4 or higher (an expansion of this term is given in Appendix G), and all mixed moments of $\mathbf{v}$ and f are of order 3 or higher.

Expansion of (53)–(54) leads to

$$\frac{\partial \Gamma_{ic}^{(l)}(\mathbf{x},\mathbf{y},t,s)}{\partial t} + \frac{\partial}{\partial x_k}\left[\sum_{n=0}^{l}\langle v_k^{(n)}(\mathbf{x},t)\rangle_c \Gamma_{ic}^{(n-l)}(\mathbf{x},\mathbf{y},t,s)\right]$$
$$= -\frac{\partial \kappa_{ikc}^{(l)}(\mathbf{x},\mathbf{y},t,s)}{\partial x_k} - \sum_{n=0}^{l}\frac{\partial Q_{kc}^{(n)}(\mathbf{x},t)}{\partial x_k}[\langle \tilde{v}_i'(\mathbf{x},t)f'(\mathbf{y},s)\rangle_c]^{(l-n)} -$$
$$-\sum_{n=0}^{l}[\langle v_k'(\mathbf{x},t)\tilde{v}_i'(\mathbf{x},t)f'(\mathbf{y},s)\rangle_c]^{(n)}\frac{\partial \langle c^{(l-n)}(\mathbf{x},t)\rangle_c}{\partial x_k} -$$
$$-\sum_{n=0}^{l}[\langle \tilde{v}_i'(\mathbf{x},t)f'(\mathbf{y},s)f'(\mathbf{x},t)\rangle_c]^{(n)}\langle c^{(l-n)}(\mathbf{x},t)\rangle_c -$$
$$-[\langle \tilde{v}_i'(\mathbf{x},t)f'(\mathbf{y},s)g'(\mathbf{x},t)\rangle_c]^{(l)}, \quad l \geqslant 3 \tag{F.7}$$

subject to the initial condition

$$\Gamma_{ic}^{(l)}(\mathbf{x},\mathbf{y},t,s) = [\langle \tilde{v}_i'(\mathbf{x},t)f'(\mathbf{y},s)C_0'(\mathbf{x})\rangle_c]^{(l)} \quad t=0, \quad \mathbf{x}\in\Omega_\infty \tag{F.8}$$

which vanishes because we consider the initial concentration to be statistically independent $\mathbf{v}$ and f. Here $\kappa_{ijc}(\mathbf{x},\mathbf{y},t,s)$ is of order 4 or higher (an expansion of this term is given in Appendix G), and all mixed moments of $\mathbf{v}$ and f are of order 2 or higher.

Expansion of (56)–(57) gives

$$\frac{\partial \Theta_{ic}^{(l)}(\mathbf{x},\mathbf{y},t,s)}{\partial t} + \frac{\partial}{\partial x_k}\left[\sum_{n=0}^{l}\langle v_k^{(n)}(\mathbf{x},t)\rangle_c \Theta_{ic}^{(l-n)}(\mathbf{x},\mathbf{y},t,s)\right]$$
$$= -\frac{\partial \vartheta_{ikc}^{(l)}(\mathbf{x},\mathbf{y},t,s)}{\partial x_k} - \sum_{n=0}^{l}\frac{\partial Q_{kc}^{(n)}(\mathbf{x},t)}{\partial x_k}[\langle \tilde{v}_i'(\mathbf{x},t)g'(\mathbf{x},t)g'(\mathbf{y},s)\rangle_c]^{(l-n)} -$$
$$-\sum_{n=0}^{l}[\langle v_k'(\mathbf{x},t)\tilde{v}_i'(\mathbf{x},t)g'(\mathbf{y},s)\rangle_c]^{(n)}\frac{\partial \langle c^{(l-n)}(\mathbf{x},t)\rangle_c}{\partial x_k} -$$
$$-\sum_{n=0}^{l}[\langle \tilde{v}_i'(\mathbf{x},t)g'(\mathbf{y},s)f'(\mathbf{x},t)\rangle_c]^{(n)}\langle c^{(l-n)}(\mathbf{x},t)\rangle_c -$$
$$-[\langle \tilde{v}_i'(\mathbf{x},t)g'(\mathbf{y},s)g'(\mathbf{x},t)\rangle_c]^{(l)} \quad l \geqslant 3, \tag{F.9}$$

subject to the initial condition

$$\Theta_{ic}^{(l)}(\mathbf{x},\mathbf{y},t,s) = [\langle \tilde{v}_i'(\mathbf{x},t)C_0'(\mathbf{x})g'(\mathbf{y},s)\rangle_c]^{(l)} \quad t=0, \quad \mathbf{x}\in\Omega_\infty \tag{F.10}$$

which vanishes because we consider the initial concentration to be statistically independent $\mathbf{v}$ and g. Here $\vartheta_{ijc}(\mathbf{x},\mathbf{y},t,s)$ is of order 4 or higher (an expansion of this term is given in Appendix G), and all mixed moments of $\mathbf{v}$ and f are of order 2 or higher.

Expansion of (59)–(60) yields

$$\frac{\partial Q_{ic}^{(l)}(\mathbf{x},\mathbf{y},t,s)}{\partial s}+\frac{\partial}{\partial y_k}\left[\sum_{n=0}^{l}\langle v_k^{(n)}(\mathbf{y},s)\rangle_c Q_{ic}^{(l-n)}(\mathbf{x},\mathbf{y},t,s)\right]$$
$$=-\frac{\partial\Delta_{ikc}^{(l)}(\mathbf{y},\mathbf{x},s,t)}{\partial y_k}-\sum_{n=0}^{l}[\langle v_i'(\mathbf{x},t)v_k'(\mathbf{y},s)\rangle_c]^{(n)}\frac{\partial\langle c^{(l-n)}(\mathbf{y},s)\rangle_c}{\partial y_k}-$$
$$-\sum_{n=0}^{l}[\langle v_i'(\mathbf{x},t)f'(\mathbf{y},s)\rangle_c]^{(n)}\langle c^{(l-n)}(\mathbf{y},s)\rangle_c,\quad l\geqslant 2 \tag{F.11}$$

subject to the initial condition

$$Q_{ic}^{(l)}(\mathbf{x},\mathbf{y},t,s)=[\langle v_i'(\mathbf{x},t)C_0'(\mathbf{y})\rangle_c]^{(l)}\quad s=0,\ \ \mathbf{y}\in\Omega_\infty \tag{F.12}$$

which vanishes because $\mathbf{v}$ and C_0 are taken to be independent. Setting $l=2$ reduces this to (72)–(73).

Expansion of (61)–(62) gives

$$\frac{\partial C_{fc}^{(2)}(\mathbf{x},\mathbf{y},t,s)}{\partial s}+\frac{\partial}{\partial y_k}[\langle v_k^{(0)}(\mathbf{y},s)\rangle_c C_{fc}^{(2)}(\mathbf{x},\mathbf{y},t,s)]$$
$$=-[\langle v_k'(\mathbf{y},s)f'(\mathbf{y},s)\rangle_c]^{(2)}\frac{\partial\langle c^{(0)}(\mathbf{y},s)\rangle_c}{\partial y_k}-$$
$$-[\langle f'(\mathbf{x},t)f'(\mathbf{y},s)\rangle_c]^{(2)}\langle c^{(0)}(\mathbf{y},s)\rangle_c,\quad l=2 \tag{F.13}$$

and

$$\frac{\partial C_{fc}^{(l)}(\mathbf{x},\mathbf{y},t,s)}{\partial s}+\frac{\partial}{\partial y_k}\left[\sum_{n=0}^{l}\langle v_k^{(n)}(\mathbf{y},s)\rangle_c C_{fc}^{(l-n)}(\mathbf{x},\mathbf{y},t,s)\right]$$
$$=-\frac{\partial\Gamma_{kc}^{(l)}(\mathbf{y},\mathbf{x},s,t)}{\partial y_k}-\sum_{n=0}^{l}[\langle v_k'(\mathbf{y},s)f'(\mathbf{y},s)\rangle_c]^{(n)}\frac{\partial\langle c^{(l-n)}(\mathbf{y},s)\rangle_c}{\partial y_k}-$$
$$-\sum_{n=0}^{l}[\langle f'(\mathbf{x},t)f'(\mathbf{y},s)\rangle_c]^{(n)}\langle c^{(l-n)}(\mathbf{y},s)\rangle_c,\quad l\geqslant 2 \tag{F.14}$$

subject to the initial condition

$$C_{fc}^{(l)}(\mathbf{x},\mathbf{y},t,s)=[\langle f'(\mathbf{x},t)C_0'(\mathbf{y})\rangle_c]^{(l)}\quad s=0,\ \ \mathbf{y}\in\Omega_\infty,\ \ l\geqslant 2 \tag{F.15}$$

which vanishes because f and C_0 are taken to be independent.

Expansion of (63)–(64) results in

$$\frac{\partial C^{(2)}_{gc}(\mathbf{x},\mathbf{y},t,s)}{\partial s} + \frac{\partial}{\partial y_k}[\langle v^{(0)}_k(\mathbf{y},s)\rangle_c C^{(2)}_{gc}(\mathbf{x},\mathbf{y},t,s)] = -\langle g'(\mathbf{x},t)g'(\mathbf{y},s)\rangle \quad l=2 \tag{F.16}$$

and

$$\frac{\partial C^{(l)}_{gc}(\mathbf{x},\mathbf{y},t,s)}{\partial s} + \frac{\partial}{\partial y_k}\left[\sum_{n=0}^{l}\langle v^{(n)}_k(\mathbf{y},s)\rangle_c C^{(l-n)}_{gc}(\mathbf{x},\mathbf{y},t,s)\right] = -\frac{\partial \Theta^{(l)}_{kc}(\mathbf{y},\mathbf{x},s,t)}{\partial y_k} \quad l>2 \tag{F.17}$$

subject to the initial condition

$$C^{(l)}_{gc}(\mathbf{x},\mathbf{y},t,s) = [\langle g'(\mathbf{x},t)C'_0(\mathbf{y})\rangle_c]^{(l)} \quad s=0,\ \mathbf{y}\in\Omega_\infty,\ l\geqslant 2 \tag{F.18}$$

which vanishes because g and C_0 are taken to be independent.

Appendix G

Expansion of (65) in terms of σ leads to the following 4th order approximation (the first integral on the right hand side of (65) is of order 5 or higher)

$$\begin{aligned}\eta^{(4)}_{ijc}(\mathbf{x},\mathbf{y},t,s) &= [\langle v'_i(\mathbf{x},t)v'_j(\mathbf{y},s)c'(\mathbf{y},s)c'(\mathbf{x},t)\rangle_c]^{(4)}\\ &= -\iint_0^t \langle\tilde{c}^{(0)}(\mathbf{x},t|\chi,\tau)\rangle_c[\langle\tilde{v}'_i(\mathbf{x},t)v'_i(\mathbf{y},s)v'_m(\chi,\tau)c'(\mathbf{y},s)\rangle_c]^{(4)}\times\\ &\quad\times\frac{\partial}{\partial\chi_m}\langle c^{(0)}(\chi,\tau)\rangle_c\,\mathrm{d}\tau\,\mathrm{d}\chi -\\ &\quad -\iint_0^t \langle\tilde{c}^{(0)}(\mathbf{x},t|\chi,\tau)\rangle_c[\langle\tilde{v}'_i(\mathbf{x},t)v'_j(\mathbf{y},s)c'(\mathbf{y},s)f'(\chi,\tau)\rangle_c]^{(4)}\times\\ &\quad\times\langle c^{(0)}(\chi,\tau)\rangle_c\,\mathrm{d}\tau\,\mathrm{d}\chi +\\ &\quad +\iint_0^t \langle\tilde{c}^{(0)}(\mathbf{x},t|\chi,\tau)\rangle_c[\langle\tilde{v}'_i(\mathbf{x},t)v'_j(\mathbf{y},s)c'(\mathbf{y},s)g'(\chi,\tau)\rangle_c]^{(4)}\,\mathrm{d}\tau\,\mathrm{d}\chi +\\ &\quad +\int\langle\tilde{c}^{(0)}(\mathbf{x},t|\chi,\tau)\rangle_c[\langle\tilde{v}'_i(\mathbf{x},t)v'_j(\mathbf{y},s)c'(\mathbf{y},s)C'_0(\chi)\rangle_c]^{(4)}\mathrm{d}\chi.\end{aligned} \tag{G.1}$$

The fourth mixed moments under the integrals can be obtained in the manner of Appendix E. In particular, the last integral is obtained upon writing (A.2) in terms of (**y**, s), multiplying by $v'_i(\mathbf{x},t)v'_j(\boldsymbol{\xi},\sigma)C'_0(\boldsymbol{\xi})$, taking conditional ensemble mean, and evaluating $v'_j(\boldsymbol{\xi},\sigma)$ at $(\mathbf{y},s)$ to yield

$$\begin{aligned}
&\langle\tilde{v}'_i(\mathbf{x},t)\tilde{v}'_j(\mathbf{y},s)C'_0(\boldsymbol{\xi})c'(\mathbf{y},s)\rangle_c\\
&\quad=\iint_0^t\langle\tilde{c}(\mathbf{y},s|\chi,\tau)\tilde{v}'_i(\mathbf{x},t)\tilde{v}'_j(\mathbf{y},s)C'_0(\boldsymbol{\xi})\rangle_c\frac{\partial}{\partial\chi_m}Q_{mc}(\chi,\tau)\,\mathrm{d}\tau\,\mathrm{d}\chi-\\
&\qquad-\iint_0^t\langle\tilde{c}(\mathbf{y},s|\chi,\tau)\tilde{v}'_i(\mathbf{x},t)\tilde{v}'_j(\mathbf{y},s)C'_0(\boldsymbol{\xi})v'_m(\chi,\tau)\rangle_c\frac{\partial}{\partial\chi_m}\langle c(\chi,\tau)\rangle_c\,\mathrm{d}\tau\,\mathrm{d}\chi+\\
&\qquad+\iint_0^t\langle\tilde{c}(\mathbf{y},s|\chi,\tau)\tilde{v}'_i(\mathbf{x},t)\tilde{v}'_j(\mathbf{y},s)C'_0(\boldsymbol{\xi})f'(\chi,\tau)\rangle_c\langle c(\chi,\tau)\rangle_c\,\mathrm{d}\tau\,\mathrm{d}\chi-\\
&\qquad-\iint_0^t\langle\tilde{c}(\mathbf{y},s|\chi,\tau)\tilde{v}'_i(\mathbf{x},t)\tilde{v}'_j(\mathbf{y},s)C'_0(\boldsymbol{\xi})g'(\chi,\tau)\rangle_c\,\mathrm{d}\tau\,\mathrm{d}\chi+\\
&\qquad+\int\langle\tilde{c}(\mathbf{y},s|\chi,\tau)\tilde{v}'_i(\mathbf{x},t)\tilde{v}'_j(\mathbf{y},s)C'_0(\boldsymbol{\xi})C'_0(\chi)\rangle_c\,\mathrm{d}\chi.
\end{aligned}\tag{G.2}$$

If we take C_0 to be statistically independent of $\mathbf{v}$, f, and g, all but the last integral vanish, and

$$\begin{aligned}
&\langle\tilde{v}'_i(\mathbf{x},t)\tilde{v}'_j(\mathbf{y},s)C'_0(\boldsymbol{\xi})c'(\mathbf{y},s)\rangle_c\\
&\quad=+\int\langle\tilde{c}(\mathbf{y},s|\chi,\tau)\tilde{v}'_i(\mathbf{x},t)\tilde{v}'_j(\mathbf{y},s)\rangle_c\langle C'_0(\boldsymbol{\xi})C'_0(\chi)\rangle\,\mathrm{d}\chi.
\end{aligned}\tag{G.3}$$

The corresponding fourth order approximation is

$$\begin{aligned}
&[\langle\tilde{v}'_i(\mathbf{x},t)\tilde{v}'_j(\mathbf{y},s)C'_0(\boldsymbol{\xi})c'(\mathbf{y},s)\rangle_c]^{(4)}\\
&\quad=\int\langle\tilde{c}^{(0)}(\mathbf{x},t|\chi,\tau)\rangle_c[\langle\tilde{v}'_i(\mathbf{x},t)\tilde{v}'_j(\mathbf{y},s)\rangle_c]^{(2)}\langle C'_0(\boldsymbol{\xi})C'_0(\chi)\rangle_c\,\mathrm{d}\chi.
\end{aligned}\tag{G.4}$$

Expansion of (66) gives to 4th order (the first integral on the right hand side of (66) is of order 5 or higher)

$$\begin{aligned}
&\varepsilon^{(4)}_{ijkc}(\mathbf{x},\mathbf{y},t,s)=[\langle v'_i(\mathbf{x},t)v'_j(\mathbf{y},s)v'_k(\mathbf{x},t)c'(\mathbf{x},t)\rangle_c]^{(4)}\\
&\quad=-\iint_0^t\langle\tilde{c}^{(0)}(\mathbf{x},t|\chi,\tau)\rangle_c[\langle\tilde{v}'_i(\mathbf{x},t)v'_j(\mathbf{y},s)\tilde{v}'_k(\mathbf{x},t)v'_m(\chi,\tau)\rangle_c]^{(4)}\times\\
&\qquad\times\frac{\partial}{\partial\chi_m}\langle c^{(0)}(\chi,\tau)\rangle_c\,\mathrm{d}\tau\,\mathrm{d}\chi,\\
&\qquad-\iint_0^t\langle\tilde{c}^{(0)}(\mathbf{x},t|\chi,\tau)\rangle_c[\langle\tilde{v}'_i(\mathbf{x},t)v'_j(\mathbf{y},s)\tilde{v}'_k(\mathbf{x},t)\times\\
&\qquad\times f'(\chi,\tau)\rangle_c]^{(4)}\langle c^{(0)}(\chi,\tau)\rangle_c\,\mathrm{d}\tau\,\mathrm{d}x+\\
&\qquad+\iint_0^t\langle\tilde{c}^{(0)}(\mathbf{x},t|\chi,\tau)\rangle_c[\langle\tilde{v}'_i(\mathbf{x},t)v'_j(\mathbf{y},s)\tilde{v}'_k(\mathbf{x},t)g'(\chi,\tau)\rangle_c]^{(4)}\,\mathrm{d}\tau\,\mathrm{d}\chi+\\
&\qquad+\int\langle\tilde{c}^{(0)}(\mathbf{x},t|\chi,\tau)\rangle_c[\langle\tilde{v}'_i(\mathbf{x},t)v'_j(\mathbf{y},s)\tilde{v}'_k(\mathbf{x},t)C'_0(\chi)\rangle_c]^{(4)}\,\mathrm{d}\chi,
\end{aligned}\tag{G.5}$$

where the last integral vanishes because $\mathbf{v}$ and C_0 are taken to be independent.

Expansion of (67) results in the 4th order approximation (the first integral on the right hand side of (67) is of order 5 or higher)

$$
\begin{aligned}
\kappa_{ikc}^{(4)}(\mathbf{x},\mathbf{y},t,s) &= [\langle v_i'(\mathbf{x},t)v_k'(\mathbf{x},t)f'(\mathbf{y},s)c'(\mathbf{x},t)\rangle_c]^{(4)} \\
&= -\iint_0^t \langle\tilde{c}^{(0)}(\mathbf{x},t|\chi,\tau)\rangle_c[\langle\tilde{v}_i'(\mathbf{x},t)\tilde{v}_k'(\mathbf{x},t)f'(\mathbf{y},s)v_m'(\chi,\tau)\rangle_c]^{(4)} \times \\
&\quad \times \frac{\partial}{\partial\chi_m}\langle c^{(0)}(\chi,\tau)\rangle_c \,\mathrm{d}\tau\,\mathrm{d}\chi - \\
&\quad - \iint_0^t \langle\tilde{c}^{(0)}(\mathbf{x},t|\chi,\tau)\rangle_c[\langle\tilde{v}_i'(\mathbf{x},t)\tilde{v}_k'(\mathbf{x},t)f'(\mathbf{y},s)f'(\chi,\tau)\rangle_c]^{(4)} \times \\
&\quad \times \langle c^{(0)}(\chi,\tau)\rangle_c \,\mathrm{d}\tau\,\mathrm{d}\chi + \\
&\quad + \iint_0^t \langle\tilde{c}^{(0)}(\mathbf{x},t|\chi,\tau)\rangle_c[\langle\tilde{v}_i'(\mathbf{x},t)\tilde{v}_k'(\mathbf{x},t)f'(\mathbf{y},s)g'(\chi,\tau)\rangle_c]^{(4)} \,\mathrm{d}\tau\,\mathrm{d}\chi + \\
&\quad + \int \langle\tilde{c}^{(0)}(\mathbf{x},t|\chi,\tau)\rangle_c[\langle\tilde{v}_i'(\mathbf{x},t)\tilde{v}_k'(\mathbf{x},t)f'(\mathbf{y},s)C_0'(\chi)\rangle_c]^{(4)} \,\mathrm{d}\chi,
\end{aligned}
\tag{G.6}
$$

where the last integral vanishes because C_0 is taken to be independent of **v** and f.

Expansion of (68) gives the 4th order expression (the first integral on the right hand side of (68) is of order 5 or higher)

$$
\begin{aligned}
\vartheta_{ikc}^{(4)}(\mathbf{x},\mathbf{y},t,s) &= [\langle v_i'(\mathbf{x},t)v_k(\mathbf{x},t)g'(\mathbf{y},s)c'(\mathbf{x},t)\rangle_c]^{(4)} \\
&= -\iint_0^t \langle\tilde{c}^{(0)}(\mathbf{x},t|\chi,\tau)\rangle_c[\langle\tilde{v}_i'(\mathbf{x},t)\tilde{v}_k'(\mathbf{x},t)g'(\mathbf{y},s)v_m'(\chi,\tau)\rangle_c]^{(4)} \times \\
&\quad \times \frac{\partial}{\partial\chi_m}\langle c^{(0)}(\chi,\tau)\rangle_c \,\mathrm{d}\tau\,\mathrm{d}\chi - \\
&\quad - \iint_0^t \langle\tilde{c}^{(0)}(\mathbf{x},t|\chi,\tau)\rangle_c[\langle\tilde{v}_i'(\mathbf{x},t)\tilde{v}_k'(\mathbf{x},t)g'(\mathbf{y},s)f'(\chi,\tau)\rangle_c]^{(4)} \times \\
&\quad \times \langle c^{(0)}(\chi,\tau)\rangle_c \,\mathrm{d}\tau\,\mathrm{d}\chi + \\
&\quad + \iint_0^t \langle\tilde{c}^{(0)}(\mathbf{x},t|\chi,\tau)\rangle_c[\langle\tilde{v}_i'(\mathbf{x},t)\tilde{v}_k'(\mathbf{x},t)g'(\mathbf{y},s)g'(\chi,\tau)\rangle_c]^{(4)} \,\mathrm{d}\tau\,\mathrm{d}\chi + \\
&\quad + \int \langle\tilde{c}^{(0)}(\mathbf{x},t|\chi,\tau)\rangle_c[\langle\tilde{v}_i'(\mathbf{x},t)\tilde{v}_k'(\mathbf{x},t)g'(\mathbf{y},s)C_0'(\chi)\rangle_c]^{(4)} \,\mathrm{d}\chi,
\end{aligned}
\tag{G.7}
$$

where the last integral vanishes because C_0 is taken to be independent of **v** and g.

Acknowledgement

This work was supported by the US Nuclear Regulatory Commission under contract NRC-04-97-056.

References

Di Federico, V. and Neuman, S. P.: 1997, Scaling of random fields by means of truncated power variograms and associated spectra, *Water Resour. Res.* **33**(5), 1075–1085.

Di Federico, V. and Neuman, S. P.: 1998a, Flow in multiscale log conductivity fields with truncated power variograms, *Water Resour. Res.* **34**(5), 975–987.

Di Federico, V. and Neuman, S. P.: 1998b, transport in multiscale log conductivity fields with truncated power variograms, *Water Resour. Res.* **34**(5), 963–973.

Di Federico, V. and Neuman, S. P.: 1999, Anisotropy, lacunarity and effective conductivity of multiscale log conductivity fields with truncated power variograms, *Water Resour. Res.* in press.

Guadagnini, A. and Neuman, S. P.: 1997, Nonlocal and Localized Finite Element Solution of Conditional Mean Flow in Randomly Heterogeneous Media, *Tech. Rep. HWR-97-100*, Department of Hydrology and Water Resources, The University of Arizona, Tucson.

Guadagnini, A. and Neuman, S. P.: 1998, Deterministic solution of stochastic groundwater flow equations by nonlocal finite elements, in: V. N. Burganos, G. P. Karatzas, A. C. Payatakes, C. A. Brebbia, W. G. Gray, and G. F. Pinder, (eds), *Computational Methods in Water Resources XII: Computational Methods in Contamination and Remediation of Water Resources*, Computational Mechanics Publications, Boston. pp. 347–354.

Guadagnini, A. and Neuman, S. P.: 1999a, Nonlocal and localized analyses of conditional mean steady state flow in bounded, randomly nonuniform domains, 1, theory and computational approach, *Water Resour. Res.* **35**(10), 2999–3018.

Guadagnini, A. and Neuman, S. P.: 1999b, Nonlocal and localized analyses of conditional mean steady state flow in bounded, randomly nonuniform domains, 2, computational examples, *Water Resour. Res.* **35**(10), 3019–3039.

Neuman, S. P.: 1990, Universal scaling of hydraulic conductivities and dispersivities in geologic media, *Water Resour. Res.* **26**(8), 1749–1758.

Neuman, S. P.: 1993, Eulerian-Lagrangian theory of transport in space-time nonstationary velocity fields: Exact nonlocal formalism by conditional moments and weak approximation, *Water Resour. Res.* **29**(3), 633–645.

Neuman, S. P.: 1995, On advective dispersion in fractal velocity and permeability fields, *Water Resour. Res.* **31**(6), 1455–1460.

Neuman, S. P.: 1997, Stochastic approach to subsurface flow and transport: A view to the future, in: G. Dagan, and S. P. Neuman (eds), *Subsurface Flow and Transport: A Stochastic Approach*, Cambridge University Press, Cambridge, United Kingdom, pp. 231–241.

Neuman, S. P. and Di Federico, V.: 1998, Correlation, flow and transport in multiscale permeability fields, in: G. Sposito (ed.) *Scale Dependence and Scale Invariance in Hydrology*, Cambridge University Press, New York, pp. 354–397.

Neuman, S. P. and Orr, S.: 1993, Prediction of steady state flow in nonuniform geologic media by conditional moments: Exact nonlocal formalism, effective conductivities, and weak approximation, *Water Resour. Res.* **29**(2), 341–364.

Neuman, S. P. and Guadagnini, A.: 1999, A new look at traditional deterministic flow models and their calibration in the context of randomly heterogeneous media, *Proc. of the International Conf. on Calibration and Reliability in Groundwater Modelling, Coping with uncertainty*, Zurich, Switzerland, vol. 1, pp. 189–197.

Neuman, S. P., Tartakovsky, D. M., Wallstrom, T. C. and Winter, C. L.: 1996, Correction to 'Prediction of steady state flow in nonuniform geologic media by conditional moments: Exact nonlocal formalism, effective conductivities, and weak approximation', *Water Resour. Res.* **32**(5), 1479–1480.

Tartakovsky, D. M. and Neuman, S. P.: 1998, Transient flow in bounded randomly heterogeneous domains: 1. Exact conditional moment equations and recursive approximations, *Water Resour. Res.* **34**(1), 1–12.

Tartakovsky, D. M. and Neuman, S. P.: 1999, Extension of 'Transient flow in bounded randomly heterogeneous domains: 1. Exact conditional moment equations and recursive approximations', *Water Resour. Res.* **35**(6), 1921–1925.

Tartakovsky, D. M., Neuman, S. P. and Lu, Z.: 1999, Conditional stochastic averaging of steady state unsaturated flow by means of Kirchhoff transformation, *Water Resour. Res.* **35**(3), 731–745.

Zhang, D. and Neuman, S. P.: 1995a, Eulerian–Lagrangian analysis of transport conditioned on hydraulic data: l. Analytical-numerical approach, *Water Resour. Res.* **31**(1), 39–51.

Zhang, D. and Neuman, S. P.: 1995b, Eulerian–Lagrangian analysis of transport conditioned on hydraulic data: 2. Effect of log transmissivity and hydraulic head measurements, *Water Resour. Res.* **31**(1), 53–63.

Zhang, D. and Neuman, S. P.: 1995c, Eulerian–Lagrangian analysis of transport conditioned on hydraulic data: 3. Spatial moments, travel time distribution, mass flow rate and cumulative release across a compliance surface, *Water Resour. Res.* **31**(1), 65–75.

Zhang, D. and Neuman, S. P.: 1995d, Eulerian-Lagrangian analysis of transport conditioned on hydraulic data: 4. Uncertain initial plume state and non-Gaussian velocities *Water Resour. Res.* **31**(1), 77–88.

Zhang, D. and Neuman, S. P.: 1996, Effect of local dispersion on solute transport in randomly heterogeneous media, *Water Resour. Res.* **32**(9), 2715–2723.

Transport in Porous Media **42:** 69–83, 2001.

The Relative Dispersion and Mixing of Passive Solutes in Transport in Geologic Media

ALDO FIORI
Dipartimento di Scienze dell'Ingegneria Civile, Universita' degli Studi di Roma Tre, Italy, e-mail: aldo@cidra.dsic.uniroma3.it

(Received: 29 December 1998; in final form: 1 October 1999)

Abstract. The spreading of a contaminant in a heterogeneous aquifer depends on the scales of variability effectively explored by the plume. In particular, we observe two major contributions of the fluctuating velocity field in the contaminant movement: (i) the spreading caused by velocity variations of scales lesser than that of the plume size, which we will call 'relative' spreading, and (ii) the meander-like movement of the plume as a whole caused by velocity variations of scale larger than that of the plume size. The aim of this work is to consider the effects of the finite size of the contaminant plume on the local concentration moments $\langle C \rangle$ and σ_C. In particular a 'relative' concentration, which depends on the scales of variability effectively explored by the plume, is defined. First, the mathematical formulation of the problem is developed along the Lagrangian framework. In particular, the expressions for the relative mean concentration and its variance are presented. Then, the methodology is applied to the regional transport problem, where the influence of the size of the plume and the pore-scale dispersion are quantitatively assessed.

Key words: aquifer transport, groundwater quality, solute dilution, heterogeneous porous formation.

1. Introduction

Spatial variability of hydraulic conductivity plays a major role in transport of solutes by groundwater. A solute plume injected in an aquifer is spreading because of the space variability of the velocity field, caused by permeability distribution. At the same time pore-scale dispersion causes diffusion of the advected blobs in the surrounding field and its action is mainly local. The spreading caused by large scale variability has been captured in the past by modeling the permeability and the dependent fields (head H, velocity $\mathbf{V}$, concentration C) as random space functions of the coordinate $\mathbf{x}$. In this manner the concentration can be characterized by its statistical moments, like the ensemble mean of the local concentration $\langle C \rangle$ and its variance σ_C^2. The calculation of such moments has been the subject of several works in the past (e.g. Dagan, 1982, 1989; Gelhar and Axness, 1983; Neuman, Winter and Newman, 1987; Chu and Sposito, 1980). More recently the influence of pore-scale dispersion on these moments has been assessed (Graham and McLaughlin, 1989; Kapoor and Gelhar, 1994; Zhang and Neuman, 1996; Dagan and Fiori, 1997; Andricevic, 1998); its impact depends mainly on the Peclet numbers, the most important being $\mathrm{Pe} = I_v/\alpha_T$, where I_v is the vertical correlation scale of logpermeability and α_T is the transverse dispersivity.

The spreading of a contaminant depends on the scales of variability explored by the plume. In particular, there are two major contributions of the fluctuating velocity field to the contaminant movement:

(1) the spreading caused by velocity variations of scales lesser than that of the plume size, which we will call *relative* spreading, and

(2) the meander-like movement of the plume as a whole caused by velocity variations of scale larger than that of the plume size.

When the plume is large enough to sample all the velocity fluctuations in the aquifer, heterogeneity is fully contributing to spreading, and the second effect above is negligible. In such a condition, the transport if referred as *ergodic* (e.g. Dagan, 1989).

The 'relative' spreading of the plume has been analyzed in the past with reference to global measures like spatial or temporal moments of the concentration field (e.g. Kitanidis, 1988; Dagan, 1989, 1991; Rajaram and Gelhar, 1993; Selroos, 1995). Less has been done about this effect on the local concentration moments, and in particular the concentration variance σ_C^2. The aim of this work is to consider the effect of the finite size of the contaminant plume on the local concentration moments $\langle C \rangle$ and σ_C. In particular a 'relative' concentration, which depends on the scales of variability effectively explored by the plume, is defined. A similar concept was investigated for the solute flux by Andricevic and Cvetkovic (1998), and in other fields, like turbulence diffusion, by a few authors (e.g. Richardson, 1926; Chatwin and Sullivan, 1979). It can be tackled by considering the limit case of a point-like solute spot of concentration C_0 injected in a heterogeneous field with the neglect of pore-scale dispersion – no matter how large is the travel time, in each realization of the random field the small plume will move erratically without dispersing. The expected concentration at the center of the spot is $\langle C \rangle \cong C_0$ and the variance σ_C^2 is equal to zero everywhere, except on the surface separating the advected blobs from the surrounding fluid. This limit case suggest that for real life plumes the concentration field predicted by the usual 'ergodic' approach may be in some cases quite unrealistic; in particular, the maximum concentrations in the field can be underestimated. Conversely, the concentration field computed through the relative approach is the closest to the observed one in a realization, that is, in the field. In fact, in the interpretation of the field tests, the concentration spatial moments are calculated with respect to the plume 'real' center of mass, implicitly conditioning in the one-realization centroid.

The work is based on a recent study (Dagan and Fiori, 1997) where a Lagrangean theoretical framework was developed to compute σ_C with the inclusion of pore-scale dispersion. The approach has been applied to analyze the Cape Cod and Borden Site concentration fluctuations, and the agreement with measurements was good (Fiori and Dagan, 1999).

2. Mathematical Statement of the Problem

We consider an aquifer of stationary, lognormally distributed hydraulic conductivity K; the logarithm of the conductivity $Y = \ln K$ is therefore completely characterized by the constant mean $\langle Y \rangle$ and by the two-point autocovariance $C_Y(\mathbf{x} - \mathbf{y})$ where $\mathbf{x}$ and $\mathbf{y}$ are the coordinate vectors of two points in the space. Flow is driven by a mean uniform hydraulic head gradient $\nabla\langle H \rangle = -\mathbf{J}$ resulting in a random velocity field $\mathbf{V} = \mathbf{U} + \mathbf{u}(\mathbf{x})$; the mean velocity $\mathbf{U} = \langle \mathbf{V} \rangle$ is constant and the fluctuation $\mathbf{u}(\mathbf{x})$ is characterized by the two-point covariance function $u_{ij}(\mathbf{x} - \mathbf{y})$ $(i, j = 1, 2, 3)$.

If we inject at time $t = 0$ in such a formation a conservative solute of initial concentration $C_0(\mathbf{a})$, where $\mathbf{x} = \mathbf{a}$ is the coordinate vector of the points in the injected volume V_0, transport is caused by advection by the random velocity field $\mathbf{V}(\mathbf{x})$ and by pore-scale dispersion, which we will assume of constant dispersion coefficients $D_{\mathrm{d}i} = \alpha_i|\mathbf{U}|$. The problem has been treated in a Lagrangean framework (Dagan and Fiori, 1997), by regarding the plume as a collection of elementary subparticles of mass $nC_0\mathbf{da}$ within the initial volume V_0 moving in the aquifer along trajectories $\mathbf{X}_{\mathrm{t}}(t; \mathbf{a}) = \mathbf{X}(t; \mathbf{a}) + \mathbf{X}_{\mathrm{d}}(t)$, where $\mathbf{X} = \int_0^t \mathbf{V}[\mathbf{X}_{\mathrm{t}}]\mathrm{d}t$ is the advective component of trajectories and $\mathbf{X}_{\mathrm{d}}$ is the displacement caused by pore-scale dispersion, modeled as a Brownian motion. The final result for the concentration was the following (Dagan and Fiori, 1997; eq. 28):

$$C(\mathbf{x}, t) = \int_{V_0} C_0(\mathbf{a})\phi\left[\mathbf{x} - \mathbf{X}(t; \mathbf{a})\right] \mathbf{da}, \tag{1}$$

where $\phi(\mathbf{x}) = \phi(x_1)\phi(x_2)\phi(x_3)$ and

$$\phi(x_i) = \frac{1}{2(\pi D_{\mathrm{d}i}t)^{1/2}} \exp\left[-\frac{x_i^2}{4D_{\mathrm{d}i}t}\right]. \tag{2}$$

The function ϕ accounts for the pore-scale dispersion mechanism. Equation (1) was the starting point in (Dagan and Fiori, 1997) for the derivation of the first two moments of the concentration field, that is, the mean concentration $\langle C \rangle$ and its variance σ_C^2. In absence of pore-scale dispersion ($\mathrm{Pe} = \infty$) $\phi(\mathbf{x}) \to \delta(\mathbf{x})$.

To account for the relative dispersion around the plume center of mass, we follow the plume spreading along its movement through the aquifer in each realization. Toward this aim, we change the coordinate system into one attached to the plume center of mass $\mathbf{R}$, defined as

$$\mathbf{R}(t) = \frac{1}{M_0}\int_{V_0} C_0(\mathbf{a})\mathbf{X}_{\mathrm{t}}(t; \mathbf{a})\, \mathbf{da}, \tag{3}$$

with M_0 the injected mass. Switching into the new coordinates $\boldsymbol{\xi} = \mathbf{x} - \mathbf{R}$, substitution into (1) leads to the definition of what will be called the *relative* concentration;

$$C_r(\boldsymbol{\xi}, t) = \int_{V_0} C_0(\mathbf{a})\phi[\boldsymbol{\xi} + \mathbf{R}(t) - \mathbf{X}(t; \mathbf{a})]\, \mathbf{da}. \tag{4}$$

For a fixed $\boldsymbol{\xi}$, C_r depends on the random variable $\mathbf{X}-\mathbf{R}$ which precisely spans the scale of variability smaller than the plume size. Hence, the difference between (4) and the previous formulation (1) is that we follow in each realization the plume as it wanders in the aquifer, concentrating on the relative spreading around its center of mass $\mathbf{R}(t)$, which is a random variable itself. Doing so we explicitly separate the effects of heterogeneity which contribute to spreading and those which cause the solute plume meander in the aquifer like a whole, directly related to $\mathbf{R}(t)$. The relative concentration $C_r(\boldsymbol{\xi}, t)$ can be also viewed as the concentration at a point $\mathbf{x} = \boldsymbol{\xi} + \mathbf{R}$ in different realizations after superimposing the plumes centroids.

With the aid of (4) we are now able to compute the first and second order statistical moments of the concentration by simply averaging (4) over $\mathbf{R}$ and $\mathbf{X}$ to obtain $\langle C_r\rangle$, and squaring (4) and averaging over $\mathbf{X}, \mathbf{Y}, \mathbf{R}$ to obtain $\langle C_r^2\rangle$. The results are as follows:

$$\begin{aligned}\langle C_r(\boldsymbol{\xi}, t)\rangle &= \iiint_{V_0} C_0(\mathbf{a})\phi[\boldsymbol{\xi} + \mathbf{R}'(t) - \mathbf{X}'(t;\mathbf{a}) + \bar{\mathbf{a}} - \mathbf{a}] \times \\ &\quad \times f(\mathbf{X}', \mathbf{R}'; t, \mathrm{Pe})\, \mathrm{d}\mathbf{a}\, \mathrm{d}\mathbf{X}'\, \mathrm{d}\mathbf{R}',\end{aligned} \tag{5}$$

$$\begin{aligned}\langle C_r^2(\boldsymbol{\xi}, t)\rangle &= \iiint \int_{V_0}\int_{V_0} C_0(\mathbf{a})C_0(\mathbf{b})\phi[\boldsymbol{\xi} + \mathbf{R}'(t) - \mathbf{X}'(t;\mathbf{a}) + \bar{\mathbf{a}} - \mathbf{a}] \times \\ &\quad \times \phi[\boldsymbol{\xi} + \mathbf{R}'(t) - \mathbf{Y}'(t;\mathbf{b}) + \bar{\mathbf{b}} - \mathbf{b}] \times \\ &\quad \times f(\mathbf{X}', \mathbf{Y}', \mathbf{R}'; t, \mathbf{a}, \mathbf{b}, \mathrm{Pe})\, \mathrm{d}\mathbf{a}\, \mathrm{d}\mathbf{b}\, \mathrm{d}\mathbf{X}'\, \mathrm{d}\mathbf{Y}' \mathrm{d}\mathbf{R}', \\ \sigma_{C_r}^2(\boldsymbol{\xi}, t) &= \langle C_r^2(\boldsymbol{\xi}, t)\rangle - \langle C_r(\boldsymbol{\xi}, t)\rangle^2.\end{aligned} \tag{6}$$

In the previous expressions we have substituted $\mathbf{X} = \mathbf{a} + \mathbf{U}t + \mathbf{X}'$, $\mathbf{R} = \bar{\mathbf{a}} + \mathbf{U}t + \mathbf{R}'$, and the prime indicate fluctuations around the mean; $\bar{\mathbf{a}}$ is the center of mass of the initial plume, and $f(\mathbf{X}', \mathbf{R}')$, $f(\mathbf{X}', \mathbf{Y}', \mathbf{R}')$ are the joint pdfs of the pertinent variables. Thus $f(\mathbf{X}', \mathbf{Y}', \mathbf{R}')$ is the joint pdf of the fluctuations of a trajectory originating at $\mathbf{a}$, of another trajectory originating at $\mathbf{b}$ and of the centroid. Expressions (5, 6) permit in principle to compute the first two statistical moments of C_r, that is the mean concentration $\langle C_r\rangle$ and its variance $\sigma_{C_r}^2 = \langle C_r^2\rangle - \langle C_r\rangle^2$. The same results could be obtained by operating with the differences $\mathbf{R}'-\mathbf{X}'$, $\mathbf{R}'-\mathbf{Y}'$ as random variables.

The particular case of zero pore-scale dispersion can be obtained by (5, 6) considering that the function ϕ tends for $D_{\mathrm{dii}} \to 0$ to $\phi \to \delta$. Substituting in (5, 6) we get

$$\langle C_r(\boldsymbol{\xi}, t)\rangle = \iint_{V_0} C_0(\mathbf{a}) f(\boldsymbol{\xi} + \mathbf{R}, \mathbf{R}; t, \mathrm{Pe})\, \mathrm{d}\mathbf{a}\, \mathrm{d}\mathbf{R}, \tag{7}$$

$$\left\langle C_r^2(\boldsymbol{\xi}, t)\right\rangle = \iint_{V_0} C_0^2(\mathbf{a}) f(\boldsymbol{\xi} + \mathbf{R}, \mathbf{R}; t, \mathrm{Pe})\, \mathrm{d}\mathbf{a}\, \mathrm{d}\mathbf{R}, \tag{8}$$

after employing the relationship $f(\mathbf{X}, \mathbf{X}, \mathbf{R}) = f(\mathbf{X}, \mathbf{R})\,\delta(\mathbf{a}-\mathbf{b})$, which stems from fluid continuity (Dagan and Fiori, 1997); in words, the probability of two particles to be at the same point at time t is zero unless $\mathbf{a} = \mathbf{b}$, that is, the particle is the same. The result (7, 8) is similar to the one obtained in the past for the ergodic case and for Pe $= \infty$ (Dagan, 1982); for C_0 constant in space, it becomes $\langle C_\mathrm{r}^2 \rangle = C_0 \langle C_\mathrm{r} \rangle$.

Two simple and interesting limits can be obtained from the expressions (5, 6). The first one is for $V_0 \to \infty$. If the logconductivity random field is of finite integral scales, it has been proved (e.g. Dagan, 1989) that the plume centroid moves like a straight line of trajectory $\mathbf{R} = \langle \mathbf{R} \rangle = \mathbf{U}t$ (the initial position of the center of mass $\bar{\mathbf{a}}$ has been set for simplicity, but without loss of generality, as $\bar{\mathbf{a}} = 0$). For this case the fluctuations of the centroid $\mathbf{R}'$ are identically zero and the trajectories pdfs modify as follows:

$$f(\mathbf{X}', \mathbf{R}') \to f(\mathbf{X}')\delta(\mathbf{R}'),$$
$$f(\mathbf{X}', \mathbf{Y}', \mathbf{R}') \to f(\mathbf{X}', \mathbf{Y}')\delta(\mathbf{R}'). \tag{9}$$

Substituting (9) in (5, 6) we obtain

$$\langle C_\mathrm{r}(\boldsymbol{\xi}, t) \rangle = \int_{V_0} \int C_0(\mathbf{a}) \phi[\boldsymbol{\xi} - \mathbf{X}'(t; \mathbf{a}) - \mathbf{a}] f(\mathbf{X}')\, \mathrm{d}\mathbf{X}'\, \mathrm{d}\mathbf{a}, \tag{10}$$

$$\langle C_\mathrm{r}^2(\boldsymbol{\xi}, t) \rangle = \int_{V_0} \int_{V_0} \iint C_0(\mathbf{a}) C_0(\mathbf{b}) \phi \left[\boldsymbol{\xi} - \mathbf{X}'(t; \mathbf{a}) - \mathbf{a}\right] \times$$
$$\times \phi \left[\boldsymbol{\xi} - \mathbf{Y}'(t; \mathbf{b}) - \mathbf{b}\right] f\left(\mathbf{X}', \mathbf{Y}'\right) \mathrm{d}\mathbf{X}'\, \mathrm{d}\mathbf{Y}'\, \mathrm{d}\mathbf{a}\, \mathrm{d}\mathbf{b}, \tag{11}$$

which are precisely the solutions given in Dagan and Fiori, 1997 (eqs. 29, 33) after substituting $\xi = \mathbf{x} - \langle \mathbf{R} \rangle = \mathbf{x} - \mathbf{U}t$; we will regard this solution as the *ergodic* one.

The other limit is for $V_0 \to 0$, that is, a 'Darcy-size' solute plume. In this case it can be shown that the cross-moments $\langle X_i' Y_i' \rangle$, $\langle X_i' R_i' \rangle$, $\langle Y_i' R_i' \rangle$, $\langle R_i'^2 \rangle$ are approximately equal (the result can be obtained by the use of expression (A1), for $V_0 \to 0$). As a consequence, the crosscorrelation between the two relative displacements $\mathbf{X} - \mathbf{R}$ and $\mathbf{Y} - \mathbf{R}$ is approximately equal to zero, and they can be assumed as uncorrelated. Thus, it is possible to write

$$f(\mathbf{X}', \mathbf{Y}', \mathbf{R}') \to f(\mathbf{X}', \mathbf{R}') f(\mathbf{Y}', \mathbf{R}'), \tag{12}$$

which substituted in (6) give

$$\langle C_r^2(\xi, t) \rangle = \langle C_\mathrm{r}(\xi, t) \rangle^2, \qquad \sigma_{C_\mathrm{r}}^2 = \langle C_\mathrm{r}^2 \rangle - \langle C_\mathrm{r} \rangle^2 = 0. \tag{13}$$

It will be shown in Section 4 that the mean relative concentration for a small particle diffuse according to pore-scale dispersion for $t D_\mathrm{d}/I^2 \ll 1$; for $t D_\mathrm{d}/I^2 \gg 1$ the spreading is ruled by the macro-dispersion coefficients. Conversely, the variance (13) is identically equal to zero, indicating that the heterogeneous velocity field is not felt by the plume as far as spreading is concerned. The consequence is that the ergodic solution will systematically underpredict the expected maximum concentration and overpredict its variance. In the following application this behavior will be investigated for the two-dimensional transport case.

3. Application to Two-Dimensional Transport

In the following we consider transport of an instantaneous rectangular injection of dimensions L_1, L_2 and constant concentration C_0 in a two-dimensional isotropic aquifer. The mean flow is aligned with the x_1 direction, with the center of the initial mass located at the origin. The logconductivity autocovariance is assumed as exponential, that is, $C_Y = \sigma_Y^2 \exp[-|\mathbf{r}|/I]$ with $\mathbf{r}(r_1, r_2)$ the distance between two point in space, I the integral scale, and σ_Y^2 the logconductivity variance. To simplify matters and approach, the first order analysis in σ_Y^2 of the trajectories statistical moments will be employed. The above assumptions are customary in the Lagrangean theories of transport (e.g., Dagan 1989); the impact of violation of these assumptions on the results is not discussed here.

The examined configuration may be viewed as resembling the problem of solute transport at a regional scale, where the flow field is usually considered to be two dimensional, and where the correlation lengths of the transmissivity field are often much larger than the characteristic dimensions of the plume, leading to nonergodic transport. In such cases, the contribution of D_d should be viewed as the one determined by heterogeneity at the local scale, rather than at the pore-scale, and the contribution of the latter can be neglected in the analysis.

Even for this relatively simple case, the computation of the concentration statistical moments with the aid of (5, 6) is a formidable task. For this reason we simplify the analysis by additional assumptions which are described in the next sections, where we distinguish between the infinite Peclet solution and the finite Peclet one.

3.1. INFINITE PECLET CASE

With C_0 = const, the basis for the calculation of the concentration statistical moments with the neglect of local-scale dispersion is (7), the concentration variance being given by $\sigma_{C_r}^2 = \langle C_r^2 \rangle - \langle C_r \rangle^2 = \langle C_r \rangle (C_0 - \langle C_r \rangle)$. We simplify the analysis by the few additional assumptions following:

(1) The trajectories pdfs in (7, 8) are assumed to be normal and completely defined by the first order moments $\langle \mathbf{X} \rangle = \mathbf{U}t + \mathbf{a}$, and by the second order ones $X_{ij} = \langle X_i' X_j' \rangle$, $R_{ij} = \langle R_i' R_j' \rangle$, $W_{ij} = \langle X_i' R_j' \rangle$. This approximation has been proved to be a valid one (e.g. Dagan and Fiori, 1997; Figure 1). The calculation of the moments for the exponential logpermeability covariance is given in the Appendix, through a first-order analysis. From a theoretical point of view the normality assumption can be justified at large times invoking the central limit theorem. It applies at any time if we adopt a small perturbation approximation,

as we do, since it leads to a linear relationship between $\mathbf{X}$ and $\ln K$. As shown in Dagan and Fiori (1997), numerical simulations tend to support such an assumption for one particle trajectories, while it has not yet been proven for two particles.

(2) Evaluation of the two particles trajectory covariance has showed that they are strongly correlated in the longitudinal direction. Moreover, such correlation increases with time. This effect has already been observed by Dagan (1991) and more recently by Zhang *et al.* (1996) and it is more pronounced if the initial plume longitudinal dimension is small compared to the logconductivity integral scale. For these reasons the dependence of the statistical moments on the longitudinal coordinates will be neglected and we will assume a relatively thin initial injection zones.

With these approximations, $\langle C_r \rangle$ and $\sigma^2_{C_r}$ are given by the following expressions:

$$\frac{\langle C_r(\xi, t)\rangle}{C_0} = \int_{-L_2/2}^{L_2/2} \frac{1}{2(2\pi\Lambda_2)^{1/2}} \exp\left[-\frac{(\xi_2 - a_2)^2}{2\Lambda_2}\right] \times \\ \times \left\{\operatorname{erf}\left[\frac{\xi_1 + L_2/2}{(2\Lambda_1)^{1/2}}\right] - \operatorname{erf}\left[\frac{\xi_1 - L_2/2}{(2\Lambda_1)^{1/2}}\right]\right\} \mathrm{d}a_2, \tag{14}$$

$$\frac{\sigma^2_{C_r}(\xi, t)}{C_0^2} = \frac{\langle C_r\rangle}{C_0}\left(1 - \frac{\langle C_r\rangle}{C_0}\right) \tag{15}$$

with $\Lambda_i = X_{ii}(t) + R_{ii}(t) - 2W_{ii}(t; a_2)$.

A useful observation is that the terms $W_{11}(a_2)$, $W_{22}(a_2)$ depend weakly on a_2 for $L_2/I \ll 1$ and $L_2/I \gg 1$. We may, therefore replace these terms by their space average over L_2 to get (see Equation A1)

$$W_{ij}(t, a_2) \cong \frac{1}{L_2}\int_{-L_2/2}^{L_2/2} W_{ij}(t, a_2)\,\mathrm{d}a_2 = R_{ij}(t). \tag{16}$$

With this approximation, (14) simplifies considerably and becomes

$$\frac{\langle C_r(\xi, t)\rangle}{C_0} = \frac{1}{4}\left[\operatorname{erf}\left(\frac{\xi_1 + L_1/2}{\sqrt{2(X_{11} - R_{11})}}\right) - \operatorname{erf}\left(\frac{\xi_1 - L_1/2}{\sqrt{2(X_{11} - R_{11})}}\right)\right] \times \\ \times \left[\operatorname{erf}\left(\frac{\xi_2 + L_2/2}{\sqrt{2(X_{22} - R_{22})}}\right) - \operatorname{erf}\left(\frac{\xi_2 - L_2/2}{\sqrt{2(X_{22} - R_{22})}}\right)\right]. \tag{17}$$

The previous result, which can be extended to the three-dimensional case, is identical to the ergodic mean concentration with replacing the single-particle moments X_{ii} by the difference $X_{ii} - R_{ii}$. The procedure is similar to that for the concentration spatial moments (e.g. Dagan, 1989, 1991) and may serve as a simple tool to approximately assess the concentration field under non-ergodic conditions; the procedure has also been employed by Andricevic (1998). It is pointed out that

(17) is an approximation, and the influence of the term W_{ii} can be important, especially for initial plumes of the order of the logconductivity integral scale.

Expression (17) is the solution of the diffusion equation for $\langle C_r(\xi, t)\rangle$ with moments given by $X_{ij} - R_{ij}$. This is not the case for the general solution (5), from which (14) derives. The reason is that with the approximation (16) the argument of ϕ in (5) does not depend on **a** anymore, and the approximated relative mean concentration (17) satisfies a diffusion equation. However, in both cases (14, 17) the concentration second order spatial moments are identical and equal to $X_{ij}(t) - R_{ij}(t; \mathbf{L})$.

3.2. FINITE PECLET CASE

The computation of the relative concentration moments $\langle C_r\rangle, \sigma^2_{C_r}$ is achieved through expressions (5, 6), the joint pdf $f(\mathbf{X}, \mathbf{Y}, \mathbf{R})$ being needed. Consistent with the previous case, we adopt a multivariate normal distribution, and neglect the dependence on the longitudinal coordinates of the trajectories/centroid spatial moments. To define the two-particles pdf, we also need the crosscorrelation between the trajectories $X_i(t; \mathbf{a})$, $Y_j(t; \mathbf{b})$, that is, $Z_{ij}(t; \mathbf{a} - \mathbf{b}) = \langle X'_i Y'_j\rangle$.

Under such approximations, the concentration statistical moments can be readily computed by the use of expressions (5, 6). The mean relative concentration is simply given by expression (14) once the following substitution is made: $\Lambda_i = X_{ii} + R_{ii} - 2W_{ii}(a_2) + 2D_{\mathrm{d}i}t$. For the term $\langle C_r^2\rangle$, we get after a few analytical integrations the following result:

$$\frac{\langle C_r^2\rangle}{C_0^2} = \int_{-L_1/2}^{L_1/2}\int_{-L_2/2}^{L_2/2}\int_{-L_1/2}^{L_1/2}\int_{-L_2/2}^{L_2/2} \Theta_1\Theta_2 \, \mathrm{d}a_1 \, \mathrm{d}a_2 \, \mathrm{d}b_1 \, \mathrm{d}b_2, \tag{18}$$

where

$$\begin{aligned}
\Theta_i = {} & \frac{1}{2\pi}[4D_{\mathrm{d}i}^2t^2 + 4D_{\mathrm{d}i}tA_3 - 4D_{\mathrm{d}i}tA_4 - A_4^2 - 4D_{\mathrm{d}i}tA_5 + 2A_4A_5 - A_5^2 + \\
& + 4D_{\mathrm{d}i}tA_1 + 2A_3A_1 - 2A_4A_1 - 2A_5A_1 + A_1^2 - 2A_3A_2 + 2A_4A_2 + \\
& + 2A_5A_2 - A_2^2]^{-1/2} \cdot \exp[(2a_i^2D_{\mathrm{d}i}t + 2b_i^2D_{\mathrm{d}i}t - 4a_i\xi_iD_{\mathrm{d}i}t - \\
& - 4b_i\xi_iD_{\mathrm{d}i}t + 4\xi_i^2D_{\mathrm{d}i}t + a_i^2A_3 - 2a_ib_iA_3 + b_i^2A_3 + 2a_ib_iA_4 - 2b_i^2A_4 - \\
& - 2a_i\xi_iA_4 + 2b_i\xi_iA_4 - 2a_i^2A_5 + 2a_ib_iA_5 + 2a_i\xi_iA_5 - 2b_i\xi_iA_5 + \\
& + a_i^2A_1 + b_i^2A_1 - 2a_i\xi_iA_1 - 2b_1\xi_iA_1 + 2\xi_i^2A_1 - 2a_ib_iA_2 + 2a_i\xi_iA_2 + \\
& + 2b_i\xi_iA_2 - 2\xi_i^2A_2)/(-8D_{\mathrm{d}i}^2t^2 - 8D_{\mathrm{d}i}tA_3 + 8D_{\mathrm{d}i}tA_4 + 2A_4^2 + \\
& + 8D_{\mathrm{d}i}tA_5 - 4A_4A_5 - 8D_{\mathrm{d}i}tA_1 - 4A_3A_1 + 4A_4A_1 + 4A_5A_1 - 2A_1^2 + \\
& + 4A_3A_2 - 4A_4A_2 - 4A_5A_2 + 2A_2^2)],
\end{aligned}$$

and $A_1 = X_{ii}(t)$, $A_2 = Z_{ii}(t, a_2 - b_2)$, $A_3 = R_{ii}(t)$, $A_4 = W_{ii}(t, a_2)$, $A_5 = W_{ii}(t, b_2)$. The moments $X_{ii}, Z_{ii}, R_{ii}, W_{ii} (i = 1, 2)$ are given in the Appendix. Evaluating (14, 18) requires two and four numerical quadratures respectively, which are carried out by the means of a Gaussian procedure; the number of points per integral scale used in the calculations is 20, which was enough for the convergence of the integrals. A simplified expression for $\langle C_r \rangle$ can be formulated also for the finite Peclet case, along the same lines as the previous chapter, reaching a formula similar to (17) but with the following substitution: $X_{ii} \rightarrow X_{ii} + 2D_{di}t$.

4. Discussion of Results and Conclusions

For the sake of illustration we compute the concentration statistical moments for an instantaneous initial plume of longitudinal dimension $L_1/I = 0.5$ and transverse dimension in the range of $L_2/I = 0.1 \div 10$. The assumed logconductivity variance is $\sigma_Y^2 = 0.5$, the Peclet number is set to $\mathrm{Pe} = I/\alpha_\mathrm{T} = \infty$ and $\mathrm{Pe} = 100$, respectively, and pore-scale dispersion is assumed to be isotropic. Considered that the order of magnitude of the parameters I, α_T one can encounter in reality at the regional scale is (in meters) about $I = \mathrm{O}(10^2) \div \mathrm{O}(10^3)$, and $\alpha_\mathrm{T} = \mathrm{O}(10^{-3}) \div \mathrm{O}(10^{-2})$, the Peclet number considered in the example is a relatively low one. The influence of local-scale dispersion on transport has already been assessed in (Dagan and Fiori, 1997) and here we want to concentrate on the effects of the finiteness of the plume and its relative dispersion around the center of mass. Results of integration of (18), at the center of the plume ($\xi_1 = \xi_2 = 0$), are displayed in Figures 1–3 where the following quantities are represented:

(i) the *ergodic* mean concentration $\langle C \rangle$ and coefficient of variation $CV[C] = \sigma_C/\langle C \rangle$, that is, the concentration moments computed regardless of the location of the plume center of mass; it is representative of the total dispersion inclusive of the uncertainty of the particle trajectories; the ergodic concentration moments are given by considering $R_{ii} = W_{ii} = 0$ in the relative formulation;

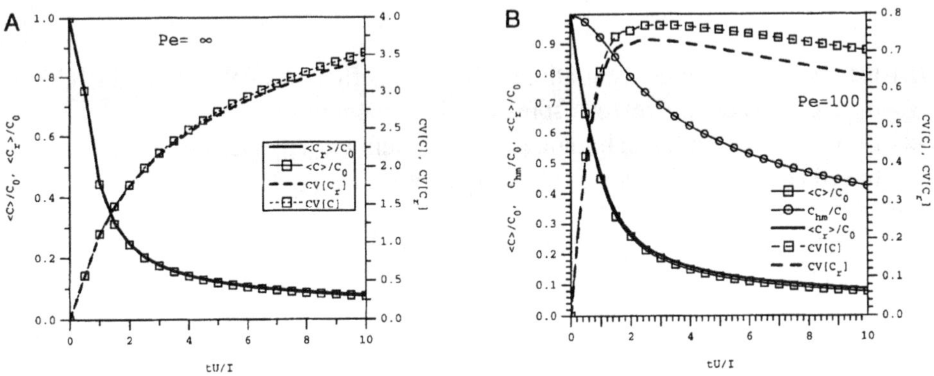

Figure 1. The concentration expected value (full lines) and coefficient of variation (dashed lines) at the center of the plume ($\xi = 0$): relative (thick line), ergodic (box) and homogeneous (circle) solution. $\sigma_Y^2 = 0.5$, $L_1/I = 0.5$, $L_2/I = 10$. (a) $\mathrm{Pe} = \infty$, (b) $\mathrm{Pe} = 100$.

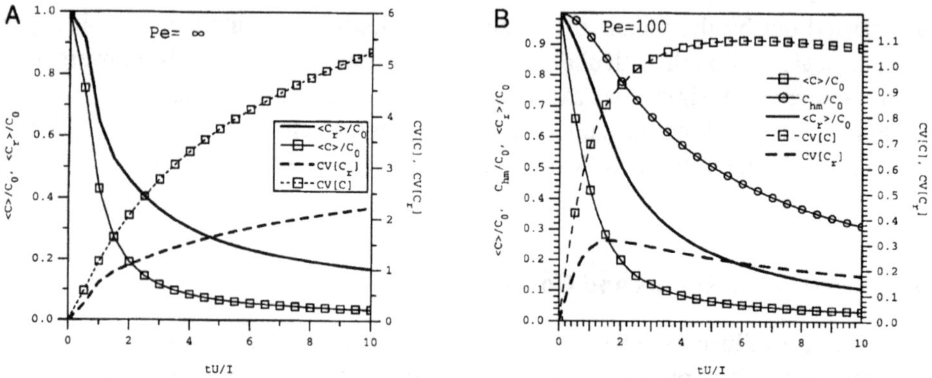

Figure 2. The concentration expected value (full lines) and coefficient of variation (dashed lines) at the center of the plume ($\xi = 0$): relative (thick line), ergodic (box) and homogeneous (circle) solution. $\sigma_Y^2 = 0.5$, $L_1/I = 0.5$, $L_2/I = 1$. (a) Pe $= \infty$, (b) Pe $= 100$.

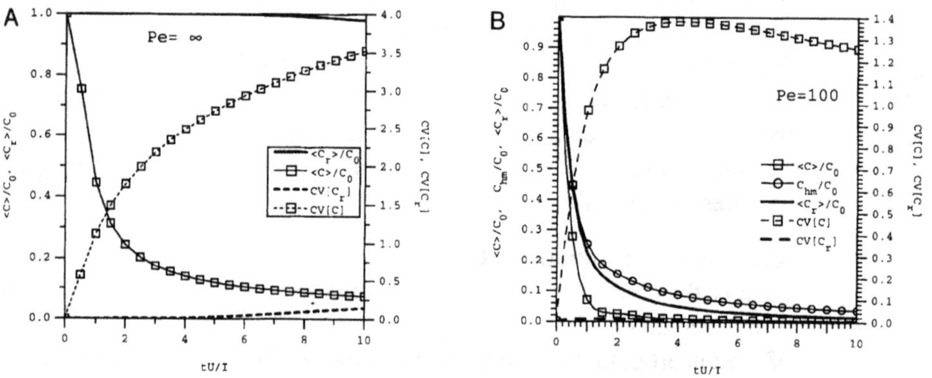

Figure 3. The concentration expected value (full lines) and coefficient of variation (dashed lines) at the center of the plume ($\xi = 0$): relative (thick line), ergodic (box) and homogeneous (circle) solution. $\sigma_Y^2 = 0.5$, $L_1/I = 0.1$, $L_2/I = 0.1$ (a) Pe $= \infty$, (b) Pe $= 100$.

(ii) the *relative* mean concentration $\langle C_r \rangle$ and coefficient of variation $CV[C_r] = \sigma_{C_r}/\langle C_r \rangle$, where the relative spreading is accounted for;

(iii) the concentration for an homogeneous medium C_{hm}, that is, the one consequent to PSD only, which has the following expression:

$$\frac{C_{hm}(\xi, t)}{C_0} = \frac{1}{4}\left[\mathrm{erf}\left(\frac{\xi_1 + L_1/2}{2\sqrt{D_{d1}t}}\right) - \mathrm{erf}\left(\frac{\xi_1 - L_1/2}{2\sqrt{D_{d1}t}}\right)\right] \times \times \left[\mathrm{erf}\left(\frac{\xi_2 + L_2/2}{2\sqrt{D_{d2}t}}\right) - \mathrm{erf}\left(\frac{\xi_2 - L_2/2}{2\sqrt{D_{d2}t}}\right)\right]. \tag{19}$$

Results for $L_2/I = 10$ are depicted in Figure 1a (Pe $= \infty$) and 1b (Pe $= 100$). In this case the plume transverse dimension is large enough for the solute to sample

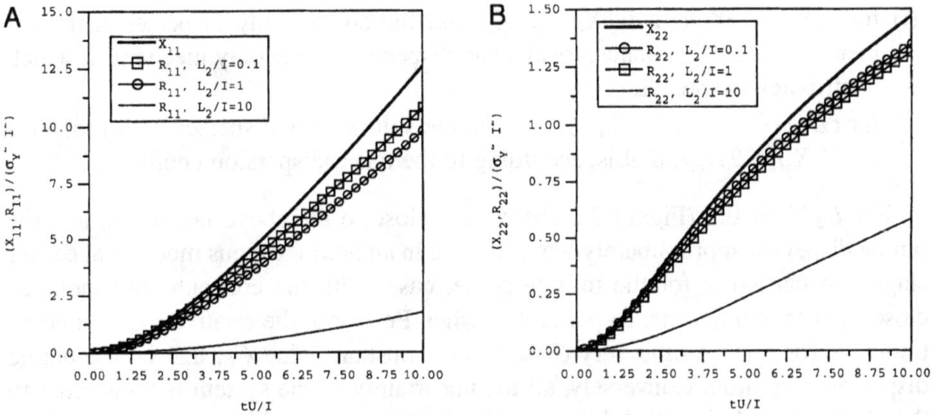

Figure 4. The dimensionless variances of the longitudinal and transverse trajectories of the center of mass R_{11} (a), R_{22} (b), for $L_2/I = 0.1, 1, 10$; Pe $= \infty$.

practically all the velocity fluctuations scales; as a consequence the expected center of mass moves approximately along a straight line and the 'ergodic' results are almost obtained, as shown in the Figure. This is the situation already discussed in the second section and expressed by Equation (10, 11). For $L_2 > 10$ (not displayed) the relative solution converges more and more toward the ergodic one. In the Figures the influence of Peclet on the coefficient of variation is also observed; in particular, $CV[C_r]$ always grows with time, while in the finite Peclet case it has a non-monotonic behavior.

For decreasing sizes of the plume L_2/I, the situation is still similar to the previous one, with the only difference that $\langle C_r \rangle$ and $CV[C_r]$ start to depart from the ergodic case, and in particular a stronger decrease of CV is observed.

Of definite interest is the case $L_2/I = 1$ (Figures 2a, 2b) for which a more pronounced departure from the ergodic case is observed. Thus, the relative mean concentration is considerably higher than the corresponding ergodic value, while the relative coefficient of variation displays a drastic reduction. This finding confirms what was already explained in the previous sections – the dispersion is caused by the heterogeneities effectively sampled by the finite size plume, while the scales of fluctuation higher than the plume size cause the plume to move as a whole and to wander throughout the heterogeneous aquifer. This picture is emphasized by Figures 4a, 4b, which shows the rapid increase of the centroid variances $R_{11}(t)$, $R_{22}(t)$ as the plume size L_2/I decreases.

For $L_2/I < 1$ the relative mean concentration shows a further increase. For Pe $= \infty$, at the limit of $L_2/I \to 0$, it is $W_{ii} \cong R_{ii} \cong X_{ii}(t)$ and $\langle C_r \rangle$ tends to C_0; the concentration variance tends to 0. Conversely, for finite Peclet and $L_2/I \to 0$, it is $W_{ii} \cong R_{ii} \cong Z_{ii}(t; 0^+)$, and $\langle C_r \rangle$ tends to expression (14) with the substitution $\Lambda_i = X_{ii}(t) - Z_{ii}(t; 0^+) + 2D_{di}t$; this result is explained by the behavior of Z_{ii} near $(a_2 - b_2) = 0$ (see Appendix). Two cases are worth considering:

(i) for $tD_{\mathrm{d}}/I^2 \ll 1$, it is $X_{ii} \cong Z_{ii}$ and the solute body disperses with moments $\Lambda_i \cong 2D_{\mathrm{d}i}t$; hence, local-scale dispersion is the only mechanism which contributes to spreading;

(ii) for $tD_{\mathrm{d}}/I^2 \gg 1$, it is $X_{ii} \gg Z_{ii}$ and the solute body disperses with moments $\Lambda_i \cong X_{ii} + 2D_{\mathrm{d}i}t$, that is, according to the macrodispersion coefficients.

For $L_2/I = 0.1$ (Figure 3a, 3b) we are close to the above behaviors, and the plume disperses approximately like if it were in an homogeneous medium at earlier stages, in particular for the infinite peclet case, with the concentration variance close to zero at any time. In particular, when $\mathrm{Pe} = \infty$, the relative mean concentration is for a large time very close to the constant value C_0, that is, the plume disperses very little. Conversely, all the uncertainty of the system is transferred to the plume meandering, and the centroid variance considerably increases, as shown in Figure 4a, 4b.

Summarizing, the relative spreading of a plume instantaneously injected in an heterogeneous aquifer depends on its size. This effect has already been shown for the second order concentration spatial moments (e.g. Kitanidis, 1988; Dagan, 1989, 1991) whereas here the point concentration moments are explored. For plumes characterized by transverse dimension much larger than the logconductivity correlation length, say about $10I$ for the examined case, all heterogeneity scales contribute to spreading and the 'ergodic' results of Dagan and Fiori, 1997, are recovered. For decreasing plume sizes, the amount of heterogeneity sampled by the plume is limited; as a consequence the spreading is less than that predicted by the ergodic analysis. In particular, the maximum concentrations is higher and the concentration variances lower than those predicted by the ergodic formulation. The ergodic result may, therefore, underpredict sometimes the maximum concentration occurring in the field; this situation is more plausible for regional transport, where high correlation lengths of logtransmissivity are often encountered. On the other hand, as the initial plume decrease, the location of its center of mass is subjected to more and more uncertainty, as shown by the centroid trajectory variances R_{11}, R_{22}. At the limit case of a very small initial plume compared to the $\ln K$ correlation length, the dispersive mechanism is similar to that for an homogeneous medium, characterized by much smaller dispersion and by less uncertainty of the concentration point values. The randomness of the system is reflected in the location of the center of mass, which is, in this case, quite uncertain. This uncertainty may be reduced by conditioning, subject outside the scope of the present work. Under non-ergodic conditions we expect the present approach to provide a more accurate estimate of the concentration fluctuations in an actual plume than previous results.

Appendix: Derivation of the Trajectories Moments

In the following are provided the expressions for the trajectories/centroid second order statistical moments by the use of the linear theory (see e.g. Dagan, 1989).

The moments R_{ij}, W_{ij} are obtained through the following definitions:

$$W_{ij}(t,a_2) = \langle X_i'(t;\mathbf{a})R_j'(t)\rangle = \frac{1}{V_0}\int_{V_0} Z_{ij}(t,\mathbf{a}-\mathbf{b})\,\mathrm{d}\mathbf{b} = \frac{1}{L_2}\int_{-L_2/2}^{L_2/2} Z_{ij}(t,a_2-b_2)\,\mathrm{d}b_2, \tag{A1}$$

$$R_{ij}(t) = \langle R_i'(t)R_j'(t)\rangle = \frac{1}{V_0^2}\int_{V_0}\int_{V_0} Z_{ij}(t,\mathbf{a}-\mathbf{b})\,\mathrm{d}\mathbf{a}\,\mathrm{d}\mathbf{b} = \frac{2}{L_2^2}\int_0^{L_2}(L-r_2)Z_{ij}(t,r_2)\,\mathrm{d}r_2, \tag{A2}$$

where we have assumed $Z_{ij}(t;a_1-b_1,a_2-b_2) \cong Z_{ij}(t;0,a_2-b_2)$ and constant C_0 (see Section 3.1). Under such approximations, it can be shown that $W_{ij} = R_{ij} = X_{ij} = Z_{ij} = 0$ for $i \neq j$.

Adopting the linear theory, the particle trajectory fluctuation $\mathbf{X}'(t;\mathbf{a})$ can be described in two-dimensional transport and for isotropic PSD by the following expression (see e.g. Dagan, 1989):

$$X_i'(t;\mathbf{a}) = \frac{1}{2\pi}\int_0^t\int \hat{u}_i(\mathbf{k})\exp[-ik_1U\tau - i\mathbf{k}\times\mathbf{a} - i\mathbf{k}\cdot\mathbf{X}_\mathrm{d}(\tau t)]\,\mathrm{d}\mathbf{k}\,\mathrm{d}\tau, \tag{A3}$$

with $\hat{\mathbf{u}}$ the Fourier transform of the velocity field.

The one and two particles second-order moments can be calculated through (A3) as follows:

$$X_{ii}(t) = \langle X_i'^2\rangle = \frac{1}{2\pi}\int_0^t\int_0^t\int \hat{u}_{ii}(\mathbf{k})\exp[-ik_1U(\tau-\tau') - i\mathbf{k}\cdot(\mathbf{a}-\mathbf{b}) - k^2D_\mathrm{d}|\tau-\tau'|]\,\mathrm{d}\mathbf{k}\,\mathrm{d}\tau\,\mathrm{d}\tau', \tag{A4}$$

$$Z_{ii}(t,\mathbf{a}-\mathbf{b}) = \langle X_i'Y_i'\rangle = \frac{1}{2\pi}\int_0^t\int_0^t\int \hat{u}_{ii}(\mathbf{k})\exp[-ik_1U(\tau-\tau') - i\mathbf{k}\cdot(\mathbf{a}-\mathbf{b}) - k^2D_\mathrm{d}(\tau+\tau')]\,\mathrm{d}\mathbf{k}\,\mathrm{d}\tau\,\mathrm{d}\tau' \quad (k=|\mathbf{k}|), \tag{A5}$$

where we have made use of the statistical stationarity of the velocity covariance, that is, $\langle\hat{u}_i(\mathbf{k})\hat{u}_i(\mathbf{k}')\rangle = (2\pi)\hat{u}_{ii}(\mathbf{k})\delta(\mathbf{k}+\mathbf{k}')$, with $\hat{u}_{ii}$ the Fourier Transform of the velocity covariance; furthermore, we have averaged over the random PSD displacements $\mathbf{X}_\mathrm{d}(t)$, considering that $\mathbf{X}_\mathrm{d}$ belonging to two distinct particles are not correlated.

Equation (A5) provides the expression for Z_{ii} for $(\mathbf{a}-\mathbf{b}) \neq 0$, being $Z_{ii}(t;0) = X_{ii}(t)$. Thus, Z_{ii} presents a discontinuity in the neighbors of $(\mathbf{a}-\mathbf{b}) = 0$, being

$Z_{ii}(t; 0^+) < Z_{ii}(t; 0) = X_{ii}(t)$; however, this discontinuity does not contribute when computing the trajectories moments (A1), or the concentration ones (5, 6), since its support is zero. For Pe $= \infty$, the discontinuity is not present, and $Z_{ii}(t; \mathbf{a} - \mathbf{b})$ tends uniformly to $X_{ii}(t)$ as $(\mathbf{a} - \mathbf{b}) \to 0$.

The moments X_{ii} (A4) and Z_{ii} (A5) are calculated by the means of numerical quadratures; switching in polar coordinates, that we get after a few manipulations

$$\begin{aligned}\frac{X_{ii}(t)}{\sigma_Y^2 I^2} &= \frac{4}{\pi}\int_0^\infty \int_0^{\pi/2} \frac{F_i}{\rho(\rho^2/\mathrm{Pe}^2 + u^2)^2(1+\rho^2)^{3/2}} \left\{ \frac{\rho^2}{\mathrm{Pe}^2}\left(\frac{\rho^2 t'}{\mathrm{Pe}} - 1\right) + \right.\\ &\quad + u^2\left(\frac{\rho^2 t'}{\mathrm{Pe}} + 1\right) + e^{-\rho^2 t'/\mathrm{Pe}}\left[\cos\left(\rho u t'\right)\left(\frac{\rho^2}{\mathrm{Pe}^2} - u^2\right) - \right.\\ &\quad \left.\left. - \frac{2\rho u}{\mathrm{Pe}} \sin\left(\rho u t'\right)\right]\right\} \mathrm{d}\rho\, \mathrm{d}\psi, \qquad \text{(A6)}\end{aligned}$$

$$\begin{aligned}\frac{Z_{ii}(t; r_2)}{\sigma_Y^2 I^2} &= \frac{2}{\pi}\int_0^\infty \int_0^{\pi/2} F_i \frac{\cos[\rho r_2 \sin\psi]}{\rho(\rho^2/\mathrm{Pe}^2 + u^2)(1+\rho^2)^{3/2}} \times \\ &\quad \times \{1 + e^{-2\rho^2 t'/\mathrm{Pe}} - 2\cos(\rho u t') e^{-\rho^2 t'/\mathrm{Pe}}\}\, \mathrm{d}\rho\, \mathrm{d}\psi \qquad \text{(A7)}\end{aligned}$$

where $u = \cos\psi$, $\mathrm{Pe} = UI/D_\mathrm{d}$, $t' = tU/I$, and

$$\begin{aligned} F_1 &= \sin^4\psi, \\ F_2 &= (\sin\psi\cos\psi)^2 . \end{aligned}$$

Acknowledgements

The author wish to thank Gedeon Dagan at the Tel Aviv University for stimulating discussion and suggestions, and Vladimir Cvetkovic at the Royal Institute of Technology in Stockholm for reviewing an earlier version of the manuscript. The anonymous reviewers are also acknowledged for their thorough reading of the paper and their constructive criticism.

References

Andricevic, R.: 1998, Effects of local dispersion and sampling volume on the evolution of concentration fluctuations in aquifers, *Water Resour. Res.* **34**(5), 1115–1129.

Andricevic, R. and Cvetkovic, V.: 1998, Relative dispersion for solute flux in aquifers, *J. Fluid Mech.* **361**, 145–174.

Chatwin, P.C. and Sullivan, P. J.: 1979, The relative diffusion of a cloud of passive contaminant in incompressible turbulent flow, *J. Fluid Mech.* **91**, 337–355.

Chu, S.Y. and Sposito, G.: 1980, A derivation of the macroscopic solute transport equation for homogeneous, saturated porous media, *Water Resour. Res.* **16**, 542–546.

Dagan, G.: 1982, Stochastic modeling of groundwater flow by unconditional and conditional probabilities, 2. The solute transport, *Water Resour. Res.* **18**(4), 813–833.

Dagan, G.: 1984, Solute transport in heterogeneous porous formations, *J. Fluid Mech.* **145**, 151–177.

Dagan, G.: 1989, *Flow and Transport in Porous Formations*, Springer-Verlag, 465 p.

Dagan, G.: 1991, Dispersion of a passive solute in non-ergodic transport by steady velocity fields in heterogeneous formations, *J. Fluid Mech.* **233**, 197–210.

Dagan, G. and Fiori, A.: 1997, The influence of pore-scale dispersion on concentration statistical moments in transport through heterogeneous aquifers, *Water Resour. Res.* **33**, 1595–1606.

Fiori, A.: 1996, Finite-Peclet extensions of Dagan's solution to transport in anisotropic heterogeneous formations, *Water Resour. Res.* **32**, 193–198.

Fiori, A. and Dagan, G.: 1999, Concentration fluctuations in transport by groundwater: comparison between theory and field experiments, *Water Resour. Res.* **35**(1), 105–112.

Gelhar, L. W. and Axness, C. L.: 1983, Three-dimensional stochastic analysis of macrodispersion in aquifers, *Water Resour. Res.* **19**, 161–180.

Graham, W. and Mclaughlin, D.: 1989, Stochastic analysis of nonstationary subsurface solute transport. 1. Unconditional moments, *Water Resour. Res.* **25**, 215–232.

Kapoor, V. and Gelhar, L. W.: 1994, Transport in three-dimensional heterogeneous aquifers: 1.Dynamics of concentration fluctuations, *Water Resour. Res.* **6**, 1775–1788.

Kitanidis, P. K.: 1988, Prediction by the methods of moments of transport in heterogeneous formation, *J. Hydrol.* **102**, 453–473.

Neuman, S. P., Winter, C. L. and Newman, C. M.: 1987, Stochastic theory of field-scale Fickian dispersion in anisotropic porous media, *Water Resour. Res.* **23**, 453–466.

Rajaram, H. and Gelhar, L. W.: 1993, Plume-scale dependent dispersion in heterogeneous aquifer: 2. Eulerian analysis and three-dimensional aquifers. *Water Resour. Res.* **29**(9), 3261–3276.

Richardson, L. F.: 1926, Atmospheric diffusion shown on a distance-neighour graph, *Proc. Roy. Soc.* Series A, **110**: 709.

Selroos, J. O.: 1995, Temporal moments for non-ergodic solute transport in heterogeneous aquifers, *Water Resour. Res.* **31**, 1705–1712.

Zhang, D. and Neuman, S. P.: 1996, Effect of local dispersion solute transport in randomly heterogeneous media, *Water Resour. Res.* **32**(9), 2715–2724.

Zhang, Y. K., Zhang, D. and Lin, L.: 1996, Non-ergodic solute transport in three-dimensional heterogeneous isotropic aquifers, *Water Resour. Res.* **32**(9), 2955–2963.

Transport in Porous Media **42:** 85–107, 2001.

Comparison of Observations from a Laboratory Model with Stochastic Theory: Initial Analysis of Hydraulic and Tracer Experiments

S. E. SILLIMAN and L. ZHENG*

Department of Civil Engineering and Geological Sciences, University of Notre Dame, Notre Dame, Indiana, U.S.A.

(Received: 19 January 1999; in final form: 23 November 1999)

Abstract. Hydraulic and tracer tests were conducted in a flow cell containing a mixture of sediments designed to mimic a two-dimensional, log-normally distributed, second-order stationary, exponentially correlated random conductivity field. With 60 integral scales in the direction of mean flow and 25 integral scales perpendicular to this direction, behavior of flow and transport in the interior of the flow cell can be compared directly with stochastic solutions for flow and transport. Using 144 piezometers and 361 platinum electrodes, the distribution of hydraulic head and the concentrations of an ionic tracer could be monitored in substantial detail. The present discussion presents the details of the experimental equipment. Results and initial analysis of hydraulic measurements and characterization of a two-dimensional tracer plume are also presented. Analysis using first-order hydraulic theory shows that the flow through the medium was consistent with an effective conductivity equal to the geometric mean of the conductivity distribution. Further, the semivariogram of head increments as observed in the experimental results was consistent with the semivariogram predicted by theory. The chemical transport experiments are here compared with the early solutions presented by Dagan (1984, 1987). The observed rate of longitudinal spread of two tracer plumes was slightly less than that predicted using this theory. Further, the spread in the transverse dimension was observed to decline from the initial plume dimensions and then remain constant or increase slightly, but at a rate lower than predicted by the theory. The difference between the hydraulic and transport results is believed to be related to the fact that the hydraulic results were averaged over a very large portion of the flow cell such that ergodic conditions could be assumed. In contrast, the initial geometry of the plume covered only approximately five integral scales in the transverse direction such that the validity of the assumption of ergodic conditions must be questioned in the analysis of results for the chemical transport.

Key words: groundwater, transport, stochastic theory, experiments.

1. Introduction

Stochastic models of groundwater flow and chemical transport hold promise to be valuable tools to assist in probabilistic and deterministic prediction of flow and transport processes in heterogeneous porous media. As early as the Monte Carlo studies of Warren and Price (1965) and Freeze (1975), characterization of

* Currently with the Kansas Geological Survey.

the hydraulic conductivity as a spatial random variable has been used as the basis for investigation and prediction of the distribution of hydraulic head, the transport of chemicals, and the uncertainties inherent in the prediction of these processes.

In terms of prediction of flow processes, early work by Freeze (1975) illustrated the variability in hydraulic head that can result from spatial variation in the hydraulic conductivity. This early study led to a number of additional papers including the work of Smith and Schwartz (1979) incorporating spatial correlation into two-dimensional conductivity fields, and the work by Bakr *et al.* (1981) extending into three-dimensions. Study of hydraulics in heterogeneous fields also led to the identification of effective hydraulic parameters in two- and three-dimensional flow fields subject to mean uniform flow (see, for example, the reviews of this work contained in Gelhar (1989) and Kitanidis (1997)). This work has also been extended to theoretical, numerical and experimental studies of effective parameters in the presence of active pumping from wells (e.g., Naff, 1991; Neuman and Orr, 1993; Indelman and Abromowitz, 1994; Silliman and Caswell, 1998).

The recognition of the link between the distribution of the hydraulic conductivity (or transmissivity) and the distribution of hydraulic head has also led to the development of methods for estimating the distribution of the hydraulic conductivity (or transmissivity) from head data. Calvete (1997), for example, reviewed the inverse method as applied to heterogeneous media. More recently, Zheng (1998) and Zheng and Silliman (2000) have presented a first-order methodology whereby the semivariogram of increments in the hydraulic head can be used to estimate the integral scale and variance of the transmissivity field.

Substantial progress has also been realized in terms of transport of conservative chemicals within saturated media. Much of this work is reviewed in Rubin (1997), thus the details will not be repeated herein. Among the significant early results from this literature was the development of expressions for the spatial variation and asymptotic limits for the dispersivity and discussion of conditions under which transport will approach behavior which can be modeled appropriately using the Fickian assumption. Much of this work has been focused on ergodic conditions, with somewhat less attention being paid to nonergodic conditions (e.g., Kitanidis, 1988; Dagan, 1990; Dagan, 1991; Zhang *et al.*, 1996; Zhang, 1997). Early theoretical efforts covering ergodic conditions have been the subject of a number of field investigations, including the Borden and Cape Cod sites (e.g., Freyberg, 1986; Garabedian *et al.*, 1991). Results from the Borden site indicated relatively close agreement with theoretical expressions for the growth of the second spatial moments of the tracer plume in both the longitudinal and transverse directions. The results from the Cape Cod site showed convergence to asymptotic behavior with reasonable dispersivities in the longitudinal and horizontal transverse directions and a very small transverse dispersivity in the vertical direction. In contrast, observations collected at the MADE site (e.g., Adams and Gelhar, 1992) provided field evidence of situations in which the ergodic theory was likely to be insufficient to describe the evolution of the first and second order moments of a tracer plume.

With respect to these latter results, Berkowitz and Scher (1998) and Berkowitz *et al.* (2000) have presented a non-ergodic, continuous time random walk model which provides predictions consistent with the observed variation in the spatial moments of the tracer plume at the MADE site.

A significant number of controlled laboratory experiments have been conducted on both flow and the transport of conservative chemicals in homogeneous and heterogeneous porous media (see, for example, the review in Silliman *et al.* (1998)). However, relatively few controlled laboratory efforts have appeared in the literature in which results predicted utilizing stochastic theory have been compared directly against laboratory results. Welty and Elsner (1997) report on a series of one-dimensional experiments in which a theory for the dispersive spread of a chemical plume under non-uniform density (Welty and Gelhar, 1982) was compared directly against observed transport in a one-dimensional laboratory column packed to reproduce a correlated conductivity field. The agreement was qualitatively quite reasonable. These authors also discussed observations of head variance within these one-dimensional columns and showed consistency with first-order hydraulic theory.

More recently, Silliman and Caswell (1998) report on the distribution of hydraulic head (and, through pump test analysis, the distribution of effective transmissivities) in the presence of a pumping well within a laboratory flow cell. This latter work was based on a pair of two-dimensional conductivity fields consisting of bimodal distributions of transmissivity. One conductivity field was designed to reproduce a second-order stationary random field and another was designed to reproduce a simple scaled field. In the case of the second-order stationary random field, these authors found good qualitative and quantitative agreement with published stochastic theories. When the conductivity field was designed to deviate from the assumptions of the published theory (e.g., the scaled medium), the results were observed to deviate from the predictions generated with these theories.

The present manuscript describes the experimental conditions, results, and initial analysis of a series of laboratory experiments that are to be used to examine hydraulic and transport theories as applied to heterogeneous porous media. The hydraulic results are compared with results by Zheng and Silliman (2000), Zheng (1998), and Gelhar and Axness (1983). The transport results are compared with predictions obtained from early stochastic theories describing the first and second moments of a plume of conservative tracer (e.g., Dagan, 1984, 1987; Freyberg, 1986; Sudicky, 1986). In both cases, the laboratory results allow direct comparison of predictions obtained from stochastic theory with observations obtained under controlled experimental conditions in which the underlying hydraulic distribution is well characterized (i.e., known within the limits of our ability to pack the experimental apparatus). It is recognized that the analysis to be presented, particularly in the area of transport, is preliminary with more sophisticated models likely providing more insight into the observed behaviors. This simplified analysis is accepted for the present manuscript as the goal here is to present the experimental design and initial results and because a second series of laboratory experiments on a second realization of the heterogeneous medium is currently being completed. More advanced analysis of these results, and the second series of results, will be presented following completion of the new experiments.

2. The Experimental Apparatus

The experiments were conducted in a laboratory flow cell with dimensions for the sediment compartment of 1.6 m × 0.67 m × 0.094 m (Figure 1). The apparatus was designed to allow flow to be established with a mean gradient parallel to the 1.6-m axis of the cell. The upper and lower borders of the flow cell were maintained as no flow boundaries. Three reservoirs were utilized on the inflow end of the flow cell. For hydraulic tests, these three reservoirs were hydraulically connected. For tracer tests, the central reservoir was hydraulically isolated for short periods of time to allow introduction of a tracer plume.

Hydraulic tests were conducted by injecting water at a known volumetric flux into the three, hydraulically-connected inflow reservoirs. The outflow reservoir was maintained at a constant hydraulic head. The distribution of head within the porous medium was measured through 144 piezometers installed through the wall of the flow cell and distributed as shown in Figure 2(a). The heads in the piezometers were measured one at a time through a series of valves on a manifold. Hydraulic heads were measured with both a pressure transducer and through visual measurement in a single manometer tube (one piezometer connected to the tube at a time). Due to electronic problems, the results from the pressure transducer were considered unreliable. Hence, only the results obtained using the manometer tube are analyzed

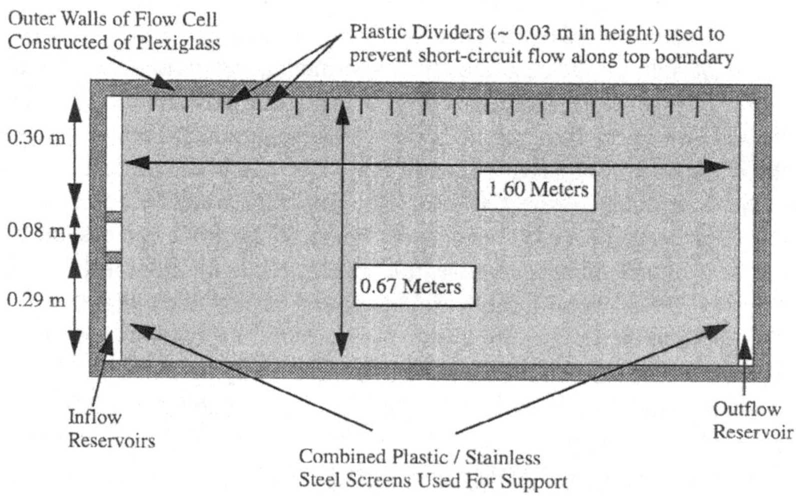

Figure 1. Schematic diagram of the flow cell used in these experiments. On inflow (left) end, the heights of the reservoirs are, from bottom to top, 0.29, 0.08, and 0.30 m.

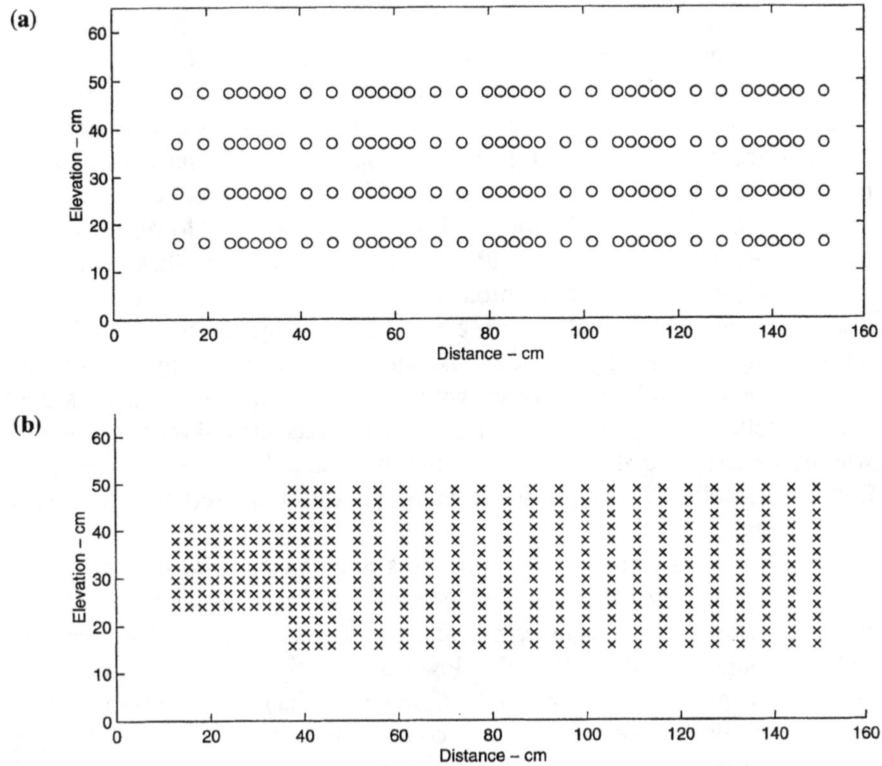

Figure 2. Internal instrumentation within flow cell: (a) 144 piezometers, and (b) 381 platinum electrodes.

below. Error in the measurement of the hydraulic heads was estimated, based on our experience with using these manometer boards, to have a standard deviation of approximately 0.05 cm. As discussed below, this estimate of error is consistent with the results obtained.

The distribution of piezometers was based on three considerations. First, as has been discussed in the literature (e.g., Rubin and Dagan, 1989), it is necessary to avoid measurements within a few integral scales of all no flow boundary conditions (so as to allow comparison of the experimental results with those derived theoretically for infinite systems). So as to be conservative, five integral scales were maintained from both of these boundaries (in the numerical work of Rubin and Dagan (1989) it was concluded that three integral scales was likely sufficient to avoid boundary effects). Second, it was considered necessary to avoid measurement in the immediate vicinity of constant head boundary conditions (Rubin and Dagan, 1988). Hence, once again, a minimum distance from the inflow and outflow boundaries of five integral scales was chosen as a conservative minimum distance. Third, based on the work of Zheng (1998), it was concluded that the first-order theory to be applied was optimal when the piezometers were spaced at distances of approximately one integral scale in the horizontal direction and at distances of at least four integral scales in the vertical direction. This led to distributing the piezometers in four horizontal lines, separated vertically by four integral scales and separated horizontally by a minimum distance of one integral scale.

The tracer tests discussed in this manuscript consisted of introducing a pulse of tracer through the central inflow reservoir. This pulse was introduced by first isolating the central reservoir (hydraulically) from the upper and lower reservoirs. This reservoir was then flushed, in the absence of flow through the flow cell, with tracer fluid. Fluid was then injected through all three reservoirs (the upper and lower reservoirs were hydraulically connected and driven by one pump while the central reservoir remained hydraulically isolated and was driven by a second pump), producing a finite volume of tracer (with an approximately rectangular shape) into the flow cell. Flow was once again terminated and the central reservoir was flushed with non-tracer fluid. The three reservoirs were then hydraulically connected and flow was established (using a single pump for all three reservoirs) to advance the tracer plume.

The movement of the tracer plume was monitored *in situ* through use of 381 electrodes (Figure 2(b)). Each electrode consisted of a twisted pair of insulated platinum wires with a diameter estimated at approximately 0.1 mm and was installed through the side wall of the flow cell. The tips of each pair of wires were clipped to expose the platinum to the fluid within the cell. The average separation between the tips of the pair of wires comprising a single electrode was substantially less than 1 mm. Hence, the measurements obtained were at a scale similar to the grain scale. The response at an electrode was recorded by passing an alternating current simultaneously through the electrode and a reference resistor, thus providing a measure of the voltage drop across the electrode. Each electrode was individually calibrated *in situ* by passing various concentrations of sodium chloride (used as tracer) through the flow cell. The resulting calibration curves showed a strong log-linear relationship between concentration and voltage drop across the electrode with minimal drift over the time periods used to complete these experiments. Within the range of concentrations used in these experiments (100–1000 mg/l NaCl), it was estimated that the error in readings from any given electrode was less than approximately 5 mg/l.

Due to the fact that the tracer was introduced through the central reservoir, the sampling for the tracer was concentrated in the central portion of the flow cell. The distribution of the electrodes is shown in Figure 2(b) (with the inflow reservoirs located at the left on this image). Near the inflow reservoir, the spacing of the electrodes parallel to the mean direction of the gradient was set equal to 0.5 times the anticipated integral scale of the conductivity field with spacing perpendicular to this direction set equal to 1.0 times the anticipated integral scale. Corresponding to the anticipated increase in the longitudinal dimension of the plume, the horizontal spacing of the electrodes was varied, increasing to a maximum of 2.0 integral scales at distance from the inflow reservoir. The distribution of electrodes represents a compromise between providing a dense sampling array with which to sample the movement of the tracer plume and the limits of our ability to install instruments through the wall of the flow cell. The limit was a function of a number of factors including the strength of the plexiglass used in constructing the flow cell, the effort involved in installing these electrodes without leaks, and our ability to sample these electrodes with reasonable sampling times.

3. The Porous Medium

Two properties were of primary interest in chosing the distribution of the sands for this experiment. The first was in establishing a realization of a log-normal, exponentially-correlated, second-order stationary, random hydraulic conductivity field. This choice was based on tying conditions within our experiments to conditions assumed in the stochastic theory used in interpreting the experimental results. The second was in establishing a correlation between the hydraulic conductivity and the sorptive capacity of the medium. The latter property was important for particle and microbial transport experiments to be reported on elsewhere.

In order to produce these fields, two field sands were obtained with various grain size distributions. The first consists of Accumin sands, a commercially available sand used in our prior experimental efforts (see, for example, Silliman and Caswell, 1998). These sands are well rounded quartz sands with minimal surface coatings. The second consists of sands excavated from a research site in Oyster, Virginia. These sands include a far wider grain size distribution than the Accumin sands and are heavily coated with iron oxides. Both sands were sieved into various sizes to produce varying hydraulic conductivities. Ten final mixtures of sands were chosen for use in the flow cell. The source and hydraulic conductivity (determined in separate falling head permeameter tests on consolidated samples of sand with variation among repetitions of less than 5%) of each sand are provided in Table I.

In order to pack these sands in the flow cell, a random field with exponential correlation was simulated on the computer using the Turning Bands field generator (Tompson *et al.*, 1989). The random field generated had an isotropic integral scale of approximately 2.75 cm and a variance of $\ln K$ of slightly less than 0.5. The final distribution generated on the computer is shown in Figure 3(a). The actual distribution of sand within the flow cell, as viewed through the front surface of the flow cell, is shown in Figure 3(b). This image shows a slight blurring of the contact between regions of different grain size that is related to minor problems involved in packing within 1 mm of the wall of the flow cell. Figure 3(c), a picture taken during destructive sampling within the flow cell, shows the sharpness of the contact between regions with different grain sizes approximately 5 cm away from both walls of the flow cell (i.e., along the centerline of the flow cell). Table I provides the percentage of each sand present within the flow cell as predicted using the Turning Bands generator and as estimated from measurements of the volumes of each sand entering the flow cell.

Table I. Sands used in packing the flow cell

Sand classification	Source of sand	Hydraulic Conductivity (cm/s)	Simulated percent of porous medium[a]	Estimate of actual per cent of medium[b]
1	Oyster	1.61	0.34	0.05
2	Oyster	0.58	1.91	0.9
3	Accumin	0.49	5.79	4.7
4	Accumin	0.224	13.78	11.0
5	Accumin	0.15	21.97	23.0
6	Accumin	0.11	24.41	26.0
7	Accumin	0.078	18.35	22.0
8	Oyster	0.041	8.98	7.8
9	Oyster	0.030	3.24	3.9
10	Oyster	0.015	1.22	0.6

[a]Per cent of porous medium comprised of particular sand type as produced by random field generator on computer.
[b]Percent of porous medium comprised of particular sand based on measured volumes used in filling the flow cell.

(a)

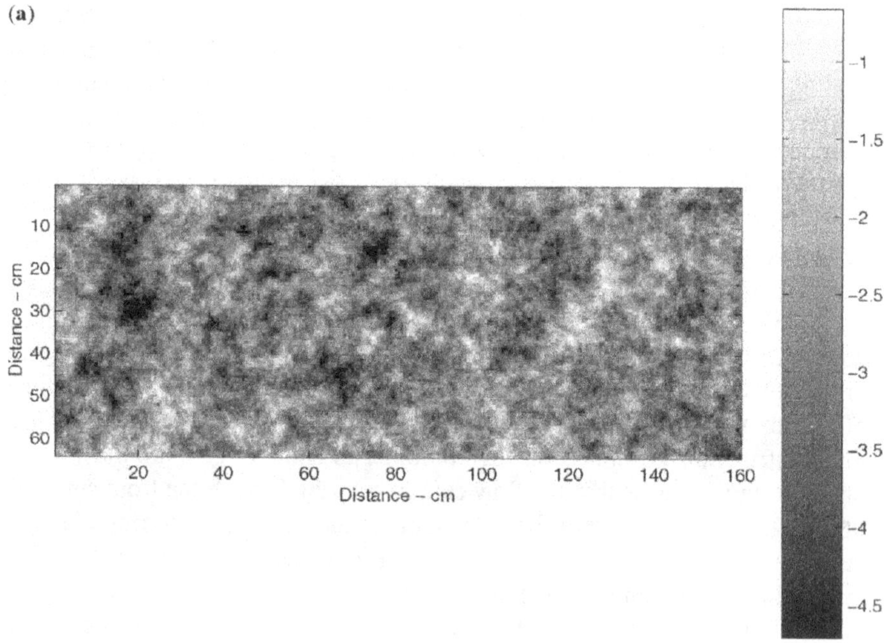

Figure 3. (a) Computer generated distribution of hydraulic conductivity subdivided into 10 classes. Gray scale based on ln(K). (b) Final distribution of sediments within the flow cell as seen through the front wall. (c) Final distribution of sediments within the flow cell as observed during the process of unpacking the flow cell. Darkest sands represent the lowest conductivities.

(b)

(c)

Figure 3. (*Continued*)

4. The Hydraulic Tests

4.1. PROCEDURE

The hydraulic test was run by injecting fluid into the inflow reservoirs (hydraulically connected) and holding the outflow reservoir at constant head. Flow rate was measured at the discharge and the system was allowed substantial time to come to steady state.

Using a network of valves, each of the 144 piezometers was individually opened to allow the water level in that piezometer to come to steady state; requiring between 2 and 10 min depending on the piezometer and sand grain size in the flow cell surrounding the tip of the piezometer. A number of the piezometers were inadvertantly plugged by the sand grains either during packing or during the initial saturation of the piezometer tubes and were unresponsive. Measurements could not be collected from these piezoemeters. Of the 144 possible water level measurements, therefore, reliable measurements were collected from 114. A measurement was considered reliable if the recorded water level at the piezometer repeatedly returned to the same level following small perturbations in the water level (created by repeatedly squeezing/release of the manometer tube).

The head measurements were utilized to estimate both the effective hydraulic conductivity within the flow cell and the semivariogram of a parameter termed the head increment. The head increment, $I(x + \Delta r)$, is defined as

$$I(x + \Delta r) = H(x) - H(x + \Delta r), \tag{1}$$

where H(x) is the head measured at position x and $H(x + \Delta r)$ is the head measured at position $(x + \Delta r)$. Zheng (1998) and Zheng and Silliman (2000) show that the semivariogram of the head increment, under mean uniform flow, is closely related to the variance and integral scale of the underlying conductivity distribution. Specifically, the semivariogram of the head increment, γ_I, in the presence of a log-normally distributed, exponentially correlated conductivity distribution is given by

$$\begin{aligned}\gamma_I(r, \theta, \Delta r) &= 2\gamma_h(\Delta r, \theta) + 2\gamma_h(r, \theta) - \gamma_h(|r - \Delta r|, \theta) - \\ &\quad -\gamma_h(r + \Delta r, \theta) + 2\sigma_e^2,\end{aligned} \tag{2}$$

where γ_h is the semivariogram of the head residual ($h(x) = H(x) - \langle H(x)\rangle$; Dagan, 1985), r is the separation between two measurements of the head increment and is here taken as being parallel to Δr, θ is the orientation of r relative to the mean gradient, σ_e^2 is the variance of the measurement error.

4.2. RESULTS

Using an r oriented parallel to the mean gradient, a magnitude equal to 11.0 cm (four times the minimum piezometer spacing), and a range of possible standard deviations in the measurement error of 0.03–0.06 cm, the theoretical semivariogram

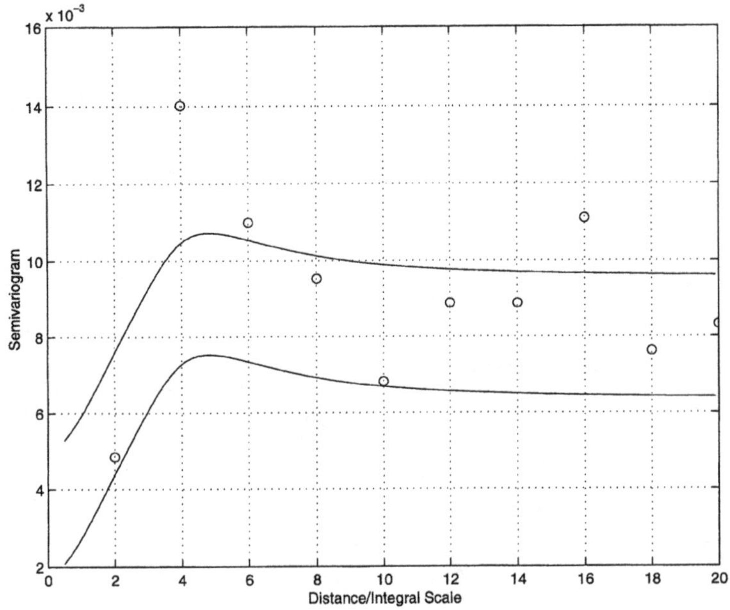

Figure 4. Sample ('o') and theoretical semivariograms for the head increments. The theoretical curves shown are for an assumed standard deviation in measurement error of 0.06 cm (upper curve) and 0.03 cm (lower curve).

of the head increment parallel to the mean gradient was determined. Figure 4 shows the theoretical semivariograms of the head increments for both the largest (0.06 cm) and smallest (0.03 cm) measurement errors considered. This figure also shows the sample semivariogram determined from measurements collected in the flow cell. It is noted that this figure was generated without calibration as all parameters required in (2) with the exception of the measurement error were known *a priori* (the integral scale and variance of ln K generated on the computer and r selected from the geometry of the piezometer locations).

The agreement between the theoretical curves and the experimental results shown in Figure 4 is considered quite strong as the shape and vertical displacement of the theoretical curve is quite sensitive to both the variance and the integral scale of the underlying conductivity field. This is illustrated in Figures 5(a) and (b). Figure 5(a) shows the theoretical curves using a measurement error of 0.05 cm, the correct variance, and assumed values of the integral scale of 0.275, 2.75 and 27.5 cm. These values for the integral scale represent an order of magnitude smaller than the true integral scale, equal to the true integral scale, and an order of magnitude larger than the true integral scale. Figure 5(b) shows the theoretical curves using a measurement error of 0.05 cm, the correct integral scale, and assumed variances of 0.1, 0.5 and 2.5. These values for the variance represent 1/5th the true value, the true value, and 5 times the true value. Based on the results shown in Figures 4, 5(a) and 5(b), it is concluded that the theory is appropriately applied to this random

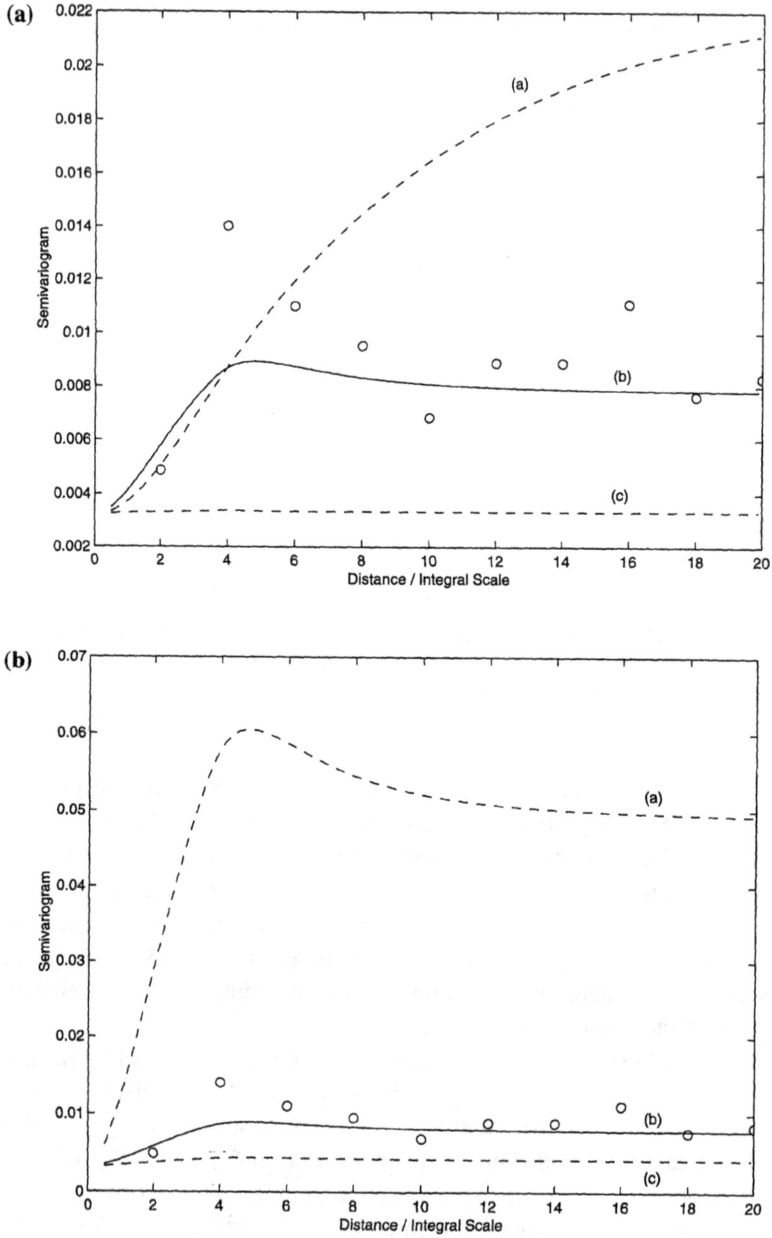

Figure 5. Bounds on fit between theory and observations with an assumed measurement error of 0.05 cm: (**a**) Curves with design variance in the conductivity, but using integral scales of 27.5 cm (a), 2.75 cm (b), and 0.275 cm (c). (**b**) Curves with design integral scale, but using variances of the ln(conductivity) of 2.5 (a), 0.5 (b), and 0.1 (c).

medium and that the experimental procedures were sufficiently rigorous to provide a reliable semivariogram of the head increment.

The data were also used to estimate the effective hydraulic conductivity via measurement of the volumetric flux through this medium with estimation of the large-scale mean hydraulic gradient based on measurements from the piezometers. From the piezometer data, it was estimated that the mean gradient was approximately 0.033 when the flow rate was 145 cc/min. Based upon this gradient and flux, in combination with the cross sectional area of (67×9.4 cm), the effective conductivity of this medium can be estimated to be 0.116 cm/s. However, given the number of uncertainties involved in making this estimate (the mean gradient, the exact width of the tanks, the measured flow rate), it is considered more appropriate to report the measured conductivity as being in the range of 0.11–0.12 cm/s. The effective conductivity that is predicted from the geometric mean of the conductivity field in the flow cell is 0.112 cm/s. Given the experimental errors inherent in this type of experiment, this agreement is considered quite reasonable. The agreement with the genometric mean also contrasts sharply with comparison to the calculated values for the arithmetic (0.139 cm/s) or harmonic (0.092 cm/s) means of this conductivity field.

5. Tracer Tests

5.1. PROCEDURES

Two tracer tests were performed, the first with a mean velocity approximately 22 times slower than the second. In both cases, a two-dimensional plume of sodium chloride was injected through the central reservoir on the inflow end of the flow cell. The plume was allowed to travel down gradient until the entire plume was captured within the electrode array. At this point, the distribution of concentration within the plume was measured and this distribution was used as the initial condition for the tracer experiment. The subsequent motion of the plume through the flow cell was then monitored.

Figures 6(a)–(d) show the motion of the tracer plume at several measurement times during the slower of the two tracer experiments. The temporal development of the plumes from the two experiments were characterized in terms of (i) the first spatial moment of the plume (the center of mass) in the direction of the mean gradient (herein termed the x direction), (ii) the first spatial moment of the plume in the direction parallel to the 0.67 m dimension of the flow cell (herein termed the y direction), and (iii) the second centered, spatial moments of the plume in both the x and y directions.

As noted earlier, concentrations within the flow cell were monitored through the 381 electrodes (calibrated *in situ*). These measurements required two levels of processing in order to allow determination of moments. The first level of processing involved the identification of anomalous measurements. This identification involved locating electrodes that were not functioning properly, identifying

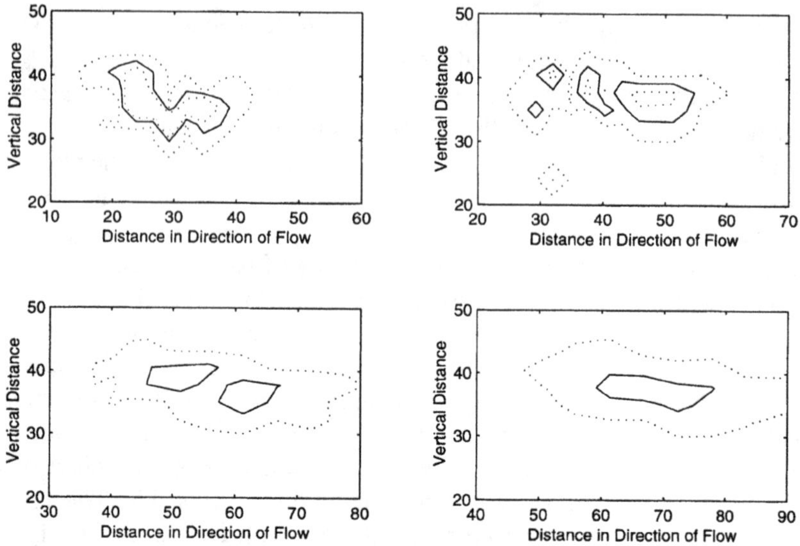

Figure 6. Movement of the tracer plume during the experiment with the lower velocity at times (after the time recorded for the initial plume) equal to: (a) 500 min, (b) 1440 min, (c) 2580 min, and (d) 3270 min. Outer dashed contour is for $C/C_0 = 0.01$. Solid contour is for $C/C_0 = 0.1$. Inner dashed contour, where present, is for $C/C_0 = 0.2$.

electrical noise which was clearly electrical error not related to the tracer plume, and determining the minimum concentration which could be considered reliable. Identifying electrodes that were not functioning properly was straight forward as these electrodes produced signals consistent either with a broken wire or an electrical short within the electrode. The second level of processing involved estimation of spatial moments from discrete spatial measurements and is discussed further below.

Of the 381 electrodes installed within the flow cell, 21 failed to produce readings. In terms of electrical noise, the fact that some electrodes were never impacted by the tracer plume allowed for identification of and correction for slight drift in the electrical readings. As one of the experiments discussed below was conducted over a period of approximately 4 days, there was a slight drift in the electronics during this experiment that was attributed to change in temperature of the reference resistor. This drift was identified through analysis of readings at electrodes not impacted by the plume and was then corrected for at all electrodes prior to analysis of the data. In addition, the contour of the final tracer concentrations led to identification of small (normalized concentration less than 0.01) patches of measured 'concentration' at random electrodes at distance from the location of the plume. Those patches which were clearly unrelated to the movement of the plume (e.g., the reading at the electrode was inconsistent from one time to the next) were eliminated prior to estimation of the spatial statistics of the plume.

The estimated reliability of the electrodes (5 mg/l), combined with the source strengths of the tracers used (500–900 mg/l), led to the conclusion that all normalized concentrations greater than 0.01 of the source strength would be considered reliably different than zero. This cutoff at normalized concentration of 0.01 also appeared consistent with the noise present within the final data sets (e.g., the background concentration within the cell appeared to have noise ranging in the normalized concentrations from slightly less than 0.0 to as high as 0.01).

Once the reliability of the data was established and the final data set identified, the second challenge was to estimate spatial moments from data collected at discrete points in space. These moments were estimated using linear interpolation within the regions enclosed by each quadruplet of sample locations, following the discussion of Garabedian *et al.* (1991). For locations where an electrode had failed, the concentration at that electrode was first estimated via linear interpolation from the four surrounding electrodes, and the moment estimation was then based on these interpolated values.

These moments, determined from the tracer test data, were then compared with the predicted variation in these moments. In terms of the first moments, the predicted variation according to first-order stochastic theory is simply a linear increase with time of the first moment parallel to the mean gradient and a constant value for the moment perpendicular to the mean gradient. For the present manuscript, the second moments are based on the solution originally developed by Dagan (1984, 1987) and presented by several authors (e.g., Freyberg, 1986; Sudicky, 1986; Sposito and Barry, 1987; Barry *et al.*, 1988) for the spread of a plume in two dimensions under ergodic conditions. It is noted that the analysis presented herein must be considered preliminary as it assumes that the initial size of the plume in the transverse direction is large as compared to the integral scale (see, for example, the discussion of ergodicity in Dagan, 1990). Further, it does not consider the impact of the original shape of the tracer plume. It is noted, for the tracer plumes realized in these experiments, that the initial transverse dimension of the plumes was on the order of five integral scales. This transverse dimension decreased with flow distance, reducing to a minimum of approximately 2.5–3.5 integral scales for some of the measurements. Additional analysis of these data, incorporating the initial geometry of the plume, the non-ergotic nature of the plume, and consideration of the variance of the predicted versus the observed concentrations, will be presented in a later manuscript.

The expressions utilized for the longitudinal (σ_{xx}^2) and transverse (σ_{yy}^2) variances are therefore given as (Freyberg, 1986; Sudicky, 1986)

$$\begin{aligned}\sigma_{xx}^2 &= 0.74(2\sigma_Y^2\lambda^2)\left\{0.75 - 1.5\mathrm{E} + \frac{L}{\lambda} + 1.5\left[\mathrm{Ei}\left(-\frac{L}{\lambda}\right) - \ln\left(\frac{L}{\lambda}\right) + \frac{\lambda}{L}\times\right.\right.\\ &\left.\left.\times \exp\left(-\frac{L}{\lambda}\right)\left(\frac{1+\lambda}{L}\right) - \frac{\lambda^2}{L^2}\right]\right\},\end{aligned}$$

$$\sigma_{yy}^2 = 0.74(2\sigma_Y^2\lambda^2)\left\{1.5\left[\frac{\lambda^2}{L^2} - \frac{\lambda}{L}\exp\left(-\frac{L}{\lambda}\right)\left(\frac{1+\lambda}{L}\right)\right] - \right.$$
$$\left. -0.5\left[\mathrm{Ei}\left(-\frac{L}{\lambda}\right) - \ln\left(\frac{L}{\lambda}\right)\right] - 0.75 + 0.5E\right\}$$

where $L =$ Velocity $\times$ time $=$ displacement of the center of mass, $\lambda =$ the integral scale of the medium, Ei is the exponential integral and E $=$ Euler's contstant. It is noted that all parameters within these equations are known *a priori* such that no calibration occurs in comparing these equations to the moments determined from the tracer test data. It is also noted that the factor, 0.74, was developed by Dagan (1987) as a means of correcting for vertical averaging of the conductivity correlation structure as an analysis is reduced from three-dimensions to two-dimensions. The magnitude of this parameter is an artifact of Dagan (1987) assuming that the vertical depth of averaging was equal to one vertical integral scale. In our situation, the covariance for the two-dimensional analysis is predetermined (assuming minimal variation in conductivity in the third direction). Hence, an argument can be made that the 0.74 should be replaced with unity. As noted below, this modification of the equation would result in further deviation between the theory and the experimental results. The theoretical results shown below are therefore based on the use of 0.74 with the recognition that a larger coefficient would lead to an increase in the estimates of both variances.

5.2. RESULTS

The two transport experiments were run with mean velocities estimated at approximately 0.015 cm/min and 0.33 cm/min. The total mass injected in the two experiments was estimated, based on the plume concentrations and geometry, to be approximately the same. Due to the manner in which the plumes were injected into the flow cell, absolute mass balances on the tracer could not be calculated (the mass of tracer could only be estimated from the measured plume concentrations). It is noted that there was significant variation in this mass (up to $+/- \sim 15\%$ of the mean estimated mass) for different measurement sets (different times) in both experiments. This variation in mass is attributed to the errors introduced by interpolating mass between measurement points in calculating the moments and measurement error (particularly at low concentrations). For further discussion of the impact of interpolation on moments, the reader is referred to Garabedian *et al.* (1991).

Figures 7(a) and 7(b) show the first moments in the x and y directions (parallel and perpendicular to the mean direction, respectively) for the two experiments performed. To allow direct comparison of these two results, the time axis for the experiment run at the higher velocity has been adjusted (multiplied) by the ratio of the volumetric fluxes used in these two experiments ($= 22.0$). As expected, the center of mass moved in the x direction approximately linearly with time (with

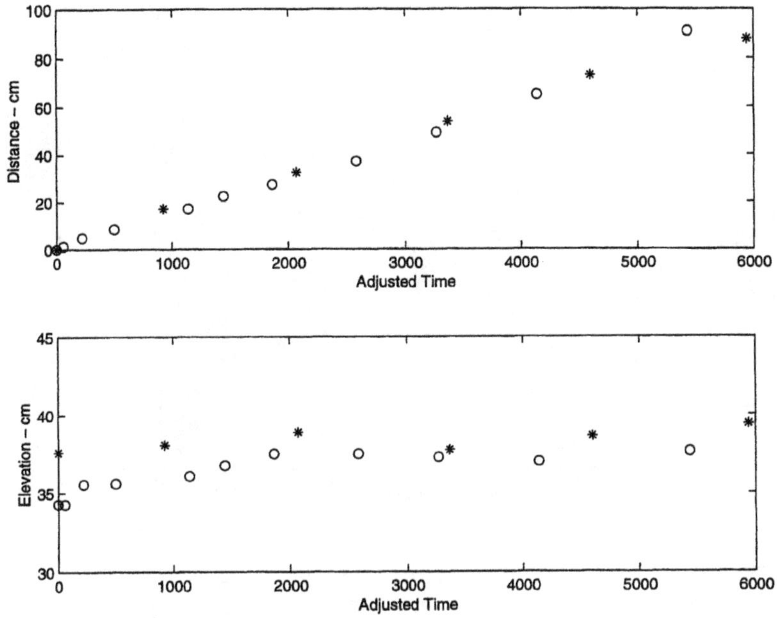

Figure 7. Variation of the first spatial moment with time for the longitudinal, or x, direction (upper plot) and the transverse, or y, direction (lower plot). The slow experiment is represented by 'o's and the fast experiment is represented by '$*$'s.

some slight variations around this mean behavior) and the center of mass in the y direction approached an approximately constant elevation after an initial period of adjustment. The fact that the two plumes were initially located at slightly different locations within the flow field is indicated by the difference in the elevation (y coordinate) of the center of mass throughout the experiment. It is interesting to note that despite the initial difference in the elevation of the center of mass, the center of mass for the two plumes converged to very similar elevations within the porous medium. It was not surprising to note that these latter elevations coincided with a region of high hydraulic conductivity.

The second spatial moments in the x and y directions are shown in Figures 8(a) and 8(b). These plots show the value of the second moments σ_{xx}^2 and σ_{yy}^2 as functions of the dimensionless quantity D/λ (dimensionless distance) such that results from the two experiments can be directly compared. For the spatial moments in the x direction, the initial value of the sample variance of each experiment has been subtracted from the calculated variances for all later measurements. Thus these results are directly comparable with the theoretical values for σ_{xx}^2. As the second sample moments in the y direction showed an initial decline in magnitude, no correction for the initial size of the plume was pursued.

It will be noted that both experiments provide very similar behavior in terms of both the longitudinal and transverse spread with the exception of the last data points. While the deviation in the last data points is somewhat troubling, it is noted

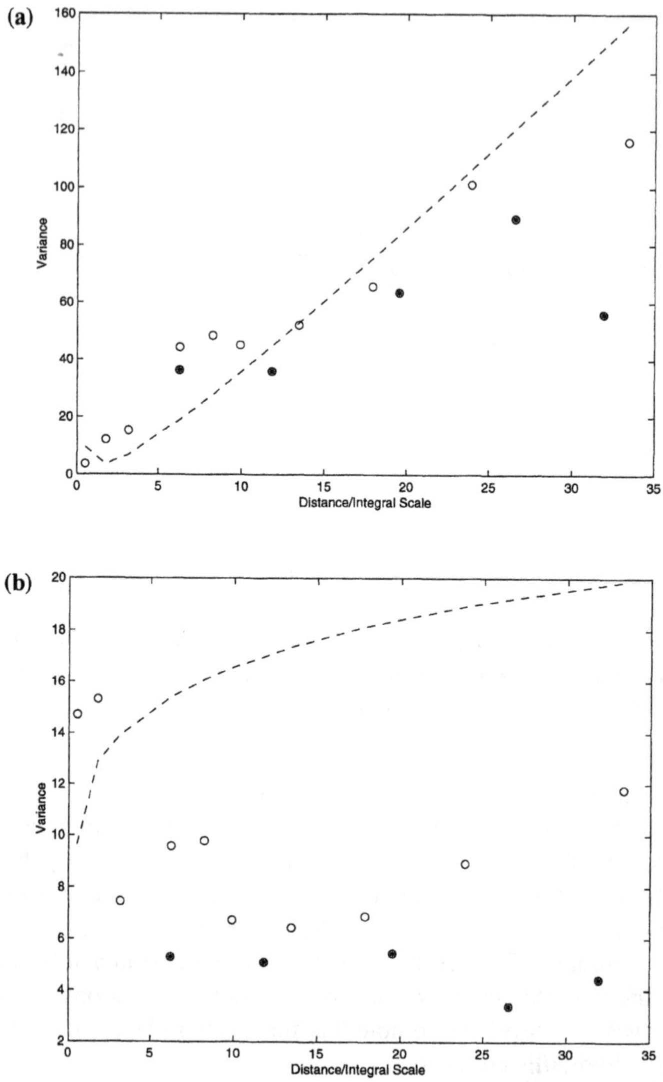

Figure 8. Variation of the second spatial moments as functions of normalized distance of travel of the center of mass. In these plots, the symbol 'o' represents the slow experiment while the symbol '∗' represents the fast experiment. The dashed line represents the theoretical prediction. The longitudinal results are shown in (a) and the transverse results are shown in (b).

that these data points are extremely sensitive to the 'cut-off' value used in analyzing the data. As noted earlier in this manuscript, it was assumed that all concentrations below $C/C_o = 0.01$ were considered unreliable and were ignored (based on our analysis of the reliability of the equipment). In Figure 9, the development of the longitudinal second moment for the slower experiment is shown for cut-off values of $C/C_o = 0.02$, 0.01 and 0.005. This figure shows that while the majority of

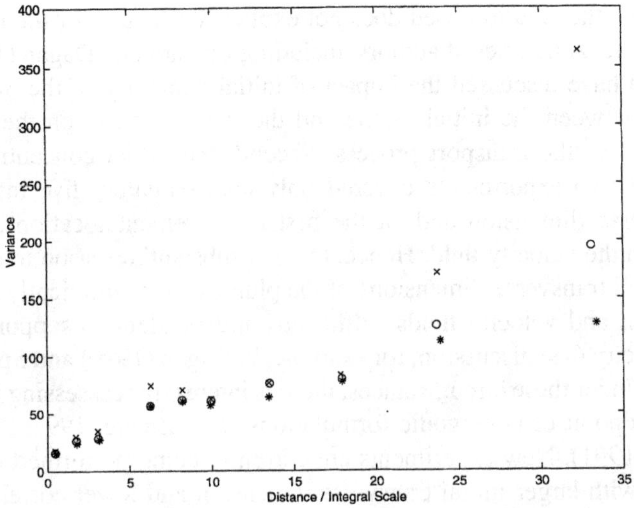

Figure 9. Sensitivity of calculated second moment to assumed cutoff value. This plot shows calculated second spatial moment in the longitudinal direction for the slow experiment assuming cutoffs of $C/C_o = 0.005$ ('*x*'), 0.01 ('o'), and 0.02 ('$*$').

the data points are relatively robust to the choice of cutoff, the last data point is extremely sensitive to this value (the plume is relatively disperse at this point and contains large regions of low concentration). Hence, we place relatively low reliability in the last data points for both experiments (see discussion for further comment on this cutoff).

The results for the longitudinal variance (Figure 8(a)) lead to two observations. First, the rate of increase in longitudinal spread is quite similar for the two experiments. This indicates that the dominant mechanism leading to the spread of the plume is likely the same in both experiments. If any difference exists, it would appear that the experiment run at the higher velocity is showing a slightly lower rate of spread. It is noted that, at the pore scale and using the mean velocity for the flow field, the Peclet number (ud/Dm = local velocity $*$ mean grain diameter/Coefficient of molecular diffusion) was on the order of 100 for the experiment run at the lower mean velocity. Hence, it is concluded that local diffusion had only minimal impact on transport in these two experiments. The resulting similarity in the longitudinal spread is therefore to be expected as mechanical dispersion is anticipated to be the dominant dispersive mechanism for both experiments.

The second observation to be derived from these results is the fact that the observed longitudinal variance of the plume, while of the correct magnitude over these travel distances, is increasing at a rate lower than the rate predicted using the solution presented above. Further, as noted above, the coefficient (0.74) used in this expression is likely leading to an underestimate in the predicted rate of spread such that the difference between theory and observation is likely larger than that shown in the figure. We anticipate that there are at least two reasons for this discrepancy. First, the solution used does not explicitly account for the initial dimension of the plume. A number of authors, including Sposito and Dagan (1994) and Fiori (this issue) have discussed the impact of initial geometry of the plume, and the correlation between the initial plume and the velocity field, on the final stochastic description of the transport process. Second, the initial concentration field for each of the two experiments covered only approximately five integral scales in the transverse dimension and, at the first measurement locations, was likely correlated with the velocity field. Hence, there is substantial reason to question whether the initial transverse dimensions of the plumes were sufficiently large and the concentration and velocity fields sufficiently independent to support the assumption of ergodicity (see discussion, for example, in Dagan (1990) and Sposito and Dagan (1994)). Under these circumstances, there is interest in reassessing these results from the standpoint of nonergodic formulations (e.g., Zhang, 1997; Zhang *et al.*, 1996; Dagan, 1991). New experiments are currently being performed using concentration fields with larger initial transverse dimension and lower correlation with the velocity field. However, these extensions are beyond the scope of the present manuscript.

The results for the transverse dispersion (Figure 8(b)) indicate that there is minimal spreading in the transverse direction, with some reduction in spread from the initial plume to the plume observed in the central region of the flow field. From the data shown in Figure 8(b), it is apparent that σ_{yy}^2 is not increasing as fast as predicted by the theory presented above. Further, the results do not rule out the possibility of a zero (or possibly negative) rate of spread in the transverse direction as suggested in the early work of Dagan (1984).

These results lend support to the arguments made by other authors (e.g., Garabedian *et al.*, 1991) that field observations of transverse dispersivities which are larger than predicted by theory may be related to the impact of transport mechanisms that were not included in the formulation of the theory. Further, these results provide additional impetus to consider the impact of the finite size of the initial plumes and the correlation of the initial concentration field with the velocity field on the observed second moments. In particular, both plumes were significantly impacted by a large region of low permeability sands near the point of initial measurement. Had the initial transverse dimension of the plume been significantly larger (e.g., 10–20 integral scales), the impact on this local low permeability zone would have played a much smaller role in determining the sample variance in the transverse dimension. As a result, there is suspicion that the transverse behavior of this plume may be specific to this combination of realization of the conductivity field and location of the injection point. Further discussion of transverse dispersion is, therefore, delayed until completion of a second set of experiments (currently being performed) in a second realization of this conductivity field.

6. Discussion

These results represent, to our knowledge, one of the few attempts to directly compare stochastic theory with laboratory experiments. In this case, the medium was packed to mimic a log-normally distributed, second-order stationary random field with an exponential correlation structure. As such, the medium reproduced the conductivity structure included in several previously published stochastic theories, including those by Zheng (1998) and Dagan (1984).

Results from these experiments were consistent with the predicted effective hydraulic conductivity and showed excellent agreement with the solution for the semivariogram of head increments. It was observed in the tracer tests, however, that the experimental results showed great sensitivity (at late time) of the calculated second moments to the choice of the concentration cut-off. Further, significant deviations were observed in comparison of the transport theory considered here and the results obtained from the transport experiments. In particular, while the theory reproduced the correct order-of-magnitude of the variance for the plume, it over predicted the rate of increase of the longitudinal second moment. The theory also predicted an increase in the transverse second moment when the experimental results indicated that the moment was constant, or decreasing, over much of the flow domain.

The sensitivity of the estimate of spatial variance to the assumed cut-off for the concentration (Figure 9) illustrates a potentially significant difficulty that must be addressed in estimating the spatial moments of a tracer plume in the field. Specifically, as the plume is transported through a heterogeneous medium, the second spatial moments become increasingly sensitive to the regions of the aquifer occupied by low concentration of the tracer. As the concentration decreases, however, analytical detection limits will often limit our ability to quantify the concentration. The result, again as illustrated in Figure 9 for the current series of experiments, is that estimates of the second spatial moments will become increasingly uncertain as a greater portion of the plume is occupied by concentrations below our analytical detection limits. Further, this result implies that two tracers, with different detection limits (relative to their initial concentrations) may provide different estimates of the second spatial moments.

It is argued that sampling design may also help to explain the difference between the hydraulic results (in which theory matched experiment) and the transport results (in which theory failed to match experiment). The distribution of piezometers allowed relatively uniform sampling of the head and head increment over the central portion of the flow field. As a result, the data were sampled over a large number of integral scales in both the horizontal and vertical dimension, thus allowing confidence in the assumption that ergodic conditions had been achieved. In contrast, ergodic conditions were likely not obtained in the transport experiments. Dagan (1990) argues that the transverse dimension of the plume should be much greater than the integral scale in order to observe transport behavior that is consistent with predictions based on assumptions of ergodicity. The results presented here indicate that the initial widths of these two plumes (less than five integral scales) were insufficient to achieve such ergodic conditions.

Reassessment of these transport plumes accounting specifically for both the initial geometry of the plume and non-ergodic conditions (e.g., Zhang, 1997) are topics of continuing analysis. Further, these experiments are being repeated in a new realization of the conductivity field and with large initial plume dimension (and a different method of introducing the tracer) in an effort to minimize these non-ergodic effects.

Acknowledgement

This work was supported by a grant from the Subsurface Science Program at the U.S. Department of Energy.

References

Adams, E. E. and Gelhar, L. W.: 1992, Field study of dispersion in a heterogeneous aquifer; 2. Spatial moments analysis, *Water Resour. Res.* **28**(12), 3293–3307.

Bakr, A. A., Gelhar, L., Gutjahr, A. and MacMillan, T.: 1978, Stochastic analysis of spatial variability in subsurface flow: 1. Comparison of one- and three-dimensional flows, *Water Resour. Res.* **14**(2), 263–271.

Barry, D. A., Coves, J. and Sposito, G.: 1988, On the Dagan model of solute transport in groundwater: Application to the Borden site, *Water Resour. Res.* **24**(10), 1805–1817.

Berkowitz, B. and Scher, H.: 1998, Theory of anomalous chemical transport in fracture networks, *Phys. Rev. E* **57**(5), 5858–5869.

Berkowitz, B., Scher, H. and Silliman, S.: 2000, Anomalous trasnport in laboratory scale, heterogeneous porous media, *Water Resour. Res.* **36**(1), 149–158.

Calvete, F. J. S.: 1997, Application of geostatistics in subsurface hydrology, In: G. Dagan and S. Neuman (eds), *Subsurface Flow and Transport: A Stochastic Approach*, Cambridge University Press, 44–61.

Dagan, G.: 1990, Transport in heterogeneous porous formations: Spatial moments, ergodicity, and effective dispersion, *Water Resour. Res.* **26**(6), 1281–1290.

Dagan, G.: 1984, Solute transport in heterogeneous porous formations, *J. Fluid Mech.* **145**, 151–177.

Dagan, G.: 1987, Theory of solute transport by groundwater, *Ann. Rev. Fluid Mech.* **19**, 183–215.

Dagan, G.: 1991, Dispersion of a passive solute in non-ergodic transport by steady velocity fields in heterogeneous formations, *J. Fluid Mech.* **233**, 197–210.

Fiori, A.: 2000, The relative dispersion and mixing of passive solutes in transport in geologic media, *Transport in Porous Media* **42**(1–2), 69–83 (this issue).

Freeze, R. A.: 1975, A stochastic-conceptual analysis of one-dimensional groundwater flow in nonuniform homogeneous media, *Water Resour. Res.* **11**, 725–741.

Freyberg, D. L.: 1986, A natural gradient experiment on solute transport in a sand aquifer; 2. Spatial moments and the advection and dispersion of nonreactive tracers, *Water Resour. Res.* **22**(13), 2031–2046.

Garabedian, S. P., LeBlanc, D. R., Gelhar, L. W. and Celia, M. A.: 1991, Large-scale natural gradient tracer test in sand and gravel, Cape Cod, Massachusetts; 2. Analysis of spatial moments for a nonreactive tracer, *Water Resour. Res.* **27**(5), 911–924.

Indelman, P. and Abramovich, B.: 1994, Nonlocal properties of nonuniform averaged flows in heterogeneous media, *Water Resour. Res.* **30**, 3385–3393.

Kitanidis, P. K.: 1988, Prediction by the method of moments of transport in a heterogeneous formation, *J. Hydrol.* **102**, 453–473.

Naff, R. L.: 1991, Radial flow in heterogeneous porous media: An analysis of specific discharge, *Water Resour. Res.* **27**, 307–316.

Neuman, S. P. and Orr, S.: 1993, Prediction of steady state flow in nonuniform geologic media by conditional moments: Exact nonlocal formalism, effective conductivities, and weak approximation, *Water Resour. Res.* **29**, 341–364, 1993.

Rubin, Y.: 1997, Transport of inert solutes by groundwater: recent developments and current issues, In: G. Dagan and S. Neuman, (eds), *Subsurface Flow and Transport: A Stochastic Approach*, Cambridge University Press, pp. 115–132.

Rubin, Y. and Dagan, G.: 1988, Stochastic analysis of boundaries effects on head spatial variability in heterogeneous aquifers; 1. Constant head boundary, *Water Resour. Res.* **24**(10), 1689–1697.

Rubin, Y. and Dagan, G.: 1989, Stochastic analysis of boundaries effects on head spatial variability in heterogeneous aquifers; 2. Impervious boundary, *Water Resour. Res.* **25**(4), 707–712.

Silliman, S. E., Zheng, L. and Conwell, P.: 1998, The use of laboratory experiments for the study of conservative solute transport in heterogeneous porous media, *Hydrogeol. J.* **6**(16), 6–177.

Silliman, S. E. and Caswell, S.: 1998, Observations of measured hydraulic conductivity in two artificial, confined aquifers with boundaries, *Water Resour. Res.* **34**(9), 2203–2213.

Smith, L. and Schartz, F.: 1981, Mass transport: 1. A stochastic analysis of macroscopic dispersion, *Water Resour. Res.* **16**(2), 303–313.

Sposito, G. and Barry, D. A.: 1987, On the Dagan model of solute transport in groundwater: Foundational aspects, *Water Resour. Res.* **23**(10), 1867–1875.

Sposito, G. and Dagan, G.: 1994, Predicting solute plume evolation in heterogeneous porous formations, *Water Resour. Res.* **30**(2), 585–589.

Sudicky, E. A.: 1986, A natural gradient experiment on solute transport in a sand aquifer: Spatial variability of hydraulic conductivity and its role in the dispersion process, *Water Resour. Res.* **22**(13), 2069–2082.

Tompson, A. F. B., Ababou, R. and Gelhar, L. W.: 1989, Implementation of the three-dimensional turning band random field generator, *Water Resour. Res.* **25**(10), 2227–2243.

Warren, J. E. and Price, H. S.: 1961, Flow in heterogeneous porous media, *Soc. Petrol. Engng J.* **1**, 153–169.

Welty, C. and Elsner, M.: 1997, Constructing correlated random fields in the laboratory for observations of fluid flow and mass transport, *J. Hydrol.* **202**(1–4), 192–211.

Welty, C. and Gelhar, L.: 1992, Simulation of large-scale transport of variable density and viscosity fluids using a stochastic mean model, *Water Resour. Res.* **28**(3), 815–827.

Zhang, Y.-K.: 1997, On the variances of second spatial moments of a nonergodic plume in heterogeneous aquifers, *Water Resour. Res.* **33**(8), 1893–1900.

Zhang, Y.-K., Zhang, D. and Lin, J.: 1996, Nonergodic solute transport in three-dimensional heterogeneous isotropic aquifers, *Water Resour. Res.* **32**(9), 2955–2963.

Zheng, L.: 1998, Geostatistical characterization of the transmissivity distribution in groundwater aquifers using hydraulic head measurements, Ph.D. Dissertation, University of Notre Dame, Notre Dame, Indiana, 127 pp.

Zheng, L. and Silliman, S.: 2000, Estimating the variance and integral of the transmissivity field using head residual increments, *Water Resour. Res.* **36**(5), 1353–1358.

Transport in Porous Media **42:** 109–132, 2001.

Large-Time Spatial Covariance of Concentration of Conservative Solute and Application to the Cape Cod Tracer Test

MARILENA PANNONE* and PETER K. KITANIDIS
Department of Civil and Environmental Engineering, Stanford University, Stanford, CA

(Received: 22 January 1999; in final form: 2 October 1999)

Abstract. Most studies on conservative transport in stationary velocity fields have focused on the description of the concentration mean. In this work, we use a Lagrangian methodology to develop an analytical expression for the spatial covariance of the concentration, based on the central limit theorem and applicable to large times after injection. We use this expression to analyze the conservative tracer test data from the natural gradient experiment conducted at Cape Cod in 1985–1988, in which bromide was quickly injected into the aquifer and the concentration was measured at many locations at certain points in time. The parameters that determine the concentration mean had been estimated in previous studies. Here, the two-particle covariance matrix, needed to describe the concentration covariance function, is derived from the measurements through a maximum likelihood method. Then, the data are interpolated on a grid using simple kriging, and contour maps of the concentration estimates are plotted. The results of cross-validation indicate that the model is consistent with the field measurements and the kriging estimates appear realistic.

Key words: concentration covariance model, two-particle moments, maximum likelihood estimation, field measurements, kriging interpolation.

1. Introduction and Background

Solute transport in natural porous formations is usually described at three distinct spatial levels of resolution: pore, laboratory, and field (Bear, 1972; Dagan, 1989). The upscaling from the pore to the laboratory and from the laboratory to the field leads to conservation equations that contain constitutive parameters globally representing the unresolved variability. Consider the advection–dispersion equation describing the transport of an ideal tracer

$$\frac{\partial c}{\partial t} + \mathbf{u} \cdot \nabla c = \nabla \cdot (\mathbf{D} \nabla c) . \tag{1}$$

At the laboratory scale, $\mathbf{u}$ is the local seepage velocity vector, the macroscopic contribution of the highly irregular velocity field within the pore network; $\mathbf{D}$ is the local dispersion tensor, which represents the superposition of molecular diffusion and hydrodynamic dispersion due to velocity fluctuations that were averaged out; finally, c is the concentration defined over a volume that contains many pores.

* Now at Department of Environmental Physics and Engineering, University of Basilicata, Italy.

The seepage velocity fluctuates with the hydraulic conductivity, which is usually extremely variable in geologic formations. Since it cannot be measured everywhere, it is impossible to predict exactly the small-scale fluctuations of the solute concentration over a whole formation. The most common approach is to attempt to resolve the slowly varying components of the distribution, averaging over the microscale heterogeneities. For instance, the stochastic approach focuses on the prediction of the concentration ensemble mean as function of the statistics of the hydraulic conductivity. Under certain conditions, the ensemble mean concentration $\langle c \rangle$ satisfies an advection–dispersion equation

$$\frac{\partial \langle c \rangle}{\partial t} + \mathbf{U} \cdot \nabla \langle c \rangle = \nabla \cdot (\mathbf{D}_{\mathrm{m}} \nabla \langle c \rangle), \tag{2}$$

where $\mathbf{U}$ is the ensemble mean velocity and $\mathbf{D}_{\mathrm{m}}$ is the macrodispersion tensor. In statistically homogeneous velocity fields and for sufficiently large travel times, both $\mathbf{U}$ and $\mathbf{D}_{\mathrm{m}}$ are constant (Dagan, 1982, 1984, 1989; Gelhar and Axness, 1983; Neuman *et al.*, 1987; Rubin, 1990).

However, the ensemble mean concentration $\langle c \rangle$ is just the best estimate of c, given the statistics of the small-scale variability. A useful measure of the expected deviation is represented by the concentration variance σ_c^2, which is the mean square difference between the actual and the mean distributions (Dagan, 1982, 1984, 1989; Graham and McLaughlin, 1989; Vomvoris and Gelhar, 1990; Rubin, 1991; Kapoor and Gelhar, 1994a, 1994b; Kapoor and Kitanidis, 1996, 1998; Dagan and Fiori, 1997; Kabala, 1997; Pannone and Kitanidis, 1999).

Concentration variance and dilution kinetics have been shown to depend on the ratio between two-particle and one-particle second-order moments (Pannone and Kitanidis, 1999), defined as

$$W_{ij} = \langle X_i' Y_j' \rangle = \langle X_i Y_j \rangle - \langle X_i \rangle \langle Y_j \rangle \tag{3}$$

and

$$P_{ij} = \langle X_i' X_j' \rangle = \langle X_i X_j \rangle - \langle X_i \rangle \langle X_j \rangle, \tag{4}$$

where vectors $\mathbf{X}$ and $\mathbf{Y}$ indicate the coordinates of two particles that started at the same location; the angle brackets indicate ensemble averaging; and $X_i' = X_i - \langle X_i \rangle$. Note that two particles that start from the same position follow different trajectories even in a single realization, because of local dispersion that makes them sample different streamlines. In macroscopically Fickian transport processes, and under asymptotic conditions, P_{ij} increases linearly with time. However, the rate of increase of W_{ij} gradually diminishes (e.g., Pannone and Kitanidis, 1999). The smaller W_{ij} compared to P_{ij}, the less correlated the locations of the two particles. Small correlation indicates a high degree of dilution, as particles started from the same location occupy different locations in a bigger volume. More complete dilution also means smaller concentration variance (Kapoor and Kitanidis, 1996, 1998), as the concentration in any realization becomes smoother and closer to its ensemble mean.

The two-particle moments have already been used in the study of turbulence (Richardson, 1926; Batchelor, 1952; Fisher *et al.*, 1979) and have also been analyzed in the context of groundwater flow and transport in heterogeneous formations through a Eulerian approach (Rajaram and Gelhar, 1993) as well as a linearized Lagrangian approach (Dagan, 1989; Fiori and Dagan, 1999). The theoretical analysis of the two-particle covariance under general conditions is complicated, since the nonlinear interaction between local- and macrodispersive mechanisms must be considered. Nonlinear effects due to the deviation of the particle trajectory from its mean have been analyzed by Neuman and Zhang (1990); Zhang and Neuman (1990); and Zhang (1995), in terms of single-particle second moments.

Pannone and Kitanidis (1999), performed a qualitative analysis suggesting that the rate of increase of $\mathbf{W}$ should diminish with increasing time; additionally, they presented exact results for the simple case of stratified confined formations and Fickian regime with constant macrodispersion coefficients, indicating that the two-particle covariance tends to a constant value. The analytical derivation was possible by considering the nonstationarity of the two-particle Lagrangian velocity.

The authors derived an expression for the large-time concentration variance in a Lagrangian framework, using normality as a consequence of the central value theorem to provide closure. In this work, we extend this approach to develop expressions for the large-time concentration covariance. The problem of concentration covariance has been previoulsy studied (Graham and McLaughlin, 1989; Neuman, 1993; Zhang, 1995; Kapoor and Kitanidis, 1997). Kapoor and Kitanidis (1997), formulated the problem of concentration covariance deriving its transport equation through a leading-order approach. Their formulation included advection with the mean velocity, macrodispersion due to velocity fluctuations, and a decay term due to local dispersion. This was an extension of previous work where, however, some of the above terms were neglected.

We apply the proposed Lagrangian theory to the data from the Cape Cod tracer test (LeBlanc *et al.*, 1991; Garabedian *et al.*, 1991). The data were previously analyzed for the parameters describing the mean concentration (Garabedian *et al.*, 1991; Thierrin and Kitanidis, 1994). We develop and apply a methodology to determine the remaining parameters needed to quantify concentration variance–covariance and dilution. Once these additional parameters are determined, we perform kriging to obtain an estimate of the actual plume concentration conditional on concentration observations.

2. Large-time Concentration Covariance

The spatial covariance between the fluctuations of concentration at two different locations $\mathbf{x}$ and $\mathbf{y}$ is

$$R_c(\mathbf{x}, \mathbf{y}, t) = \left\langle c'(\mathbf{x}, t)\, c'(\mathbf{y}, t)\right\rangle = \langle c(\mathbf{x}, t)\, c(\mathbf{y}, t)\rangle - \langle c(\mathbf{x}, t)\rangle \langle c(\mathbf{y}, t)\rangle, \quad (5)$$

where $c(\mathbf{x}, t)$ indicates the concentration at location $\mathbf{x}$ and time t.

Without loss of generality, the concentration function may be written as

$$c\left(\mathbf{x}, t\right) = \int_{V_0} C_0\left(\mathbf{a}\right) \delta\left(\mathbf{x} - \mathbf{X}\left(\mathbf{a}, t\right)\right) \mathrm{d}\mathbf{a}, \tag{6}$$

where $\mathbf{X}(\mathbf{a}, t)$ is the position at time t of a particle starting at time $t = 0$ from location $\mathbf{a}$; $C_0(\mathbf{a})$ is the initial concentration; V_0 is the injection volume, and δ is the Dirac function. Then,

$$c(\mathbf{x}, t)c(\mathbf{y}, t) = \int_{V_0}\int_{V_0} C_0\left(\mathbf{a}\right) C_0\left(\mathbf{b}\right) \delta\left(\mathbf{x} - \mathbf{X}\left(\mathbf{a}, t\right)\right) \delta\left(\mathbf{y} - \mathbf{Y}\left(\mathbf{b}, t\right)\right) \mathrm{d}\mathbf{a}\, \mathrm{d}\mathbf{b}, \tag{7}$$

where $\mathbf{Y}(\mathbf{b}, t)$ represents the position at time t of a different particle that started at location $\mathbf{b}$.

The total positions $\mathbf{X}$ and $\mathbf{Y}$ can be subdivided in a mean and its fluctuation;

$$\mathbf{X}\left(\mathbf{a}, t\right) = \mathbf{a} + \mathbf{U}t + \mathbf{X}'\left(\mathbf{a}, t\right) \tag{8}$$

and

$$\mathbf{Y}\left(\mathbf{b}, t\right) = \mathbf{b} + \mathbf{U}t + \mathbf{Y}'\left(\mathbf{b}, t\right). \tag{9}$$

By virtue of the stationarity of the velocity field, the marginal statistics of the random variables $\mathbf{X}'$ and $\mathbf{Y}'$ do not depend on the initial location. However, their joint statistics depend on the separation between the two initial positions.

The statistical moments of the concentration function are obtained through averaging the actual quantities over the different possible realizations of the trajectory fluctuations. From (6),

$$\langle c\left(\mathbf{x}, t\right)\rangle = \int\int_{V_0} C_0\left(\mathbf{a}\right) \delta\left(\mathbf{x} - \mathbf{a} - \mathbf{U}t - \mathbf{X}'\right) f_{\mathbf{X}'}\left(\mathbf{X}'\right) \mathrm{d}\mathbf{a}\, \mathrm{d}\mathbf{X}'. \tag{10}$$

From (7),

$$\begin{aligned} \langle c(\mathbf{x}, t)c(\mathbf{y}, t)\rangle = & \iiint_{V_0}\int_{V_0} C_0\left(\mathbf{a}\right) C_0\left(\mathbf{b}\right) \delta\left(\mathbf{x} - \mathbf{a} - \mathbf{U}t - \mathbf{X}'\right) \times \\ & \times \delta\left(\mathbf{y} - \mathbf{b} - \mathbf{U}t - \mathbf{Y}'\right) f_{\mathbf{X}'\mathbf{Y}'}\left(\mathbf{X}', \mathbf{Y}'; \mathbf{a} - \mathbf{b}\right) \mathrm{d}\mathbf{a}\, \mathrm{d}\mathbf{b}\, d\mathbf{X}'\, \mathrm{d}\mathbf{Y}', \end{aligned} \tag{11}$$

where $f_{\mathbf{X}'}\left(\mathbf{X}'\right)$ indicates the probability density function of $\mathbf{X}'$ and $f_{\mathbf{X}'\mathbf{Y}'}(\mathbf{X}', \mathbf{Y}'; \mathbf{a} - \mathbf{b})$ is the joint probability density function of $\mathbf{X}'$ and $\mathbf{Y}'$. Then,

$$\langle c\left(\mathbf{x}, t\right)\rangle = \int_{V_0} C_0\left(\mathbf{a}\right) f_{\mathbf{X}'}\left(\mathbf{x} - \mathbf{a} - \mathbf{U}t\right) \mathrm{d}\mathbf{a} \tag{12}$$

and

$$\begin{aligned} \langle c\left(\mathbf{x}, t\right) c\left(\mathbf{y}, t\right)\rangle = & \int_{V_0}\int_{V_0} C_0\left(\mathbf{a}\right) C_0\left(\mathbf{b}\right) \times \\ & \times f_{\mathbf{X}'\mathbf{Y}'}\left(\mathbf{x} - \mathbf{a} - \mathbf{U}t, \mathbf{y} - \mathbf{b} - \mathbf{U}t; \mathbf{a} - \mathbf{b}\right) \mathrm{d}\mathbf{a}\, \mathrm{d}\mathbf{b}. \end{aligned} \tag{13}$$

Due to the stationarity of the velocity field, and by virtue of the central limit theorem, $f_{\mathbf{X}'}$ and $f_{\mathbf{X}'\mathbf{Y}'}$ are assumed to be asymptotically normal (Pannone and Kitanidis, 1999).

This means that these distributions tend to become normal as time passes and, in practical terms, that they are approximated by normal ones. Thus, they are completely described by first and second moments of $\mathbf{X}'$ and $\mathbf{Y}'$,

$$f_{\mathbf{X}'}\left(\mathbf{X}'\right) = (2\pi)^{-n/2}\,|\mathbf{P}|^{-1/2}\exp\left(-\tfrac{1}{2}\mathbf{X}'^{T}\mathbf{P}^{-1}\mathbf{X}'\right), \tag{14}$$

$$\begin{aligned} f_{\mathbf{X}'\mathbf{Y}'}\left(\mathbf{X}',\mathbf{Y}';\mathbf{a}-\mathbf{b}\right) &= f_{\mathbf{Y}'|\mathbf{X}'}\left(\mathbf{Y}'|\mathbf{X}'\right) f_{\mathbf{X}'}\left(\mathbf{X}'\right) \\ &= (2\pi)^{-n}\begin{vmatrix} \mathbf{P} & \mathbf{W} \\ \mathbf{W} & \mathbf{P} \end{vmatrix}^{-1/2} \times \\ &\times \exp\left(-\frac{1}{2}\left[\mathbf{X}'^{T}\ \mathbf{Y}'^{T}\right]\begin{pmatrix} \mathbf{P} & \mathbf{W} \\ \mathbf{W}^{T} & \mathbf{P} \end{pmatrix}^{-1}\begin{bmatrix} \mathbf{X}' \\ \mathbf{Y}' \end{bmatrix}\right), \end{aligned} \tag{15}$$

where n is the dimensionality of the flow domain; $|\mathbf{A}| = \det\mathbf{A}$; the vertical line indicates statistical conditioning, and the tensors of single- and two–particle covariances, $\mathbf{P}$ and $\mathbf{W}$, are defined by Equations (4) and (3). $\mathbf{P}$ and $\mathbf{W}$ depend on t and $\mathbf{W}$ also depends on the distance $\mathbf{a}-\mathbf{b}$.

Equations (12) through (15) apply for any initial condition. For further simplification, consider now that the solute mass is injected at a single point $\mathbf{a}_0$,

$$C_0(\mathbf{a}) = \frac{M}{\eta}\delta\left(\mathbf{a}-\mathbf{a}_0\right), \tag{16}$$

$$C_0(\mathbf{b}) = \frac{M}{\eta}\delta\left(\mathbf{b}-\mathbf{a}_0\right), \tag{17}$$

where M is the total mass and η the porosity. Then,

$$\langle c(\mathbf{x},t)\rangle = \frac{M}{\eta} f_{\mathbf{X}'}\left(\mathbf{x}-\mathbf{a}_0-\mathbf{U}t\right) \tag{18}$$

and

$$\langle c(\mathbf{x},t)c(\mathbf{y},t)\rangle = \left(\frac{M}{\eta}\right)^{2} f_{\mathbf{X}'\mathbf{Y}'}\left(\mathbf{x}-\mathbf{a}_0-\mathbf{U}t,\mathbf{y}-\mathbf{a}_0-\mathbf{U}t\right) \tag{19}$$

where, now, the two-particle statistics are for two particles starting at the same location.

In terms of spatial covariance, we obtain

$$\begin{aligned} R_c(\mathbf{x},\mathbf{y},t) &= \left(\frac{M}{\eta}\right)^{2} f_{\mathbf{X}'\mathbf{Y}'}\left(\mathbf{x}-\mathbf{a}_0-\mathbf{U}t,\mathbf{y}-\mathbf{a}_0-\mathbf{U}t\right) - \\ &- \left(\frac{M}{\eta}\right)^{2} f_{\mathbf{X}'}\left(\mathbf{x}-\mathbf{a}_0-\mathbf{U}t\right) f_{\mathbf{X}'}\left(\mathbf{y}-\mathbf{a}_0-\mathbf{U}t\right). \end{aligned} \tag{20}$$

3. Estimation of Two-particle Moments and Kriging

The concentration c at time t is modeled as a nonstationary random field with the first and second moments given in the previous section. Kriging can be used to condition the predicted concentration on the observations of c that are available at a number of locations. Before applying kriging, however, the observations can be used to determine the model parameters. The analysis from this point on is for a specific time, though we will not show the dependence on t.

Parameter estimation can be accomplished using a cross-validation approach, that is, by selecting parameters that reproduce one set of observations given another set (Kitanidis, 1991). Since the model is described through only the first and second moments, kriging predictors are reasonable. Cross-validation using kriging may be applied through a maximum likelihood approach, assuming that the concentration field is Gaussian. This procedure allows us to search for the values of the parameters that best explain the oscillations of the actual concentration field about its ensemble mean.

Consider that at a time t there are N observations z_i. If c_i is the actual concentration and e_i the measurement error,

$$z_i = c_i + e_i \qquad i = 1, ...N. \tag{21}$$

These errors are modeled as normally distributed, with zero mean and covariance function R_e. It is reasonable to assume that they are uncorrelated and that the standard deviation σ_e linearly increases with the measured concentration

$$R_e\left(\mathbf{x}_i, \mathbf{x}_j\right) = \begin{cases} \sigma_e^2\left(\mathbf{x}_i\right) & \text{if } i = j \\ 0 & \text{if } i \neq j, \end{cases} \tag{22}$$

where

$$\sigma_e\left(\mathbf{x}_i\right) = \alpha z\left(\mathbf{x}_i\right) + \beta. \tag{23}$$

In (23), $z_i = z\left(\mathbf{x}_i\right)$ indicates the observed concentration at location $\mathbf{x}_i$; for concentrations given in mg/l or ppm, $\beta = 0.1$ mg/l represents a minimum error that accounts for the uncertainty related to measurement of values near the detection limit. The coefficient α represents the relative error. We test two reasonable (G. Hopkins, M. Reinhard, personal communication) values of α: 0.05 and 0.07. Additional studies should be performed to evaluate the reliability of the data.

A priori, the observation vector $\mathbf{z}$ is Gaussian with mean $\langle\mathbf{c}\rangle$ and covariance matrix $\mathbf{\Gamma} + \mathbf{R}_e$. Here, $\langle\mathbf{c}\rangle$ is the N by 1 vector of unconditional mean values at observation locations; $\mathbf{\Gamma}$ is the N by N symmetric matrix whose ijth element is the unconditional covariance of concentration at locations i and j, and $\mathbf{R}_e$ is the N by N diagonal matrix whose iith element is the error variance at location i. Thus, the negative logarithm of the pdf of $\mathbf{z}$ given the structural parameters is, except for a constant term

$$L = \tfrac{1}{2}\ln\left|\mathbf{\Gamma} + \mathbf{R}_e\right| + \tfrac{1}{2}\left(\mathbf{z} - \langle\mathbf{c}\rangle\right)^T\left(\mathbf{\Gamma} + \mathbf{R}_e\right)^{-1}\left(\mathbf{z} - \langle\mathbf{c}\rangle\right). \tag{24}$$

The *a posteriori*, that is, given the observations, estimates of the structural parameters can then be determined by minimizing L with respect to them.

The structural parameters in this problem are the quantities M, $\mathbf{U}$, $\mathbf{P}$, and $\mathbf{W}$. The first three of them are needed to estimate the mean concentration and have thus been evaluated in previous studies (Garabedian *et al.*, 1991; Thierrin and Kitanidis, 1994). We will use these estimates and we will limit our attention to finding $\mathbf{W}$. Thus, our problem is to minimize L as function of the different W_{ij}.

Calculation of the second moments for the field solute distribution (Garabedian *et al.*, 1991) produced a non-diagonal tensor of solute variance terms. However, to make the investigation possible within a reasonable effort, we have performed all the computations in a reference frame whose axes are parallel to the plume principal directions. In this case, both $\mathbf{P}$ and $\mathbf{W}$ are diagonal matrices

$$\mathbf{P} = \begin{bmatrix} P_{11} & 0 & 0 \\ 0 & P_{22} & 0 \\ 0 & 0 & P_{33} \end{bmatrix}, \tag{25}$$

$$\mathbf{W} = \begin{bmatrix} W_{11} & 0 & 0 \\ 0 & W_{22} & 0 \\ 0 & 0 & W_{33} \end{bmatrix}. \tag{26}$$

In order to estimate the structural parameters, W_{ii}, we have chosen the Gauss–Newton method (see Kitanidis and Lane, 1985), an iterative optimization procedure

$$\boldsymbol{\theta}_{l+1} = \boldsymbol{\theta}_l - \mathbf{F}_l^{-1}\mathbf{g}_l. \tag{27}$$

In (27), l represents the step counter, $\boldsymbol{\theta} = [W_{11}\ W_{22}\ W_{33}]^T$ is the vector of the unknowns, $\mathbf{g}$ is the vector of the gradients of the negative loglikelihood function L with respect to $\boldsymbol{\theta}$, and $\mathbf{F}$ indicates the Fisher information matrix. The Fisher information matrix is an approximation of the Hessian $\mathbf{H}$, which is the matrix of the second derivatives.

Equation (27) can be easily derived by approximating the negative loglikelihood function with a quadratic polynomial in terms of vector of unknowns. Through a Taylor series expansion about the most recent guess $\boldsymbol{\theta}_l$, where $\mathbf{H}$ is replaced by $\mathbf{F}$, we get

$$L(\boldsymbol{\theta}) = L(\boldsymbol{\theta}_l) + \mathbf{g}_l^T(\boldsymbol{\theta} - \boldsymbol{\theta}_l) + \tfrac{1}{2}(\boldsymbol{\theta} - \boldsymbol{\theta}_l)^T\mathbf{F}_l(\boldsymbol{\theta} - \boldsymbol{\theta}_l) + O(\|\boldsymbol{\theta} - \boldsymbol{\theta}_l\|^3). \tag{28}$$

Then, neglecting the higher order terms and minimization of the right-hand side of (28) lead to (27).

The different components of $\mathbf{F}$ are obtained by taking the expected values of the corresponding components of $\mathbf{H}$; that is, by substituting $(\mathbf{z} - \langle\mathbf{c}\rangle)$ and $(\mathbf{z} - \langle\mathbf{c}\rangle)(\mathbf{z} - \langle\mathbf{c}\rangle)^T$ with their mean values $\mathbf{0}$ and $(\boldsymbol{\Gamma} + \mathbf{R}_e)$, respectively. Compared to Newton's method (that uses (28) without the substitution $\mathbf{H} \rightarrow \mathbf{F}$), the Gauss–Newton's method offers some important advantages – first, it avoids the cumbersome numerical calculation of the second derivatives of $\mathbf{\Gamma}$; second, it often converges faster.

It is possible to show that, in this case, $\mathbf{g}$ is given by

$$\mathbf{g} = \begin{bmatrix} \dfrac{\partial L}{\partial W_{11}} \\ \dfrac{\partial L}{\partial W_{22}} \\ \dfrac{\partial L}{\partial W_{33}} \end{bmatrix} \tag{29}$$

$$= \begin{bmatrix} \frac{1}{2} Tr\left[(\mathbf{\Gamma}+\mathbf{R}_e)^{-1}\frac{\partial\mathbf{\Gamma}}{\partial W_{11}}\right] - \frac{1}{2}(\mathbf{z}-\langle\mathbf{c}\rangle)^T\left[(\mathbf{\Gamma}+\mathbf{R}_e)^{-1}\frac{\partial\mathbf{\Gamma}}{\partial W_{11}}(\mathbf{\Gamma}+\mathbf{R}_e)^{-1}\right](\mathbf{z}-\langle\mathbf{c}\rangle) \\ \frac{1}{2} Tr\left[(\mathbf{\Gamma}+\mathbf{R}_e)^{-1}\frac{\partial\mathbf{\Gamma}}{\partial W_{22}}\right] - \frac{1}{2}(\mathbf{z}-\langle\mathbf{c}\rangle)^T\left[(\mathbf{\Gamma}+\mathbf{R}_e)^{-1}\frac{\partial\mathbf{\Gamma}}{\partial W_{22}}(\mathbf{\Gamma}+\mathbf{R}_e)^{-1}\right](\mathbf{z}-\langle\mathbf{c}\rangle) \\ \frac{1}{2} Tr\left[(\mathbf{\Gamma}+\mathbf{R}_e)^{-1}\frac{\partial\mathbf{\Gamma}}{\partial W_{33}}\right] - \frac{1}{2}(\mathbf{z}-\langle\mathbf{c}\rangle)^T\left[(\mathbf{\Gamma}+\mathbf{R}_e)^{-1}\frac{\partial\mathbf{\Gamma}}{\partial W_{33}}(\mathbf{\Gamma}+\mathbf{R}_e)^{-1}\right](\mathbf{z}-\langle\mathbf{c}\rangle) \end{bmatrix},$$

while $\mathbf{F}$ is defined by

$$\mathbf{F} = \begin{bmatrix} \left\langle\dfrac{\partial^2 L}{\partial W_{11}^2}\right\rangle & \left\langle\dfrac{\partial^2 L}{\partial W_{11}\partial W_{22}}\right\rangle & \left\langle\dfrac{\partial^2 L}{\partial W_{11}\partial W_{33}}\right\rangle \\ \left\langle\dfrac{\partial^2 L}{\partial W_{22}\partial W_{11}}\right\rangle & \left\langle\dfrac{\partial^2 L}{\partial W_{22}^2}\right\rangle & \left\langle\dfrac{\partial^2 L}{\partial W_{22}\partial W_{33}}\right\rangle \\ \left\langle\dfrac{\partial^2 L}{\partial W_{33}\partial W_{11}}\right\rangle & \left\langle\dfrac{\partial^2 L}{\partial W_{33}\partial W_{22}}\right\rangle & \left\langle\dfrac{\partial^2 L}{\partial W_{33}^2}\right\rangle \end{bmatrix}, \tag{30}$$

$$F_{ij} = \left\langle\frac{\partial^2 L}{\partial W_{ii}\partial W_{jj}}\right\rangle = \frac{1}{2} Tr\left[(\mathbf{\Gamma}+\mathbf{R}_e)^{-1}\frac{\partial\mathbf{\Gamma}}{\partial W_{ii}}(\mathbf{\Gamma}+\mathbf{R}_e)^{-1}\frac{\partial\mathbf{\Gamma}}{\partial W_{jj}}\right] \tag{31}$$

and Tr indicates matrix trace, that is, the sum of the diagonal entries.

To further improve the performance of the method, we have resorted to a type of Lavenberg–Marquardt procedure that consists of reducing the length of each step and turning the step direction slightly toward the steepest-descent direction

$$\boldsymbol{\theta}_{l+1} = \boldsymbol{\theta}_l - (\mathbf{F}_l + \nu\mathbf{I})^{-1}\,\mathbf{g}_l. \tag{32}$$

In (32), $\mathbf{I}$ is the identity matrix and ν is a multiple of the smallest eigenvalue of $\mathbf{F}_l$. This modification is especially helpful at early iterations, by preventing divergence, if the starting point is far from the optimum. However, it tends to slow down the convergence of the algorithm.

For each sampling round of the Cape Cod tracer test that we have analyzed, the iterations have been stopped when the gradient of L with respect to the unknown parameters was practically zero. In the neighborhood of the optimum $\boldsymbol{\theta}^*$, with $\mathbf{g}^* = \mathbf{0}$, we have

$$\begin{aligned} L(\boldsymbol{\theta}) &= L(\boldsymbol{\theta}^*) + \tfrac{1}{2}(\boldsymbol{\theta}-\boldsymbol{\theta}^*)^T \mathbf{H}^* (\boldsymbol{\theta}-\boldsymbol{\theta}^*) + O\left(\|\boldsymbol{\theta}-\boldsymbol{\theta}^*\|^3\right) \\ &\simeq L(\boldsymbol{\theta}^*) + \tfrac{1}{2}(\boldsymbol{\theta}-\boldsymbol{\theta}^*)^T \mathbf{F}^* (\boldsymbol{\theta}-\boldsymbol{\theta}^*). \end{aligned} \tag{33}$$

Thus, at a point $\boldsymbol{\theta}_l$ very near to the optimum $\boldsymbol{\theta}^*$, the gradient is

$$\mathbf{g}_l \simeq \mathbf{F}^* (\boldsymbol{\theta}_l - \boldsymbol{\theta}^*) \tag{34}$$

and

$$L(\boldsymbol{\theta}_l) - L(\boldsymbol{\theta}^*) \simeq \tfrac{1}{2}\mathbf{g}_l^T (\mathbf{F}^*)^{-1} \mathbf{g}_l. \tag{35}$$

The convergence criterion is then equivalent to

$$\mathbf{g}_l^T (\mathbf{F}^*)^{-1} \mathbf{g}_l \leqslant \epsilon, \tag{36}$$

where ϵ is a small dimensionless number. In the results presented later, we have used $\epsilon = 0.5$ and we have approximated $\mathbf{F}^*$ by $\mathbf{F}$ at the latest iteration.

The consistency of the model with the data can be verified after the fitting of the parameters through the analysis of the orthonormal residuals (Kitanidis, 1997). In our case, orthonormal residuals are a set of N uncorrelated random variables, expressed by

$$\varepsilon_i = \frac{\gamma_i}{d_i}, \quad i = 1, \ldots, N \tag{37}$$

where

$$\gamma_i = \{\mathbf{T}(\mathbf{z} - \langle \mathbf{c} \rangle)\}_i, \tag{38}$$

$$d_i = \left\{\left[\mathrm{diag}\left[\mathbf{T}(\boldsymbol{\Gamma} + \mathbf{R}_e)\mathbf{T}^T\right]\right]_i\right\}^{1/2} \tag{39}$$

and $\mathbf{T}$ is a matrix such that $\mathbf{T}(\boldsymbol{\Gamma} + \mathbf{R}_e)\mathbf{T}^T$ is diagonal. In (39), diag[$\mathbf{A}$] indicates the vector of the diagonal elements of $\mathbf{A}$.

In order to compare the actual performance of the model with the expected one, we have computed the following statistics:

$$Q_1 = \frac{1}{N}\sum_{i=1}^{N} \varepsilon_i \tag{40}$$

and

$$Q_2 = \frac{1}{N}\sum_{i=1}^{N} \varepsilon_i^2. \tag{41}$$

Q_1 should be a normally distributed random variable with mean 0 and variance $1/N$. Q_2, when N is large, is approximately Gaussian with mean 1 and variance $2/N$. Thus, for model validation, we checked that the ε_i distribution was at least approximately normal, and that both Q_1 and Q_2 were reasonably close to their theoretical mean values 0 and 1, respectively. Additionally, we judged the predictive ability of the model through

$$cR = Q_2 \exp\left[\frac{1}{N}\sum_{i=1}^{N} \ln\left(d_i^2\right)\right], \tag{42}$$

which is a measure of the estimation error of the residuals. The smaller the value of cR, the better the predictive ability of the model.

With the correlation structure completely defined through the variance–covariance matrix $\mathbf{\Gamma}$, it is possible to perform interpolation through kriging, in order to get the best estimate of the concentration on a regular grid as a linear weighting of the data. As before, $\mathbf{z}$ is the N by 1 vector of observations, with mean $\langle \mathbf{c} \rangle_z$ and covariance matrix $(\mathbf{\Gamma} + \mathbf{R}_e)_{zz}$; let $\mathbf{s}$ be the M by 1 vector of unknowns with mean $\langle \mathbf{c} \rangle_s$ and covariance matrix $\mathbf{\Gamma}_{ss}$. The M by N measurement-unknown cross-covariance is $\mathbf{\Gamma}_{sz} = \mathbf{\Gamma}_{zs}^T$. Then, the best estimate in this given-mean case is

$$\hat{\mathbf{s}} = \langle \mathbf{c} \rangle_s + \mathbf{\Lambda}\left(\mathbf{z} - \langle \mathbf{c} \rangle_z\right), \tag{43}$$

where $\mathbf{\Lambda}$, the kriging coefficients matrix, is found by solving the linear system

$$\mathbf{\Lambda}\,(\mathbf{\Gamma} + \mathbf{R}_e)_{zz} = \mathbf{\Gamma}_{sz}. \tag{44}$$

4. Results from Data Analysis

Summarizing the key modeling assumptions made for the analysis of the data:

1. The velocity field is time-independent and statistically homogeneous.
2. The velocity field satisfies the conditions for the application of the central limit theorem to the distribution of particle displacements.
3. Sufficient time has elapsed for the normality assumption used in the determination of the concentration ensemble mean and covariance to be valid.
4. The injection volume is very small compared to the plume size at observation times.
5. The random function that describes the concentration is approximately Gaussian. (Although this assumption is not essential to apply minimum-variance linear estimation methods, these methods work best for near-Gaussian data.)
6. The statistics of the observation errors are known or can be determined.

Since the proposed model is based on a large-time approximation, it has been applied to the analysis of the data collected during two late sampling rounds of the tracer test – 273 and 426 days after injection. The optimization of (24) as function of the two-particle covariances and the solution of (44) in terms of kriging coefficients have been carried out using all the available data (1923 at $t = 273$ days and 2002 at $t = 426$ days). The dimensions of the interpolation grid are slightly larger than the maximum size of the experimental plume – 100 m along the longitudinal direction, 20 m along the transverse horizontal direction, and 10 m along the vertical one. The grid spacing accounts for the more pronounced irregularity of the concentration in the vertical direction due to the anisotropy

$$\Delta Z = \Delta X/2 = \Delta Y/2 = 0.25\,\mathrm{m}. \tag{45}$$

Here, X, Y and Z stand for the distances measured in meters along the plume principal directions – longitudinal, transverse horizontal, and transverse vertical, respectively. The origin of the principal reference frame is located at the sea level and in correspondence of the horizontal coordinates of the injection point.

At $t = 273$ days, the known structural parameters are (Garabedian *et al.* 1991):

$$M = 4470 \quad [\mathrm{g}],$$

$$\mathbf{a}_0 + \mathbf{U}t = \begin{bmatrix} 115 \\ -2.1 \\ 6.6 \end{bmatrix} \quad [\mathrm{m}],$$

$$\mathbf{P} = \begin{bmatrix} 196 & 0 & 0 \\ 0 & 5.2 & 0 \\ 0 & 0 & 1.05 \end{bmatrix} \quad [\mathrm{m}^2].$$

The estimated effective porosity is $\eta = 0.39$.

Figures 1, 2 and 3 show three vertical longitudinal slices of the bromide plume through the centroid, and for three different models of the measurement-error covariance matrix

$$R_{e,ij}^{(A)} = R_e^{(A)}\left(\mathbf{x}_i, \mathbf{x}_j\right) = [0.07z\left(\mathbf{x}_i\right) + 0.1]^2\,\delta_{ij}, \tag{46}$$

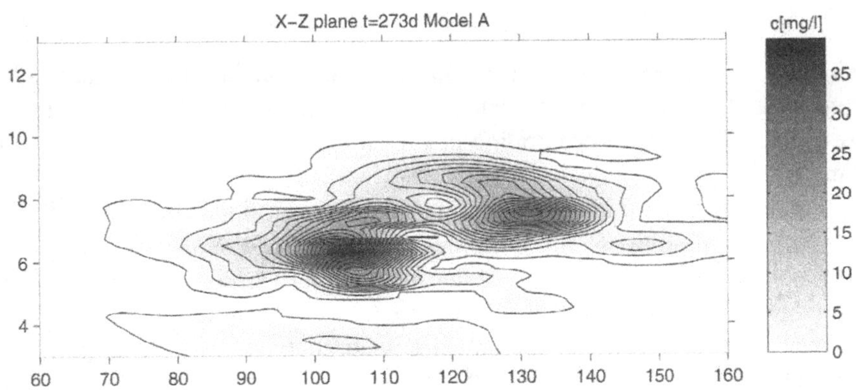

Figure 1. Vertical longitudinal slice through the centroid at $t = 273$d (Model A).

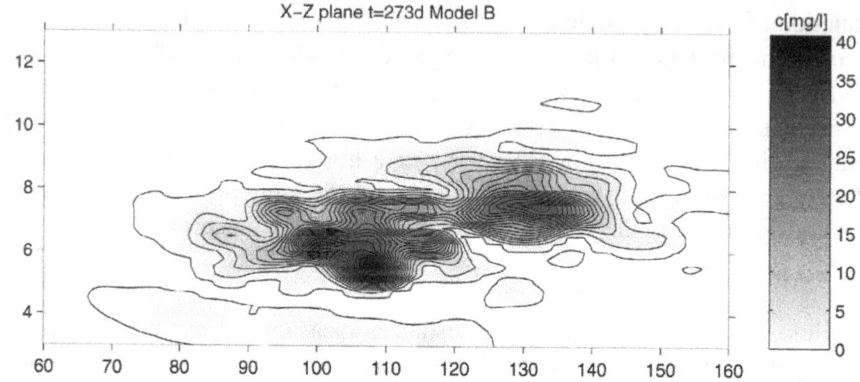

Figure 2. Vertical longitudinal slice through the centroid at $t = 273$d (Model B).

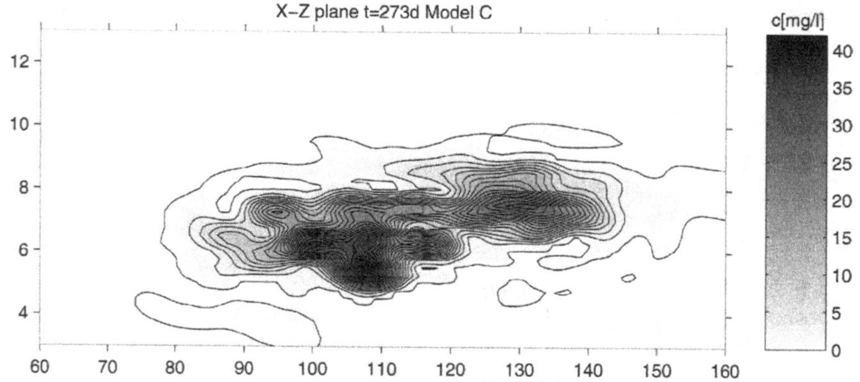

Figure 3. Vertical longitudinal slice through the centroid at $t = 273$d (Model C).

$$R_{e,ij}^{(B)} = R_e^{(B)}\left(\mathbf{x}_i, \mathbf{x}_j\right) = [0.05z\,(\mathbf{x}_i) + 0.1]^2\,\delta_{ij} \tag{47}$$

and

$$R_{e,ij}^{(C)} = R_e^{(C)}\left(\mathbf{x}_i, \mathbf{x}_j\right) = \delta_{ij}, \tag{48}$$

where δ_{ij} is Kronecker's delta. In these figures, and in all the figures showing concentration distributions, the exterior contour line delimits the area characterized by $c \geqslant 0.2$ mg/l. The concentration increment between contiguous contour lines is 2 mg/l.

Figures 4, 5 and 6 show the comparison between fitted (interpolated) and observed concentrations at the sampling points, as well as the distribution of the orthonormal residuals. The larger the measurement-error covariance matrix, $\mathbf{R}_e$, the larger the difference between observed and fitted values; however, a large difference does not mean that the model is inappropriate. If the measurement-error covariance matrix were zero, the fitted values would reproduce the observed values exactly.

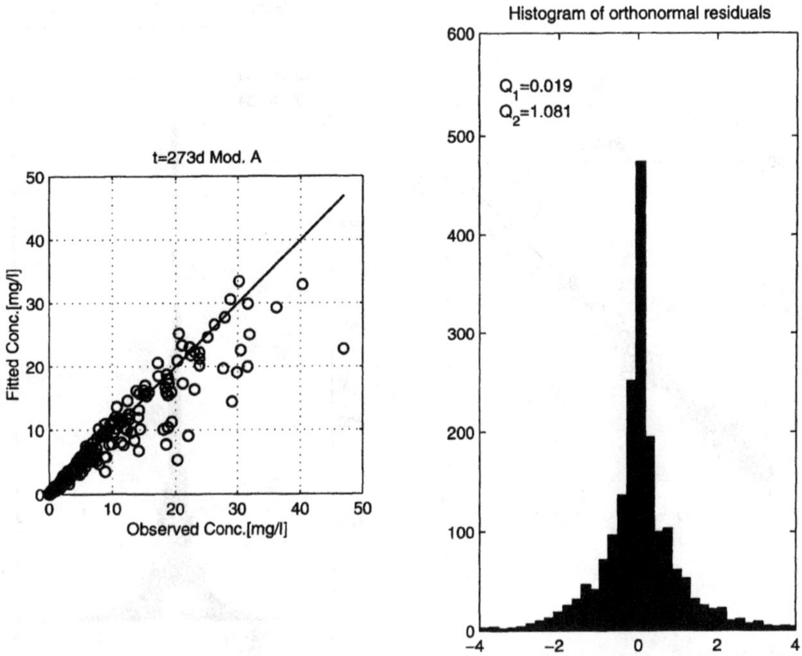

Figure 4. Comparison between fitted and observed concentrations at the sampling points, with corresponding histogram of orthonormal residuals; $t = 273\text{d}$ (Model A).

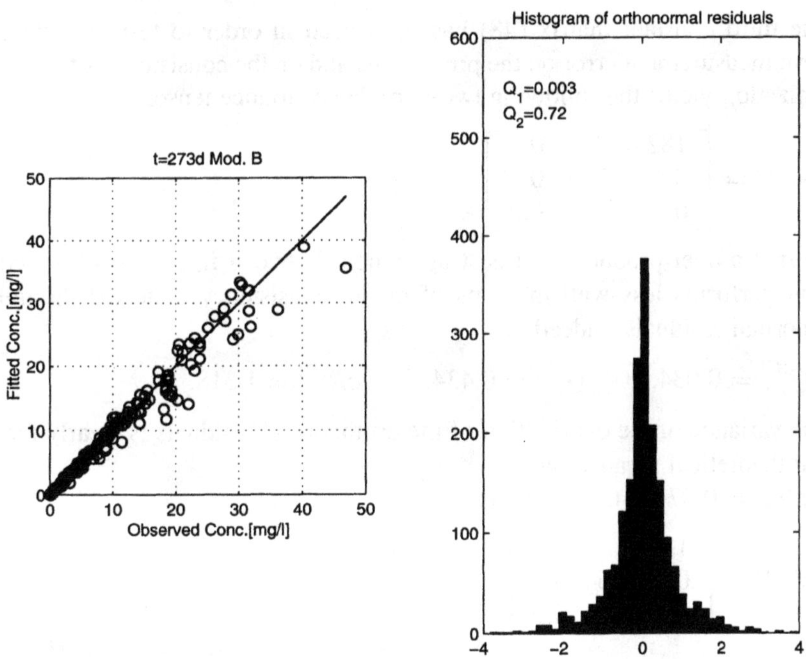

Figure 5. Comparison between fitted and observed concentrations at the sampling points, with corresponding histogram of orthonormal residuals; $t = 273\text{d}$ (Model B).

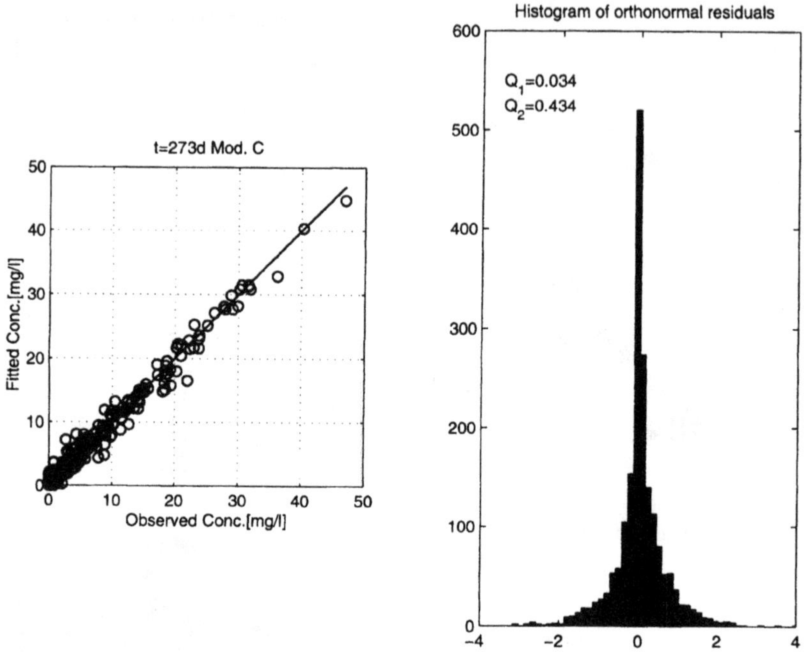

Figure 6. Comparison between fitted and observed concentrations at the sampling points, with corresponding histogram of orthonormal residuals; $t = 273$d (Model C).

The third constant matrix (48) has been used in order to test the effect of a uniform measurement error on the predictions and on the consistency of the model. Optimization yields the following two-particle covariance tensor

$$\mathbf{W}^{(C)} = \begin{bmatrix} 182.6 & 0 & 0 \\ 0 & 2.36 & 0 \\ 0 & 0 & 0.9 \end{bmatrix} \quad [\mathrm{m}^2]. \tag{49}$$

This error model produces the best agreement between fitted and observed values but performs less well in terms of model consistency as tested through the orthonormal residuals. Indeed,

$$Q_1^{(C)} = 0.034, \qquad Q_2^{(C)} = 0.434, \qquad cR^{(C)} = 1.318. \tag{50}$$

The variance of the distribution of the orthonormal residuals is clearly far from its unit theoretical mean value.

With $\alpha = 0.07$ (46), we obtain

$$\mathbf{W}^{(A)} = \begin{bmatrix} 157 & 0 & 0 \\ 0 & 3.64 & 0 \\ 0 & 0 & 0.9 \end{bmatrix} \quad [\mathrm{m}^2] \tag{51}$$

and

$$Q_1^{(A)} = 0.019, \qquad Q_2^{(A)} = 1.081, \qquad cR^{(A)} = 0.185. \tag{52}$$

In this case, the values of Q_1 and Q_2 are very close to the expected ones; plus, the shape of the frequency distribution of the orthonormal residuals is approximately Gaussian.

Finally, for $\alpha = 0.05$ (47), the results are

$$\mathbf{W}^{(B)} = \begin{bmatrix} 180 & 0 & 0 \\ 0 & 3.13 & 0 \\ 0 & 0 & 0.89 \end{bmatrix} \left[\mathrm{m}^2\right], \tag{53}$$

$$Q_1^{(B)} = 0.003, \qquad Q_2^{(B)} = 0.72, \qquad cR^{(B)} = 0.257. \tag{54}$$

Thus, model A, Equation (46), seems to be in better agreement with the data because its Q_1 and Q_2 statistics are closer to their optimum, cR is the smallest among the three models, and the residuals appear more normally distributed than for the other two cases. The estimates of the two-particle covariance matrix $\mathbf{W}$ depend partially on the measurement-error model. This is because the fluctuations in the observations can be attributed to either observation errors, quantified by $\mathbf{R}_e$, or fluctuations in the concentration, whose variance is affected by the choice of $\mathbf{W}$. The larger the $\mathbf{R}_e$, the smaller the estimate of $\mathbf{W}$. It is clear that the estimation of $\mathbf{W}$ from field measurements is more difficult than the estimation of the single-particle moments. This was to be anticipated, since the two-particle covariance must be evaluated from the concentration variance, which itself is difficult to identify from the data. Nevertheless, by choosing appropriate models and by testing them through cross-validation, we arrived at reasonable estimates. The results indicate that W_{11}, the diagonal element corresponding to the flow direction, is markedly smaller than P_{11}, the one-particle variance. The ratio W_{11}/P_{11}, which is the correlation between the displacement in the flow direction of the two particles at the time of the observation, is 0.80 for model A. Specifically, the difference between the two moments ($196\,\mathrm{m}^2$–$157\,\mathrm{m}^2 = 39\,\mathrm{m}^2$) cannot be attributed to the purely local dispersive contribution (the estimated field local dispersivity was about $1.5 \div 2$ mm) and may indicate nonlinear synergies between advection and local dispersion.

Figures 1, 2 and 3 should be compared with Figure 7 that depicts the unconditional mean and compares its predictions at observation points with the actual measurements. The unconditional mean is the prediction in the absence of data or if the data are noninformative; it can be obtained formally when $\mathbf{W}$ is very small in relation to $\mathbf{P}$. The unconditional mean is a Gaussian function that underestimates the highest concentration values. Note that the contour plots of Figures 1, 2, 3, and 7 show concentration distributions characterized by the same zeroth, first, and second moments. Their pronounced differences are due to different values of $\mathbf{W}$, which affects the concentration variance and the dilution index, a global measure of concentration nonuniformity (Kitanidis, 1994). Larger $\mathbf{W}$ signifies larger concentration variance and lower dilution index (i.e., less complete dilution).

Figures 8, 9 and 10 show horizontal slices through the centroid for the three different measurement-error models. The horizontal slices are generally less irregular

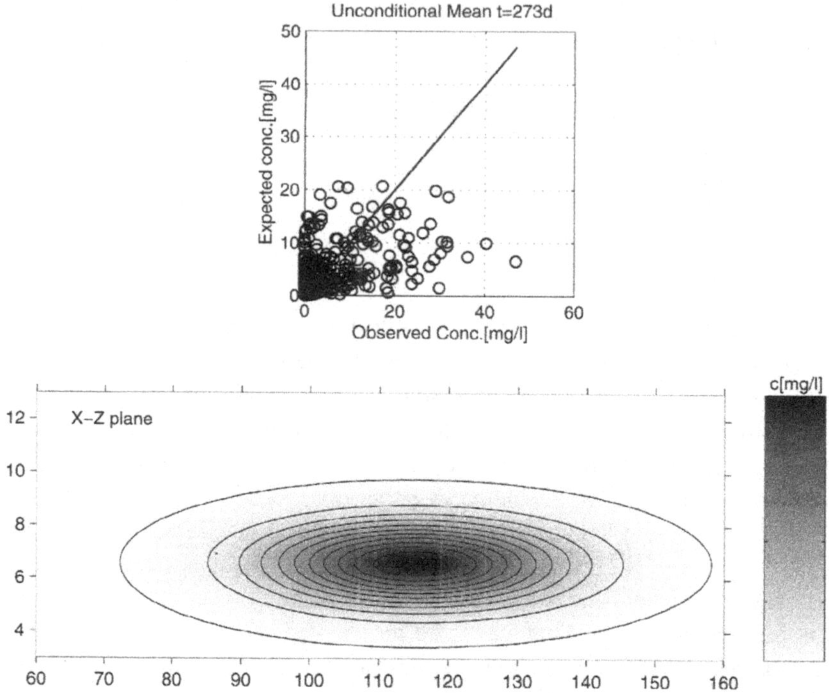

Figure 7. Unconditional mean distribution at $t = 273$d: Comparison between expected and observed concentrations, with vertical longitudinal slice through the centroid.

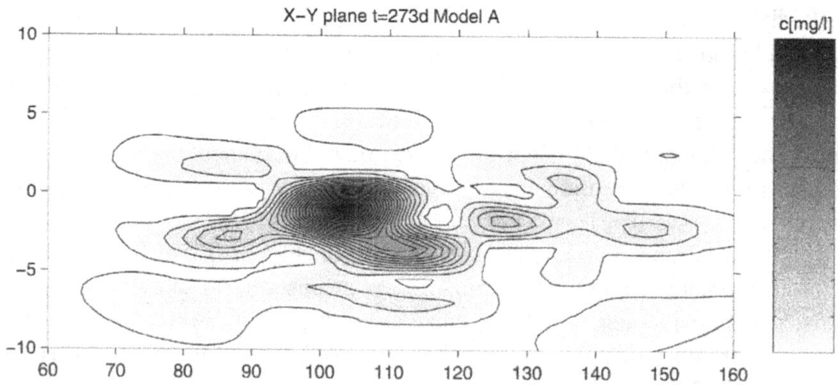

Figure 8. Horizontal slice through the centroid at $t = 273$d (Model A).

than the vertical distributions. Model $\mathbf{R}_e^{(A)}$ predicts a peak of concentration very close to the measured one (46.9 mg/l), and practically at the same location. Figures 11 and 12 compare two parallel vertical longitudinal slices taken at the same horizontal distance from the centroid, for $\alpha = 0.05$. These plots, which stress that our model describes variability of three-dimensional concentration, show a clearly asymmetric distribution and increased nonuniformity away from the centroid.

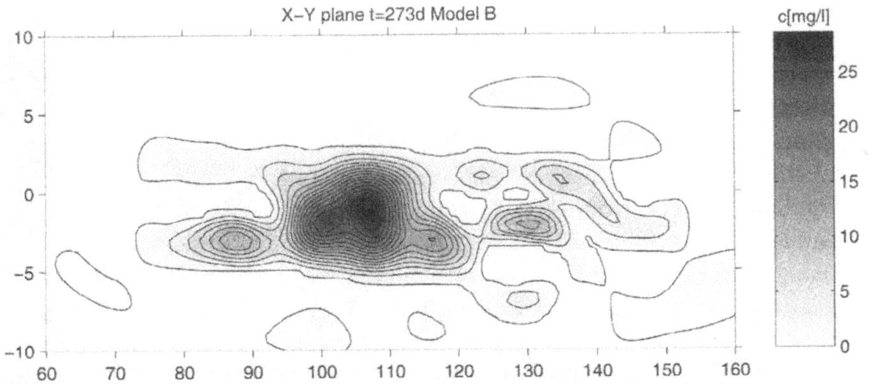

Figure 9. Horizontal slice through the centroid at t = 273d (Model B).

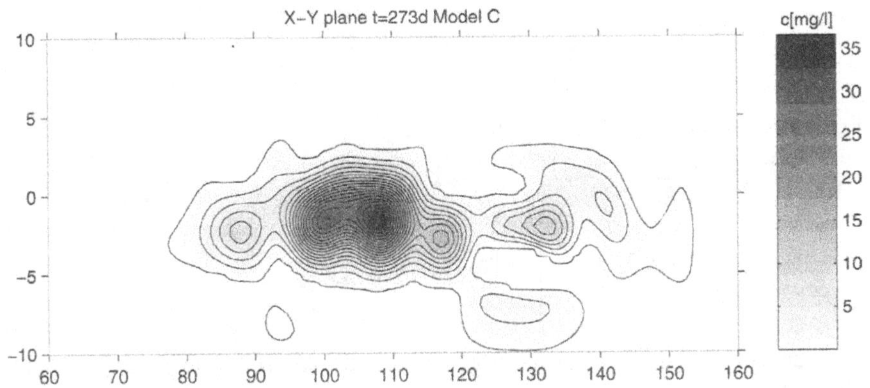

Figure 10. Horizontal slice through the centroid at t = 273d (Model C).

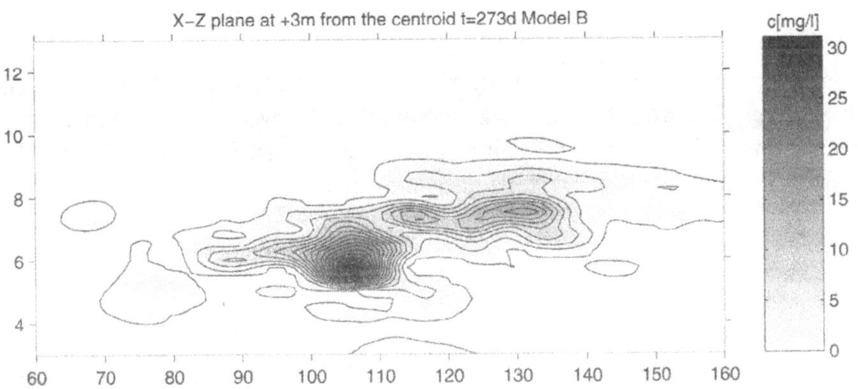

Figure 11. Vertical longitudinal slice at +3 m from the centroid; t = 273d (Model B).

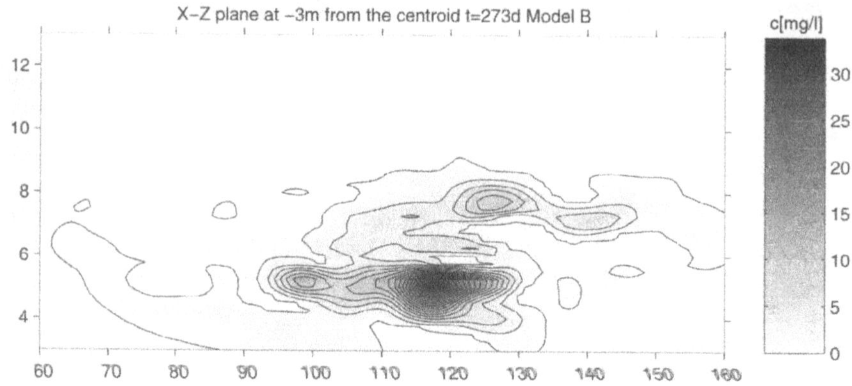

Figure 12. Vertical longitudinal slice at -3 m from the centroid; $t = 273$d (Model B).

At $t = 426$ days, the known structural parameters are (Garabedian *et al.*, 1991)

$$M = 4940 \quad [\mathrm{g}],$$

$$\mathbf{a}_0 + \mathbf{U}t = \begin{bmatrix} 181 \\ -21.2 \\ 5.4 \end{bmatrix} \quad [\mathrm{m}],$$

$$\mathbf{P} = \begin{bmatrix} 334 & 0 & 0 \\ 0 & 7.7 & 0 \\ 0 & 0 & 0.81 \end{bmatrix} \quad [\mathrm{m}^2].$$

The optimization of the negative loglikelihood function using Model A produces the following two-particle covariance tensor

$$\mathbf{W}^{(A)} = \begin{bmatrix} 226.9 & 0 & 0 \\ 0 & 7.13 & 0 \\ 0 & 0 & 0.5 \end{bmatrix} \quad [\mathrm{m}^2]. \tag{55}$$

Figures 13 and 14 show the vertical and the horizontal slice through the centroid, respectively. Figure 15 illustrates the comparison between fitted and measured concentrations at the observation points, as well as the distribution of the orthonormal residuals. The latter is characterized by a mean and a variance given by

$$Q_1^{(A)} = 0.014, \tag{56}$$

$$Q_2^{(A)} = 0.79. \tag{57}$$

The concentration vertical field shown in Figure 13 is much more diluted than the corresponding distribution at $t = 273$ days (Figure 1). The maximum detected value decreases from ~ 40 mg/l to ~ 25 mg/l; furthermore, the shape of

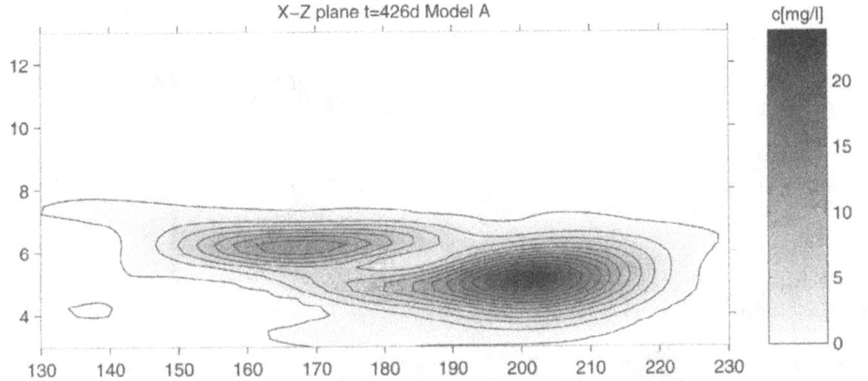

Figure 13. Vertical longitudinal slice through the centroid at t = 426d (Model A).

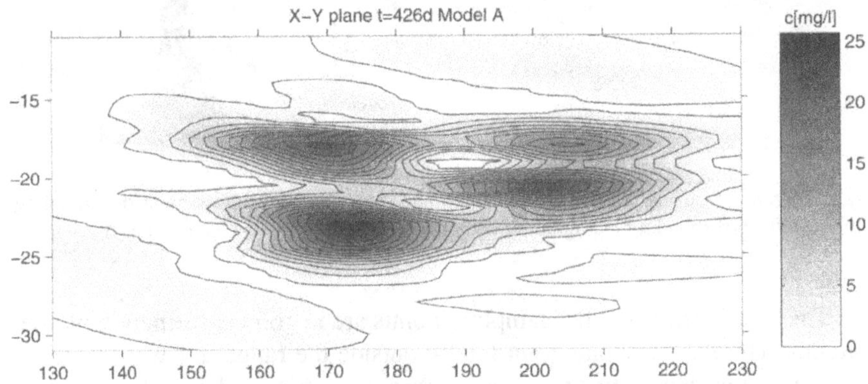

Figure 14. Horizontal slice through the centroid at t = 426d (Model A).

the iso-concentration lines is clearly more regular. The correlation coefficients $\rho_i = W_{ii}/P_{ii}$ are (0.68, 0.93, 0.62) compared to (0.80, 0.70, 0.86) at time 273. We anticipated that the correlation coefficient would decrease over time, indicating that **W** increases more slowly than **P** as a consequence of local dispersion and perhaps other dispersive processes enhancing dilution. The correlation coefficient corresponding to the longitudinal and the vertical transverse direction indeed dropped from day 273 to day 426. However, the coefficient corresponding to the horizontal transverse direction increased significantly. Compared to the horizontal slice at t = 273 days (Figure 8), the concentration distribution of Figure 14 shows that the plume has spread out in the horizontal transverse direction, Y, and the peak concentration has been reduced; however, the concentration in the Y direction seems to be more nonuniform than it was at the earlier time.

Such a result is hard to explain from a purely theoretical standpoint and may be indicative of experimental errors or difficulties associated with trying to estimate the two-particle covariance from a tracer test.

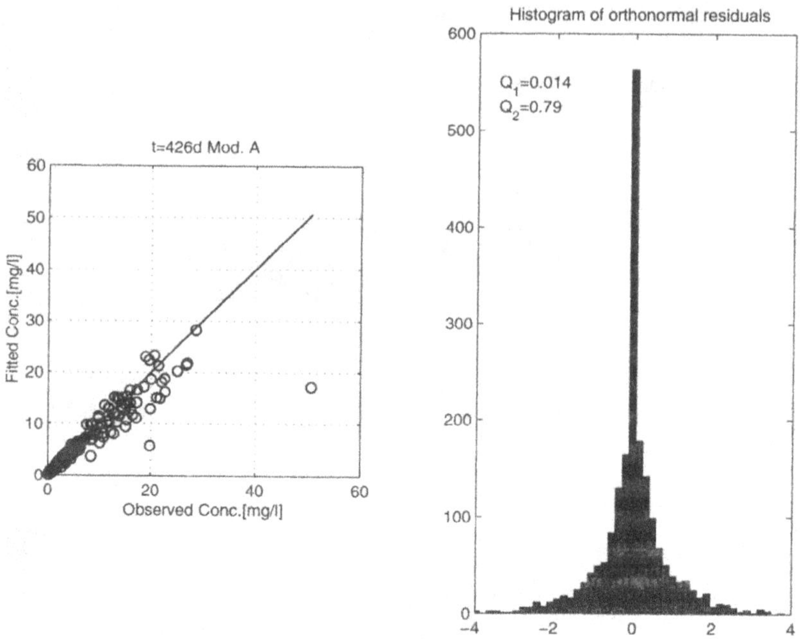

Figure 15. Comparison between fitted and observed concentrations at the sampling points, with corresponding histogram of orthonormal residuals; $t = 426$d (Model A).

The best estimates at the sampling points are in good agreement with the observations. There is only one point totally outside the range, for which $c \simeq 50$ mg/l. However, this value is two times larger than the one immediately lower; thus, it may have been affected by a particularly large measurement error. The total recovered mass, 4940 g, is quite close to the injected one (4900 g).

5. Discussion and Conclusions

This work includes the development of an analytic expression for the large-time covariance of the concentration of a conservative solute and the analysis of data from the Cape Cod field experiment.

In the first part, we extended the analysis of Pannone and Kitanidis (1999) to derive an expression for the concentration covariance at a given time t, which was shown to depend on the pdf of the single-particle trajectory as well as on the joint pdf of the trajectories of two distinct particles. Then, we assumed that, at large times, these pdf are both Gaussian; as a consequence, the concentration covariance depends on one-particle and two-particle first and second moments only. The one-particle moments had already been analyzed in previous studies. The two-particle covariances are constitutive variables introduced by Pannone and Kitanidis (1999) as global measures of solute nonuniformity.

Since the normality assumption is key to the derivation of the analytical expression, it behooves us to discuss its validity. The normality of the distribution of particle displacement is predicated on the validity of the central limit theorem, considering the total displacement as the sum of many successive steps. In its simplest form, the central value theorem states that the sum of many independent and identically distributed random variables tends to become Gaussian as the sum increases (Parzen, 1960). The theorem can be extended so that, according to Benjamin and Cornell (1970, p. 251), "Under very general conditions, as the number of variables in the sum becomes large, the distribution of the sum of random variables will approach the normal distribution". In our case, the theorem applies if each step of the particle displacement is correlated with only a limited number of other steps, as when the velocity is a stationary random field. The relation of large-time transport theory and the central limit theorem has been studied (e.g., Bhattacharya, 1985; Bhattacharya *et al.*, 1989), for the statistics of a single particle.

For the joint probability density function of the displacements of two particles to be normal, it is sufficient that the marginal pdf and also the pdf of their difference are normal. The marginal pdf, that is, the pdf of a single-particle displacement, is accepted to generally tend to a Gaussian (Dagan, 1989). This conclusion has been directly verified in Dagan and Fiori (1997), on the basis of a 2D numerical experiment performed by A. Bellin. Using the same computational techniques as in Bellin *et al.* (1992), they generated and analyzed data from over 1500 Monte Carlo realizations, for stationary log-conductivity distribution with exponential covariance function, and showed that the single-particle displacement is normal except at very short times. The normality of the single-particle displacement has been explicitly used by Neuman and Zhang (1990) in order to get a quasi-linear solution for the second moments of the plume. Under the same conditions, the pdf of the distance between the two particles tends to become normal because the distance is the sum of many successive increments and each increment is correlated only with a limited number of the others, due to the finite correlation of the velocity field.

Thus, the normality assumption applies to cases in which the central limit theorem is valid, which is considered as the normative case. However, in some circumstances, the validity of the central limit theorem cannot be taken for granted and alternative models have been proposed. Berkowitz and Scher (1998) have used a continuous time random walk approach to describe the anomalous transport taking place in fractured systems and in highly heterogeneous porous formations. In such cases, the flow domain turns out to be made of high velocity channels surrounded by large areas of very slow flow. The concentration distribution becomes increasingly asymmetric and the moment of inertia of the plume increases at the same rate as the square of the displacement of the centroid. Our normality assumption would be inconsistent with such a model.

Next, we analyzed the evolution of the spatial structure of the bromide distribution from an experiment at Cape Cod (Garabedian *et al.*, 1991; LeBlanc *et al.*, 1991). The concentration mean was already modeled in previous studies as normal.

Adding the spatial covariance function, developed in this study, we could perform simple kriging in order to interpolate between observation points. The parameters needed were the total mass, the ensemble mean trajectory, the tensor of one-particle covariances, and the tensor of two-particle covariances. The first three parameters were known from previous studies (Garabedian *et al.*, 1991) but the two-particle covariance needed to be estimated from the data. The estimates obtained using a maximum likelihood method (Kitanidis and Lane, 1985) depended on the model of the accuracy of the observations. We presented the results of three such models for data from sampling period 273 days. In two of the models the standard observation error increased linearly with the measured value and in the third model the standard error was constant. To evaluate which of the three models was the most reliable, we used cross-validation with orthonormal residuals (Kitanidis, 1991). On the basis of cross-validation, we found that the model with the larger observation error appeared to be more consistent with the data.

Using the estimated parameters, we resorted to simple kriging to interpolate the data on a regular grid which was then used for plotting contour maps of the concentration. A particularly interesting conclusion is that the contour maps of the best estimates depend on the assumed accuracy of the observations and the two-particle covariance tensor, for constant solute mass, centroid location, and moments of inertia. The results of kriging give reasonable representations of plumes, with values vanishing away from the centroid, larger nonuniformity at areas of steeper slope, etc. Thus, the model seems to have worked very well as interpolation tool. However, it is difficult to make the case that the estimated two-particle covariance tensors represent faithfully conditions at the Cape Cod site, due to model and data limitations. The data may need to be revisited after more analysis and simulations of transport in domains with known parameters.

Acknowledgements

Funding for this study was provided through a post-doctoral scholarship by University of Basilicata (Potenza) - Italy, and by NSF under projects EAR-9523922, ‘Constitutive Relations of Solute Transport and Transformations’, and EAR-9522651, ‘Factors Affecting Solute Dilution in Heterogeneous Formations’. We thank K. M. Hess, S. P. Garabedian and D. R. Leblanc for providing the data and useful discussions; O. Cirpka and three anonymous reviewers for their comments, that have greatly improved the manuscript.

References

Batchelor, K.: 1952, Diffusion in a field of homogeneous turbulence, 2. The relative motion of particles, *Proc. Cambridge Philos. Soc.* **48**, 345–362.

Bear, J.: 1972, *Dynamics of Fluids in Porous Media*, Elsevier, New York, pp. 764.

Bellin, A., Salandin, P. and Rinaldo, A.: 1992, Simulation of dispersion in heterogeneous formations: Statistics, first-order theories, convergence of computations, *Water Resour. Res.* **28**(9), 2211–2227.

Benjamin, J. R., and Cornell, C. A.: 1970, *Probability, Statistics, and Decision for Civil Engineers*, MacGraw-Hill, NY, pp. 684.

Berkowitz, B. and Scher, H.: 1998, Theory of anomalous chemical transport in random fracture networks, *Physical Rev. E* **57**(5), 5858–5869.

Bhattacharya, R.: 1985, A central limit theorem for diffusions with periodic coefficients, *The Annals of Probab.* **13**(2), 385–396.

Bhattacharya, R. N., Gupta, V. K. and Walkers, H. F.: 1989, Asymptotics of solute dispersion in periodic porous media, *Siam J. Appl. Math.* **49**(1), 86–98.

Dagan, G.: 1982, Stochastic modeling of groundwater flow by unconditional and conditional probabilities, 2. The solute transport, *Water Resour. Res.* **18**(4), 835–848.

Dagan, G.: 1984, Solute transport in heterogeneous porous formations, *J. Fluid Mech.* **145**, 151–177.

Dagan, G.: 1989, *Flow and Transport in Porous Formations*, Springer-Verlag, New York, pp. 465.

Dagan, G. and Fiori, A.: 1997, The influence of pore-scale dispersion on concentration statistical moments in transport through heterogeneous aquifers, *Water Resour. Res.* **33**(7), 1595–1605.

Fiori, A. and Dagan, G.: 1999, Concentration fluctuations in transport by groundwater: Comparison between theory and field experiments, *Water Resour. Res.* **35**(1), 105–112.

Fisher, H. B., List, E. J., Koh, R. C. Y., Imberger, J. and Brooks, N. H.: 1979, *Mixing in Inland and Coastal Waters*, Academic Press, NY, pp. 483.

Garabedian, S. P., LeBlanc, D. R., Gelhar, L. W. and Celia, M. A.: 1991, Large-scale natural gradient test in sand and gravel, Cape Cod, Massachusetts, 2: Analysis of spatial moments for nonreactive tracer, *Water Resour. Res.* **27**(5), 911–924.

Gelhar, L. W. and Axness, C. L.: 1983, Three-dimensional stochastic analysis of macrodispersion in aquifers, *Water Resour. Res.* **19**(1), 161–180.

Graham, W. D. and McLaughlin, D.: 1989, Stochastic analysis of nonstationary subsurface solute transport: 1. Unconditional moments, *Water Resour. Res.* **25**(2), 215–232.

Kabala, Z. J.: 1997, Analytical solutions for the coefficient of variation of the volume-averaged solute concentration in heterogeneous aquifers, *Stoch. Hydrol. Hydraul.* **11**(4), 331–348.

Kapoor, V. and Gelhar, L. W.: 1994a, Transport in three-dimensionally heterogeneous aquifers: 1. Dynamics of concentration fluctuations, *Water Resour. Res.* **30**(6), 1775–1788.

Kapoor, V. and Gelhar, L. W.: 1994b, Transport in three-dimensionally heterogeneous aquifers: 2. Predictions and observations of concentration fluctuations, *Water Resour. Res.* **30**(6), 1789–1801.

Kapoor, V. and Kitanidis, P. K.: 1996, Concentration fluctuations and dilution in two-dimensionally periodic heterogeneous porous media, *Transport in Porous Media* **21**(1), 91–119.

Kapoor, V. and Kitanidis, P. K.: 1997, Advection–diffusion in spatially random flows: formulation for the concentration covariance, *Stoch. Hydrol. Hydraul.* **11**(5), 397–422.

Kapoor, V. and Kitanidis, P. K.: 1998, Concentration fluctuations and dilution in aquifer, *Water Resour. Res.* **34**(5), 1181–1193.

Kitanidis, P. K. and Lane, R. W.: 1985, Maximum likelihood parameter estimation of hydrologic spatial processes by Gauss–Newton method, *J. Hydrol.* **79**, 53–71.

Kitanidis, P. K.: 1991, Orthonormal residuals in geostatistics: model criticism and parameter estimation, *Math. Geol.* **23**(5), 741–758.

Kitanidis, P. K.: 1994, The concept of the dilution index, *Water Resour. Res.* **30**(7), 2011–2026.

Kitanidis, P. K.: 1997, *Introduction to Geostatistics*, Cambridge University Press, pp. 249.

LeBlanc, D. R., Garabedian, S. P., Hess, K. M., Gelhar, L. W., Quadri, R. D., Stollenwerk, K. G. and Wood, W. W.: 1991, Large-scale natural gradient test in sand and gravel, Cape Cod, Massachusetts, 1: Experimental design and observed movement, *Water Resour. Res.* **27**(5), 895–910.

Neuman, S. P., Winter, C. L. and Newman, C. M.: 1987, Stochastic theory of field-scale Fickian dispersion in anisotropic porous media, *Water Resour. Res.* **23**(3), 453–466.

Neuman, S. P. and Zhang, Y. K.: 1990, A quasi-linear theory of non-Fickian and Fickian subsurface dispersion, 1. Theoretical analysis with application to isotropic media, *Water Resour. Res.* **26**(5), 887–902.

Neuman, S. P.: 1993, Eulerian–Lagrangian theory of transport in space-time nonstationary velocity fields: Exact nonlocal formalism by conditional moments and weak approximations, *Water Resour. Res.* **29**(3), 633–645.

Pannone, M. and Kitanidis, P. K.: 1999, Large time behavior of concentration variance and dilution in heterogeneous formations, *Water Resour. Res.* **35**(3), 623–634.

Parzen, E.: 1960, *Modern Probability Theory and Its Applications*, Wiley, NY, pp. 464.

Richardson, L. F.: 1926, Atmospheric diffusion shown on a distance-neighbour graph, *Proc. R. Soc. London, Ser. A* **110**, 709.

Rajaram, H. and Gelhar, L. W.: 1993, Plume scale-dependent dispersion in heterogeneous aquifers 2., Eulerian analysis and three-dimensional aquifers, *Water Resour. Res.* **29**(9), 3249–3260.

Rubin, Y.: 1990, Stochastic modeling of macrodispersion in heterogeneous porous media, *Water Resour. Res.* **26**(1), 133–141.

Rubin, Y.: 1991, Transport in heterogeneous porous media: prediction and uncertainty, *Water Resour. Res.* **27**(7), 1723–1738.

Thierrin, J. and Kitanidis, P. K.: 1994, Solute dilution at the Borden and Cape Cod groundwater tracer tests, *Water Resour. Res.* **30**(11), 2883–2890.

Vomvoris, E. G. and Gelhar, L. W.: 1990, Stochastic analysis of the concentration variability in a three-dimensional heterogeneous aquifer, *Water Resour. Res.* **26**(10), 2591–2602.

Zhang, Y. K. and Neuman, S. P.: 1990, A quasi-linear theory of non-Fickian and Fickian subsurface dispersion, 2. Application to anisotropic media at the Borden site, *Water Resour. Res.* **26**(5), 903–913.

Zhang, Q.: 1995, Transient behavior of mixing induced by a random velocity field, *Water Resour. Res.* **31**(3), 577–591.

Zhang, D.: 1995, Impacts of dispersion and first-order decay on solute transport in randomly heterogeneous porous media, *Transport in Porous Media* **21**, 123–144.

Transport in Porous Media **42:** 133–154, 2001.

Macrodispersivity and Large-scale Hydrogeologic Variability

DENNIS McLAUGHLIN[1] and FENG RUAN[2]
[1] *Ralph M. Parsons Laboratory, Department of Civil and Environmental Engineering, Massachusetts Institute of Technology, Cambridge, MA 02139, U.S.A.*
[2] *Applications Research, Schlumberger GeoQuest, Houston, TX 77056, U.S.A.*

(Received: 16 March 1999; in final form: 21 December 1999)

Abstract. Although groundwater velocities vary over a wide range of spatial scales it is generally only feasible to model the largest variations explicitly. Smaller-scale velocity variability must be accounted for indirectly, usually by increasing the magnitude of the dispersivity tensor (i.e. by introducing a so-called macrodispersivity). Most macrodispersion theories tacitly assume that a macrodispersivity tensor which works well when there is only small-scale velocity variability will also work well when there is larger-scale variability. We analyze this assumption in a high resolution numerical experiment which simulates solute transport through a two-scale velocity field. Our results confirm that a transport model which uses an appropriately adjusted macrodispersivity can reproduce the large-scale features of a solute plume when the velocity varies only over small scales. However, if the velocity field includes both small and large-scale components, the macrodispersivity term does not appear to be able to capture all of the effects of small-scale variability. In this case the predicted plume is more well mixed and consistently underestimates peak solute concentrations at all times. We believe that this result can be best explained by scale interactions resulting from the nonlinear transformation from velocity to concentration. However, additional analysis will be required to test this hypothesis.

Key words: macrodispersion, multi-scale, solute transport.

1. Introduction

It is now widely acknowledged that the hydrogeological properties which control solute transport in the subsurface vary over a wide range of space and/or time scales. In practical modeling studies it is convenient to make a fundamental distinction between large-scale variability that can be measured or inferred from site-specific information and smaller-scale variability that cannot be resolved. Unresolved fluctuations in hydraulic conductivity, recharge, and other quantities may have significant, but difficult to predict, effects on larger-scale solute transport. The resulting uncertainty can be accounted for explicitly with a stochastic description of solute transport which seeks to predict probability distributions, or at least ensemble moments, of the solute concentration.

Much attention has been devoted in recent years to the development of theories that predict the statistical properties of solute plumes transported through velocity fields characterized by small-scale variations about a known constant mean (Gelhar and Axness, 1983; Gelhar, 1986; Dagan, 1984, 1986, 1989, 1990; Neuman *et al.*, 1987). Many of these studies have focused on derivations of the ensemble mean of the solute concentration. The hope is that this mean will provide useful information about the behavior, not only of a hypothetical ensemble, but also of particular real-world plumes. Of course, we don't expect any specific plume to look exactly like the ensemble mean. But it would be useful if the ensemble mean correctly predicted important large-scale features of the plume, such as its spatial moments or the percentage of plume area with concentrations exceeding a given threshold.

Efforts to investigate connections between the ensemble mean concentration and the properties of particular plumes have tended to focus on the first two spatial moments of concentration. These moments provide convenient quantitative measures of the translation and spreading of particular plumes. The basis for a probabilistic description of solute variability is a hypothetical ensemble (or population) of solute plumes. This ensemble is characterized by a set of probability densities which define the first, second, ... nth-order statistics of the solute concentration at particular times and locations. The spatial moments of these plumes are random time series which have their own ensemble properties (Dagan, 1990). We can say that the plume ensemble is 'weakly ergodic' if the time-averaged derivatives of the first and second spatial moments of any replicate approach the corresponding ensemble mean quantities at large times (Papoulis, 1984). When this form of ergodicity applies the asymptotic (large time) ensemble mean plume will have the same overall size and shape as each replicate, although the smaller-scale details will generally differ. In practice it is important to know just how long it will take for a real plume to reach this asymptotic state.

Under certain special conditions, the ensemble mean concentration of a non-reactive solute satisfies a classical advection–dispersion transport equation with a constant (effective) velocity and macrodispersivity tensor. These effective properties may be derived from the velocity statistics which define the plume ensemble. If the ensemble is weakly ergodic the mean concentration obtained from the transport equation should have the same asymptotic first and second spatial moments as individual plumes. It is in this sense that we can say that macrodispersivities account for the large-scale effects of unresolved small-scale variability.

Generally speaking, the ensemble-based mean concentrations derived in stochastic transport theories provide good asymptotic approximations to individual plumes when the mean velocity is constant and the length scales of velocity fluctuations are small compared to the displacement and lateral dimensions of the plume. In such cases the plume has an opportunity to 'sample' a large range of velocity variations and is not unduly influenced by any particular anomaly. These are the somewhat idealized conditions that have applied in the field tests where macrodispersivities derived from stochastic theories have been most successful (McKay *et al.*, 1986; Freyberg *et al.*, 1986; Sudicky, 1986; Rajaram and Gelhar, 1991; Hess *et al.*, 1992). Ergodicity in field settings is usually checked by noting that the time derivatives of the spatial moments of the field plume approach constant asymptotic values which are close to the corresponding moments of the derived ensemble mean. These requirements are generally satisfied in the experiments cited above, over times ranging from a few months to a few years.

Relatively few investigators have considered how macrodispersivity concepts focused on the effects of small-scale variability might be applied in more complicated field situations where large-scale variability is also important. Large-scale velocity variations reflect the effects of hydrogeologic features such as aquifer boundaries, stratigraphic sequences, depositional gradients, faults, surface water bodies, and regional or seasonal variations in recharge. Such features are typically represented deterministically in large-scale flow models. Most stochastic theories implicitly assume that a macrodispersivity tensor that works well when there is only small-scale velocity variability will also work well when there is larger-scale variability. This hypothesis has not been adequately verified in field settings. If the transformation from velocity to concentration were linear the effects of small and large-scale variability could be superimposed. But this transformation is nonlinear and velocity fluctuations at different scales may interact, in a process similar to modulation. It is difficult to say how important this effect might be without investigating particular cases.

In this paper we examine the applicability of classical macrodispersion concepts to situations where small and large-scale velocity variations are both present. It is important to emphasize that the two scales considered here do not represent the scales of actual geological features. Rather, the large scale represents resolved velocity variations (advective velocities specified explicitly in a numerical model) and the small scale represents unresolved velocity variations (velocities accounted for with a Fickian macrodispersivity coefficient). This approach enables us to conveniently partition the effects of multi-scale geological variability into two categories. The dividing line between large and small-scale variability depends on the data available for defining the advective velocity field and on the size of the grid cells used to model solute transport.

Our analysis relies on individual velocity and concentration realizations, as would occur in a field experiment. However these replicates are generated numerically, so they can be readily manipulated. In particular, we are able to simulate transport through velocity fields with only small-scale variability, only large-scale variability, or two-scales of variability (one small and one large). Although our analysis is performed without reference to particular stochastic theories, we recognize that our results have implications for the practical application of such theories. We comment on this in the final section of the paper.

There is a growing literature on alternative methods for characterizing multi-scale hydrogeologic variability and its effects on solute transport. In stochastic studies small-scale Eulerian velocity variations are most often modeled as time-invariant spatially-correlated random deviations from a constant mean. These velocity variations are generally assumed to be caused by small-scale hydraulic conductivity fluctuations. Characteristic correlation lengths for small-scale hydraulic conductivity and velocity fluctuations are on the order of a few meters in the horizontal and a few centimeters in the vertical. In this paper we also treat small-scale velocity fluctuations as time-invariant correlated homogeneous random fields.

Most theoretical analyses of solute macrodispersion do not explicitly distinguish large-scale advective velocity fluctuations from smaller-scale dispersive fluctuations. The primary exceptions are a few studies which account for systematic (e.g. linear) log conductivity trends (Rubin and Seong, 1994; Indelman and Rubin, 1996). The large-scale effects of interest in our study are more complex and less systematic than such trends. This complexity could be captured if we obtained the large-scale velocity field from a typical deterministic regional flow simulation which includes irregular boundaries, variable soil properties and recharge rates, etc. Although such an approach has merit, it produces non-stationary fields which do not have well-defined velocity length scales or correlation structures. Here we adopt a different approach which is more compatible with the general philosophy of stochastic transport theory. Our large-scale velocity variations are also modeled as random time-invariant fluctuations about a constant mean but with much larger correlation lengths than the small-scale fields (Rajaram and McLaughlin, 1990). Typical correlation lengths for the large-scale features of interest here are on the order of tens of meters or more in the horizontal and meters in the vertical. Although the large-scale velocity fluctuations used in our experiment are generated randomly, they are assumed to be known perfectly. In this sense they represent the spatially variable but resolved portion of the velocity field that is included in the advective term of a traditional transport model.

Our numerical approach allows us not only to isolate different components of the velocity field but also to simulate transport with only local dispersion or with both local and macroscopic dispersion. These various options can be examined at much higher resolution and for much longer times and travel distances than is possible in a field experiment. Aggregate transport measures such as spatial moments or the peak concentration can be computed accurately and sensitivity analyses can be performed. Finally, extraneous factors such as temporal variations in the mean velocity can be avoided.

Of course, the results of our experiment are dependent on the assumptions and approximations we make. Since it is limited in fundamental ways this experiment does not provide a definitive assessment of the applicability of macrodispersion theory. However, the experimental results raise a number of questions which deserve further investigation. These are examined at the end of the paper.

2. The Computational Experiment

2.1. EXPERIMENTAL SETUP

Our computational experiment is intended to simulate a typical field tracer test, with solute released from an instantaneous point source into a steady flow field. We intentionally consider a single realization, rather than an ensemble, since our goal is to examine the applicability of macrodispersion concepts to field situations, rather than to confirm the ability of any particular theory to predict ensemble properties.

In order to be credible our experiment must be able to properly simulate small-scale velocity and concentration variability without introducing spurious oscillations or numerical dispersion. These are challenging requirements which can only be met with an accurate transport solver discretized on a high resolution spatial grid. Numerical accuracy comes at the cost of increased computational effort, leading to some difficult tradeoffs. For a given effort, we can choose between simulating three-dimensional transport in a relatively small spatial domain or simulating two-dimensional transport in a much larger domain. In this context, computational effort is related primarily to the number of spatial grid nodes and domain size is measured by the number of large correlation distances traveled. Also, we can choose between generating exact velocity fields from a computationally demanding flow simulator run over a small domain versus generating approximate velocity fields from a much more efficient random field generator run over a larger domain. These choices affect the types of questions we can examine as well as the generality of the results we can produce.

In this study we have given first priority to simulating transport over a large enough distance to allow the solute plume to adequately sample both the smallest and largest scales of velocity variability. As a result, we use a large two-dimensional (horizontal plane or plan view) domain with velocities obtained from a random field generator. The hydraulic conductivity in this domain is assumed to be isotropic and spatially variable. The domain extends over 200 large correlation lengths (λ_l) in the longitudinal (mean flow) direction and 12.5 large correlation lengths in the transverse direction. The left and right halves of the domain are illustrated in the upper and lower sections of Figure 1, respectively. We will examine detailed 'snapshots' of solute plumes located in the four windows indicated at time steps 10, 60, 100, and 300 (200, 1200, 2000, and 6000 days, respectively) after the instantaneous source release. These windows are defined for display purposes only – the domain used to solve the transport equation at all times is the $200\lambda_l$ by $12.5\lambda_l$ outer region shown in the figure. The release is distributed over the 20 m by 20 m square source area indicated in the figure and the mean flow is from left to right along the longitudinal (x_1) axis.

The random velocity generator used in our experiment relies on approximate velocity spectra derived from a specified log hydraulic conductivity spectrum and linearizations of Darcy's law and the groundwater flow equation (Ruan and McLaughlin, 1998). This approach is much less computationally demanding than currently

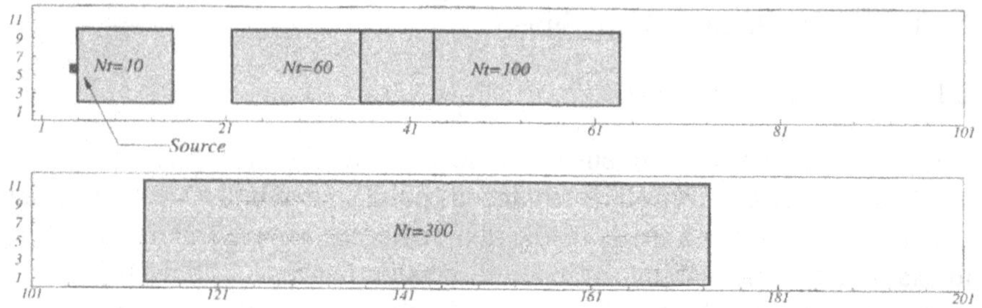

Figure 1. Left and right halves of experimental domain. Axes units in number of large log hydraulic conductivity correlation scales ($\lambda_l = 20\,\mathrm{m}$). Shaded areas indicate windows used to display results at time steps 10, 60, 100, and 300 (200, 1200, 2000, and 6000 days, respectively).

available methods for solving the groundwater flow equation with random coefficients on grids of order 10^6. Both theoretical and numerical analyses show that our linearized spectral approach provides physically reasonable divergence-free longitudinal and transverse velocities at each grid location (Graham and McLaughlin, 1988; Ruan and McLaughlin, 1998). In the simulations considered here we assume that the small and large-scale log conductivities are independent random fields, with Gaussian spectra characterized by correlation lengths λ_s and λ_l, respectively. We chose Gaussian spectra because they produce replicates with reasonably well-defined features that have length scales similar to the specified values of λ_s and λ_l.

The small and large-scale log conductivity spectra are used to derive the spectra of the corresponding linearized small and large-scale velocity fields. There is no need to actually generate random log conductivity replicates since the transport model only depends on velocity. We obtain a two-scale velocity field by summing the deviations of the small and large-scale velocity fields from their means

$$v_{it}(\mathbf{x}) = \overline{v_i} + v'_{is}(\mathbf{x}) + v'_{il}(\mathbf{x}), \tag{1}$$

where v_{it} is the two-scale velocity in direction i, v'_{is}, and v'_{il} are zero-mean small and large-scale velocity fluctuations in direction i, and $\overline{v_i}$ is the constant mean velocity in direction i. Figure 2 shows some typical longitudinal velocity realizations produced by our random field generator. The upper, middle, and lower plots are small-scale only, large-scale only, and two-scale realizations, respectively. These plots cover a small portion of our experimental domain extending over about 25 by 12.5 large correlation scales. The small-scale log conductivity correlation length is 2 m and the large-scale correlation length is 20 m, giving a scale disparity ratio of 1:10. Other inputs are summarized in Table I. Note that the longitudinal velocities are anisotropic, with longer correlation scales in the longitudinal (mean flow) direction. Also, note how the major features in the large-scale velocity field influence the two-scale field.

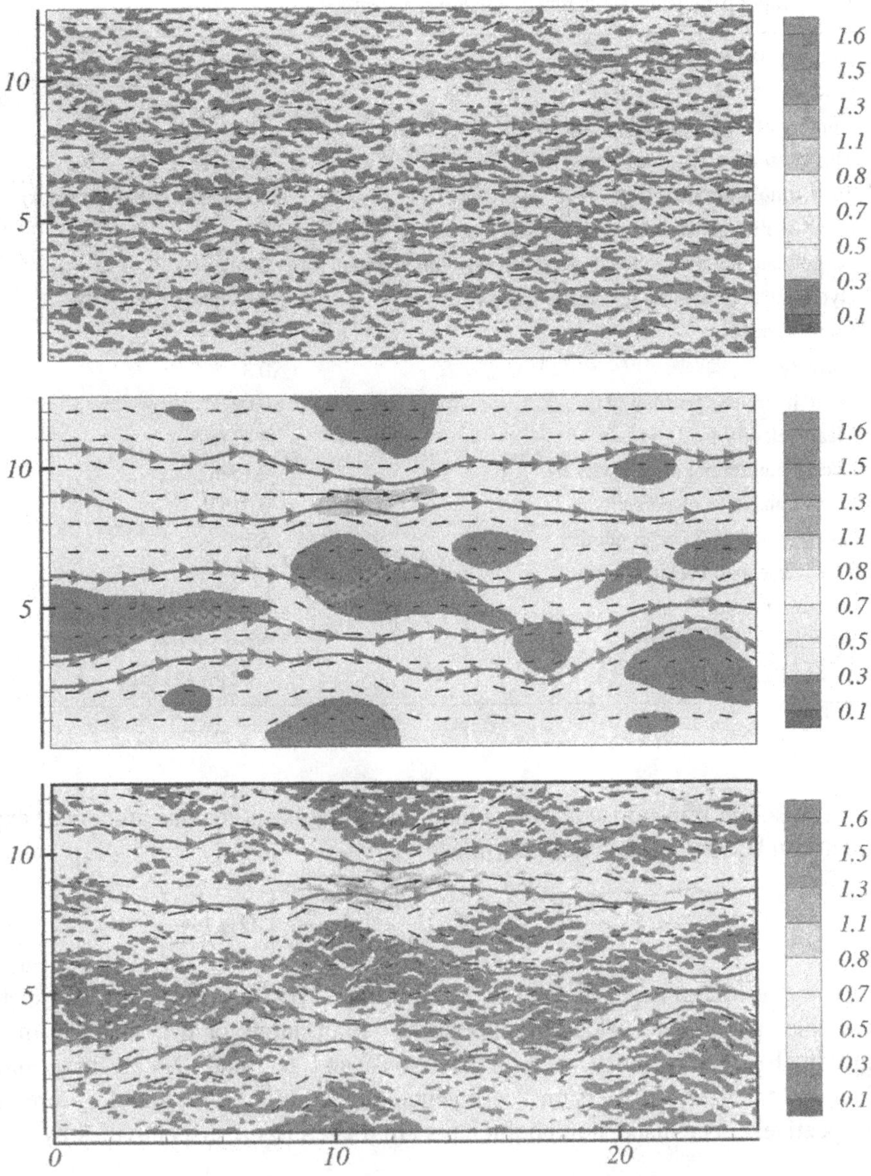

Figure 2. Synthetic realizations of longitudinal velocity for small ($\lambda_s = 2\,\text{m}$), large ($\lambda_l = 20\,\text{m}$) and two-scale log hydraulic conductivity spectra. Axis units in numbers of large correlation scales. Velocities are in m/day.

Table I. Input parameters for computational experiment

Domain size $L_1 \times L_2$	4080 m × 250 m
Grid size $\Delta x = \Delta x_1 = \Delta x_2$	1.0 m
Number of computational nodes N	1,024,331
Time step Δt	20 days
Total simulation time	8000 days (400 time steps)
Small-scale log hydraulic conductivity correlation length λ_s	2.0 m
Large-scale log hydraulic conductivity correlation length λ_l	20.0 m
Hydraulic conductivity geometric mean K_g	6.25 m/day
Log k variance σ_f^2	1.0
Porosity ρ	0.3
Mean hydraulic gradient $\|J\|$	0.02
Mean velocity $\overline{v}$	0.42 m/day
Local longitudinal dispersivity α_l	0.15 m
Local transverse dispersivity α_t	0.015 m
Longitudinal grid Peclet number	6.7
Transverse grid Peclet number	67.0
Courant number	8.2
Rectangular source size	20 m × 20 m
Source location (x_0, y_0)	(102 m, 125 m)

The solute transport equation which forms the basis for our numerical experiment can be written in generic form as

$$\frac{\partial c}{\partial t} + v_1 \frac{\partial c}{\partial x_1} + v_2 \frac{\partial c}{\partial x_2} = \frac{\partial}{\partial x_1}\left[D_{11}\frac{\partial c}{\partial x_1}\right] + \frac{\partial}{\partial x_2}\left[D_{22}\frac{\partial c}{\partial x_2}\right], \tag{2}$$

where t is time, $c(\mathbf{x}, t)$ is solute concentration, $v_1(\mathbf{x})$ and $v_2(\mathbf{x})$ are the spatially variable steady-state velocity components (small, large, or two-scale) in the longitudinal (x_1) and transverse (x_2) directions, and D_{11} and D_{22} are the longitudinal and transverse dispersion coefficients, aligned along the x_1 and x_2 directions, respectively. The dispersion coefficients are related to the dispersivities by

$$D_{11} = \overline{v}A_L, \qquad D_{22} = \overline{v}A_T, \qquad D_{12} = D_{21} = 0, \tag{3}$$

where A_{L} and A_{T} are, respectively, the longitudinal and transverse dispersivities.

We solve (2) with the cubic spline Eulerian–Lagrangian algorithm described in Ruan and McLaughlin (1999). This solver is computationally efficient and introduces little or no numerical dispersion at the grid resolutions we have used. Ruan and McLaughlin (1999), present a set of detailed quantitative tests of the cubic spline algorithm which demonstrate that it performs well in heterogeneous advection-dominated transport problems similar to the one considered here. When

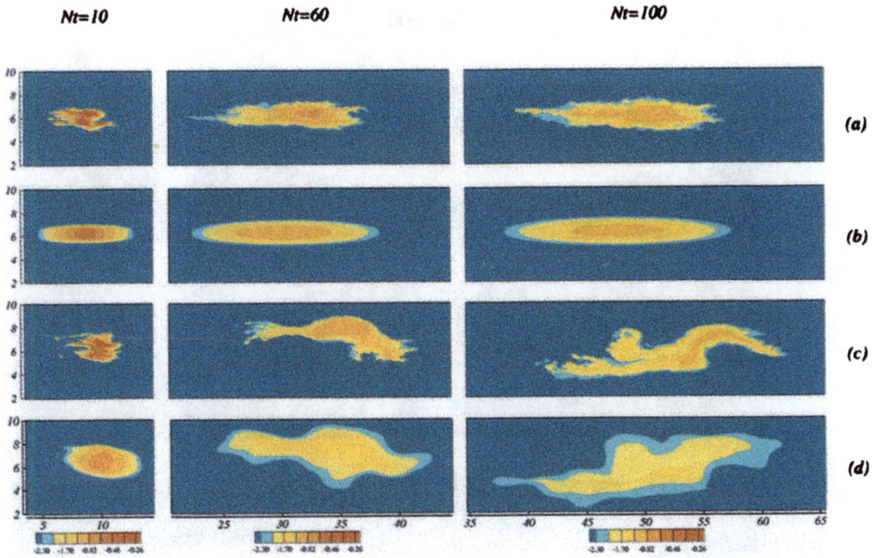

Figure 3. Solute log concentration contours at time steps 10 (left column), 60 (center column), and 100 (right column), corresponding to days 200, 1200, and 2000, respectively. First row (a) shows small-scale experimental plume (small-scale velocity with local dispersivity), second row (b) shows predicted (ensemble mean) small-scale plume (constant velocity with estimated macrodispersivity), third row (c) shows two-scale experimental plume (two-scale velocity with local dispersivity) and fourth row (d) shows predicted two-scale plume (large-scale velocity with estimated macrodispersivity). Compare (b) to (a) and (d) to (c).

compared to traditional Eulerian or Eulerian–Lagrangian methods, the cubic spline interpolator yields smaller computational errors for a given level of computational effort while allowing the use of larger Courant and Peclet numbers. The accuracy of the cubic spline transport solver is also demonstrated by its ability to resolve small-scale highly variable plume features without adding spurious oscillations (see Figures 3c and 4c).

In our experiment the concentration is specified to be zero on the left boundary of the computational domain. Outflow conditions are specified on the remaining boundaries. The outflow conditions allow streamlines to cross boundaries. They are implemented in the dispersive step of our Eulerian–Lagrangian solution algorithm by fixing the outflow boundary concentration at the value obtained from the preceding advective step (see Ruan and McLaughlin (1999) for details). The initial concentration is zero everywhere in the computational domain, except over the source region, where it is equal to the specified source concentration c_S.

2.2. INVESTIGATION OF MACRODISPERSIVE TRANSPORT

We now wish to use our experimental setup to examine the performance of a transport model which uses macrodispersivities to account for the effects of small-

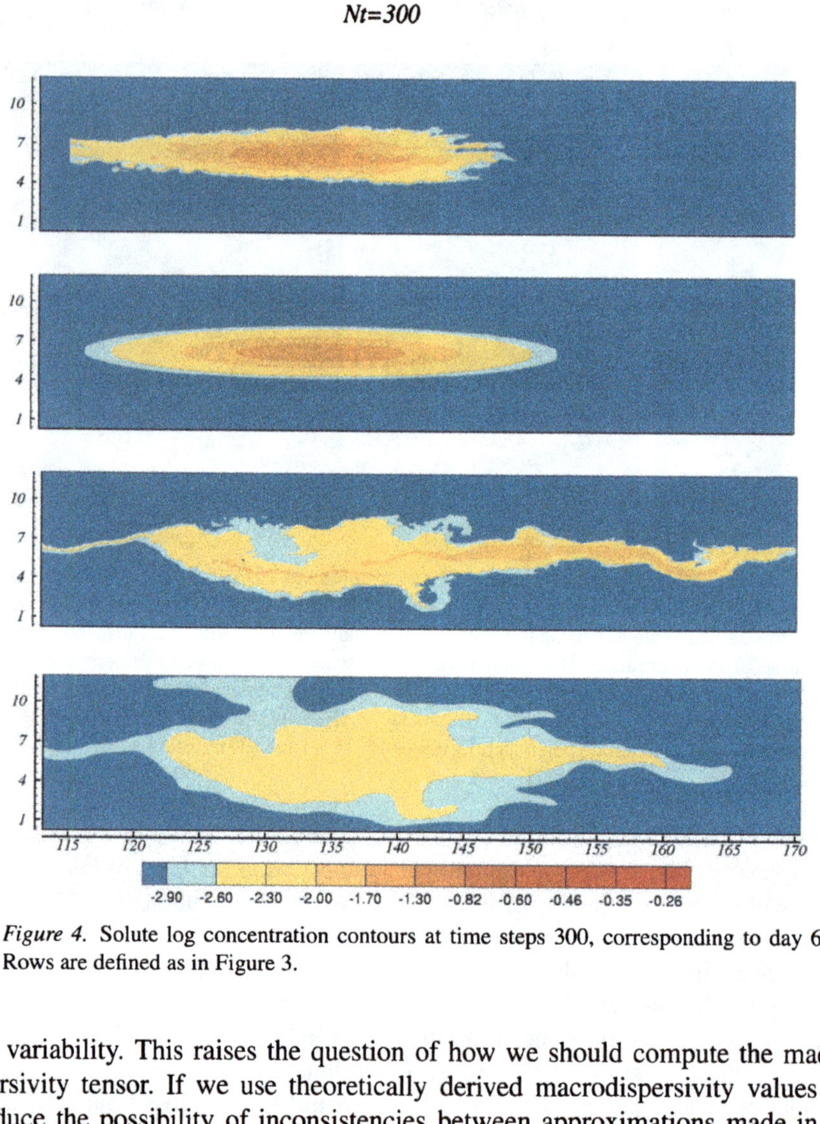

Figure 4. Solute log concentration contours at time steps 300, corresponding to day 6000. Rows are defined as in Figure 3.

scale variability. This raises the question of how we should compute the macrodispersivity tensor. If we use theoretically derived macrodispersivity values we introduce the possibility of inconsistencies between approximations made in the numerical experiment and approximations made in the theory. For example, the two-dimensional asymptotic macrodispersivities of Gelhar and Axness (1983) are based on linearizations of both the flow and transport equations while our experiment relies on a linearization of the flow equation (since the velocity random field generator is based on a linearization) but retains all the nonlinearities introduced by the transport process.

In order to avoid such inconsistencies, we adopt a different approach. The constant velocity and macrodispersivity used in the ensemble mean transport equation are related to the asymptotic spatial moments of the mean concentration (the velocity is the time derivative of the first moment while the macrodispersivity is proportional to the time derivative of the second moment). If the assumptions of the stochastic theories are correct, the weak ergodicity assumption should hold and the asymptotic spatial moments of the smooth ensemble mean plume should be essentially the same as the asymptotic moments of the irregular experimental plume. Consequently, we can estimate the desired constant velocity and macrodispersivity from the spatial moments of the experimental plume. The latter alternative is analogous to estimating effective transport properties from a tracer test in a domain where there is only small-scale variability. It has the advantage of focusing on the common objective of most of the stochastic theories without depending on the assumptions of any particular theory.

We begin by generating an experimental plume that emanates from an instantaneous point source and moves through a velocity field with only small-scale variability about a constant mean. These are the conditions considered in classical macrodispersion theory. The first rows (Case a) of Figures 3 and 4 show log concentration contour plots obtained at four different times for such a plume. Table II summarizes the velocity and dispersivity assumptions made for this and all of the other cases simulated in our experiment.

Note that only local dispersivity is used in the (Case a) simulation since small-scale velocity fluctuations are explicitly included in the advective term of the transport equation. The resulting tracer plume has a roughly Gaussian shape, with quite a bit of fine structure contributed by small-scale velocity variations. The very presence of this fine structure is a qualitative indication of the cubic spline algorithm's ability to solve difficult advection-dominated transport problems. Even a small amount of numerical dispersion tends to suppress such features.

The first and second moments of an ensemble mean plume are related to the constant effective velocity and macrodispersivity in the solute transport equations

Table II. Summary of experimental cases

Case	v_1	v_2	A	Comments
(a)	$\overline{v} + v_{s1}$	v_{s2}	Local only	Small-scale experimental plume, no large-scale variability
(b)	$\overline{v}$	0	Local + macrodispersivity	Predicted (ensemble mean) small-scale plume, small-scale variability not resolved but accounted for with macrodispersivity, no large-scale variability
(c)	$\overline{v} + v_{t1}$	v_{t2}	Local only	Two-scale experimental plume, both small and large-scale variability
(d)	$\overline{v} + v_{l1}$	v_{l2}	Local + macrodispersivity	Predicted two-scale plume, small-scale variability not resolved but accounted for with macrodispersivity, large-scale variability resolved

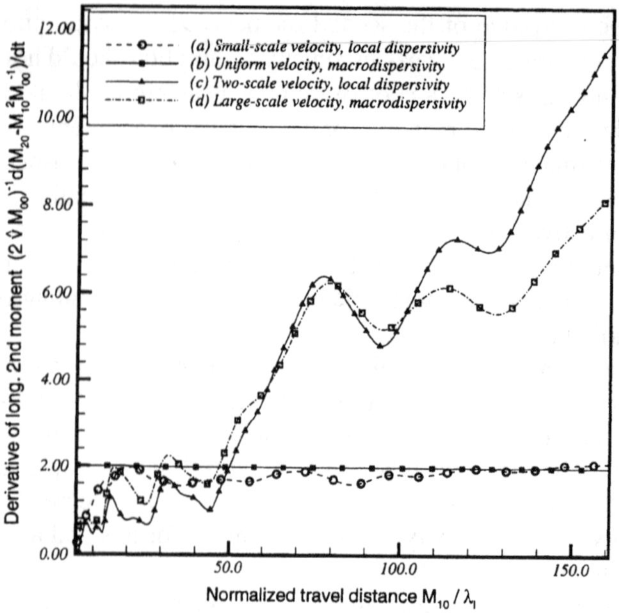

Figure 5. Normalized time derivative of longitudinal second spatial moment $(2\widehat{v}M_{00})^{1/2}d \cdot (M_{20} - M_{10}^2 M_{00}^{-1})/dt$ versus normalized travel distance (M_{10}/λ_l) for each of the cases shown in Figures 3 and 4.

as follows (Dagan, 1989; Gelhar; 1993):

$$\widehat{v} = \frac{1}{M_{00}} \frac{dM_{10}}{dt}, \tag{4}$$

$$\widehat{A}_L = \frac{1}{(2\hat{v}M_{00})} \frac{d(M_{20} - M_{10}^2 M_{00}^{-1})}{dt}, \tag{5}$$

$$\widehat{A}_T = \frac{1}{(2\hat{v}M_{00})} \frac{d(M_{02} - M_{01}^2 M_{00}^{-1})}{dt}, \tag{6}$$

where M_{ij} is the (i, j)th spatial moment, defined in discrete form as

$$M_{ij}(t) = \sum_{k=1}^{N} x_{k1}^i x_{k2}^j \, c(x_{k1}, x_{k2}, t) \, \Delta x \Delta x \tag{7}$$

and $M_{00}(t)$ is the solute mass (practically a constant for our simulations). If we substitute the nodal concentrations from the small-scale experimental plume into (7) we find that the time derivatives on the right-hand sides of (4) through (6) vary over time. The first moment time derivative fluctuates slightly about the ensemble mean velocity of 0.42 m/day used in the random field generator. Figures 5 and 6 reveal that the longitudinal and transverse central second moment time derivatives

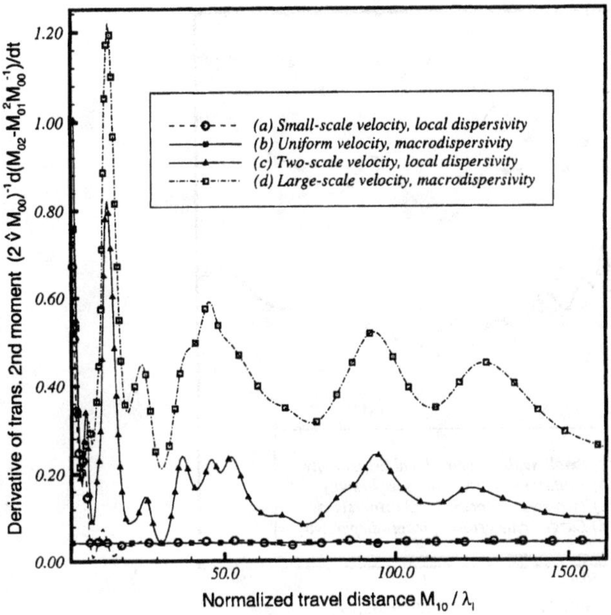

Figure 6. Normalized time derivative of transverse second spatial moment $(2\widehat{v}M_{00})^{1/2}d \cdot (M_{02} - M_{01}^2 M_{00}^{-1})/dt$ versus normalized travel distance (M_{10}/λ_l) for each of the cases shown in Figures 3 and 4.

for Case a (small circles) fluctuate initially but soon approach constant asymptotic values, a prerequisite for ergodicity. We substitute these asymptotic values into (4) through (6) to obtain the effective velocity and macrodispersivity for our ensemble mean model. Note that this forces the ergodicity condition to apply.

If we insert the asymptotic effective velocity and macrodispersivities derived from Case a into the solute transport Equation (2) the result is the predicted (ensemble mean) small-scale plume shown in the second rows of Figures 3 and 4 (Case b). As expected, the Case b plume has a Gaussian shape with the same asymptotic first and second spatial moments as the Case a plume in Figures 3(a) and 4(a). This is confirmed by the Case b second moment time derivatives plotted in Figures 5 and 6 (small filled squares).

Although the predicted plume of Figures 3(b) and 4(b) has the same first two spatial moments as the more irregular plume in Figures 3(a) and 4(a), the smaller-scale differences between these two cases are apparent. Most striking is the moderate longitudinal asymmetry (non-zero third moment) of the irregular plume. Despite minor differences between Cases a and b, it might be argued that the predicted plume provides a reasonable aggregate approximation to the irregular plume. This prediction does not resolve small-scale variations in velocity but, instead, uses a macrodispersivity coefficient to account for their aggregate effect.

Other quantitative measures of plume structure provide additional insight which can help us compare our experimental cases. One of these is a dilution (or entropy)

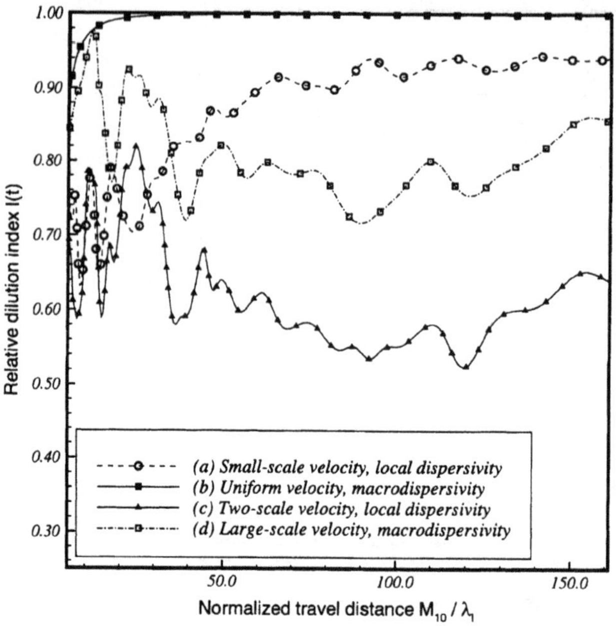

Figure 7. Relative dilution index ($I(t)$) versus normalized travel distance (M_{10}/λ_l) for each of the cases shown in Figures 3 and 4.

index based on concepts proposed by Kitanidis (1994). This index is defined as

$$I(t) = E(t)/E_{\max}(t), \tag{8}$$

where

$$E(t) = \exp\left\{-\sum_{k=1}^{N} p_k \log p_k \Delta x \Delta x\right\} \tag{9}$$

and $E_{\max}(t)$ is the maximum value of $E(t)$. The ratio $p_k(t) = c_k(t)/M_{00}(t)$ is the normalized solute concentration at time t and node k. Figure 7 plots the dilution index versus normalized travel distance for each of our experimental cases.

It can be shown that the value $E_{\max}(t)$ is attained when $c_k(t)$ corresponds to a Gaussian plume with the same second spatial moments as the plume of interest (Kitanidis, 1994). That is, the Gaussian condition is the one that provides the most dilution or mixing. The dilution index $I(t)$ measures the ratio of area (or volume) occupied by a simulated plume to the area (or volume) occupied by the corresponding Gaussian plume. When $I(t)$ approaches 1.0 the plume is as well-mixed as possible, for a given set of spatial second moment values. This interpretation is confirmed by noting that the dilution index for Case b approaches the maximum value of 1.0 as soon as velocity variations overcome the initialization effects imposed by the square solute source. Although the dilution index for Case a does not

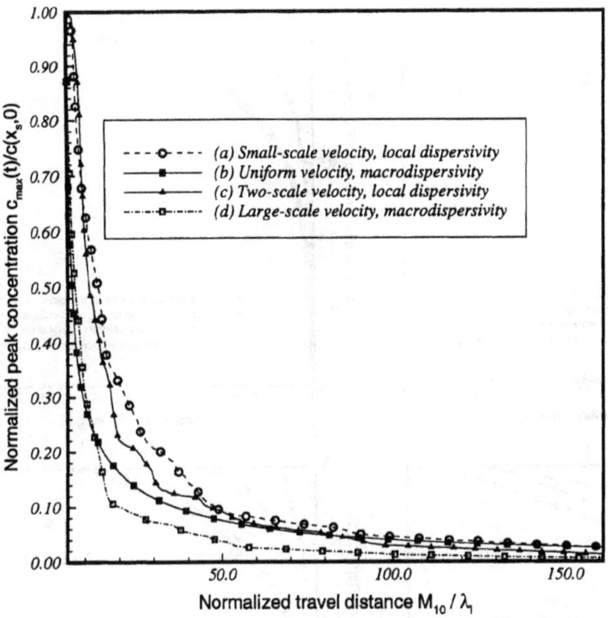

Figure 8. Normalized peak solute concentration $c_{max}(t)/c_S$ versus normalized travel distance (M_{10}/λ_l) for each of the cases shown in Figures 3 and 4.

reach 1.0 it approaches an asymptotic value above 0.90, indicating that this case produces a nearly Gaussian plume.

Another important measure of plume behavior is the maximum (or peak) concentration taken over all simulation nodes. Figure 8 plots the normalized peak concentration $c_{max}(t)/c_s$ versus normalized travel distance for each of our experimental cases. Here c_s is the initial source concentration. Note that the normalized peak decays quickly from its maximum value of 1.0 for both Cases a and b. However, the peak for Case a, which retains small-scale variability, is consistently higher than the peak for Case b, which accounts for small-scale variability only through the enhanced spreading provided by the macrodispersivity coefficients. The differences in the two peaks are greatest at intermediate times when the peak values are moderately large. At these times the peaks for Case a can be twice the values obtained for Case b.

Figure 9 provides a more global indication of the distribution of simulated concentration values. The measure plotted here is a cumulative areal distribution function of the concentration at four times after the initial solute release. At each of these times concentrations at all computational nodes with values greater than the threshold $10^{-6}c_{max}(t)$ are ranked from smallest to largest ($c_{max}(t)$ is the maximum concentration over all nodes at time t). This rank is then divided by the number of nodes N_{cmax} which have concentrations above the threshold to give a number between 0.0 and 1.0. The normalized rank and concentration are plotted for each

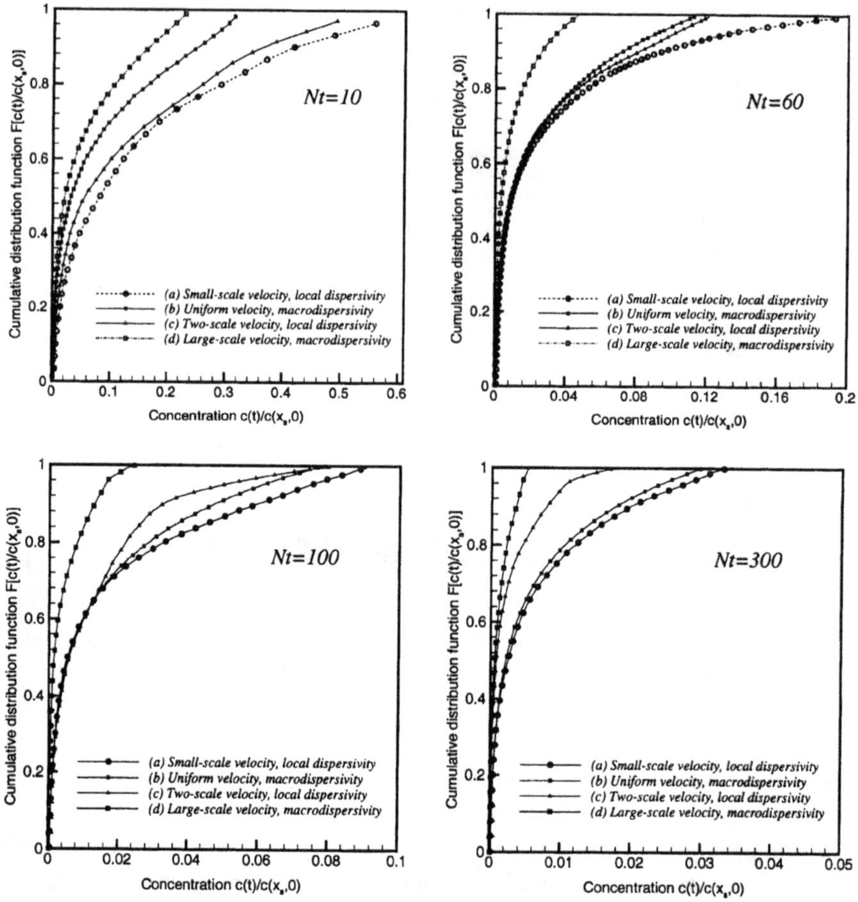

Figure 9. Cumulative areal distribution functions $F(c)$ of solute concentration versus normalized concentration for each of the cases shown in Figures 3 and 4.

of the N_{cmax} nodes, giving the curves shown in Figure 9. Since every node has an associated area Δx^2 the ordinate of each curve can be viewed as the fraction of the plume area with concentrations less than or equal to the value specified on the abscissa. It is apparent from the plot that the concentrations obtained for Case a are consistently higher than those obtained for Case b at early times but nearly the same at later times. This appears to reflect the time needed in Case a for local dispersion and small-scale velocity fluctuations to dilute the solute originally concentrated at the source.

The results discussed above for Cases a and b confirm the basic assumptions of stochastic macrodispersion theory, particularly the rapid convergence of the spatial moment time derivatives to asymptotic values and the near-Gaussian appearance of the asymptotic plume. This supports the use of macrodispersivities in situations where velocity varies only over small scales. Now we can examine the results obtained when large-scale velocity variations are also present.

The third rows (Case c) of Figures 3 and 4 show the evolution of an experimental solute plume through a two-scale velocity field with only local dispersion included. Case c is analogous to Case a except for the addition of a large-scale velocity component. We can consider this to be a more-or-less realistic case which we would like to reproduce, at least in an aggregate fashion, with a model based on effective properties. The Case c plume clearly has a less regular structure than Cases a and b, reflecting the influence of large-scale flow features. This is most apparent at time step $N_t = 100$, when cross-sections through portions of the plume are multi-modal and the area of highest concentration runs nearly perpendicular to the mean flow direction.

Some of the quantitative measures used to analyze Cases a and b behave much differently for Case c. Figure 5 indicates that the time derivative of the longitudinal spatial second moment (small triangles) does not approach an asymptotic value even after the plume has traveled 150 large correlation lengths (1500 small correlation lengths). This means that the weak ergodicity condition cannot be satisfied, at least over the time period considered in this experiment. Although the transverse component plotted in Figure 6 does vary around a constant value after about 50 large correlation scales this value is significantly larger than the ones obtained for Cases a and b.

Figure 7 shows that the Case c dilution index is much lower than the indices obtained for Cases a and b, confirming the visual observation that the two-scale plume is not Gaussian and not as well mixed as plumes which have a more Gaussian shape. The peak concentration measure of Figure 8 indicates that the Case c plume has lower peak values than Case a except at very early times. This may reflect the marked stretching of the two-scale plume that takes place starting after time step $N_t = 60$. The cumulative areal distribution functions plotted in Figure 9 confirm that Case c has consistently lower concentrations than Case a with the difference increasing over time.

Now we can consider what happens when we insert the effective dispersivities estimated from Case a into a transport equation which resolves the large-scale features of the Case c flow field. In particular, we are interested in the ability of these macrodispersivities to account for the effects of small-scale variability. The resulting plume, which can be viewed as a prediction of Case c, is shown in the fourth rows of Figures 3 and 4 (Case d)). The Case d plume reproduces the general location and shape of the Case c plume, although it is much smoother. The most noticeable qualitative difference is at the higher concentration levels, especially at later times. At $N_t = 100$ the Case c peak is higher and in a different location than the Case d peak and the region of moderately high concentrations is more extensive in Case c. The contrast is much more pronounced at $N_t = 300$, when the predicted plume has nearly disappeared and the concentrations levels along its centerline are greatly underestimated.

Table III. Summary of mass balance errors (in percent)

Case	(a)	(b)	(c)	(d)
$N_t = 10$	−3.28	−0.18	−0.80	−0.25
$N_t = 60$	−5.74	0.25	−1.36	−0.02
$N_t = 100$	−4.03	0.46	0.23	0.89
$N_t = 300$	−4.85	1.02	3.40	14.66

Additional insight can be obtained by comparing the Case c and d quantitative measures plotted in Figures 5 through 9. The longitudinal second moment time derivatives for Cases c and d are quite similar up to about 100 large correlation distances (time step $N_t = 100$). After that, the Case d time derivative increases more slowly, underestimating the longitudinal spreading of the Case c plume. The transverse second moment time derivative of the predicted plume (open squares) is significantly larger than the comparable Case c result at large times. This indicates that the Case d plume is spreading more rapidly in the transverse direction, as is apparent in Figures 3 and 4. Many of these differences between the Case c and d second moment plots are probably replicate-specific. The more general phenomenon noted here is the lack of convergence to asymptotic values, especially in the longitudinal direction, even after 150 large-scale correlation distances. This indicates that a critical prerequisite for ergodicity (moment derivative convergence) is not satisfied.

The dilution index plotted in Figure 7 indicates that the predicted Case d plume is much better mixed and 'more Gaussian' than the Case c plume, reflecting the smoothing effect of the macrodispersivity. Figures 8 and 9 confirm that this effect lowers the peak concentration and shifts the cumulative areal distribution function towards lower values at all times. Overall, the effective property approximation locates the solute reasonably well but consistently underestimates its concentration. This could have serious implications for applications where concentration levels are important, either for regulatory reasons or because of possible effects on chemical and biological transformations not considered here.

Since the plumes in Figures 3 and 4 are plotted with discrete log concentration intervals (one color for each interval) it is difficult to visually confirm that their masses are all the same. Table III summarizes computed mass balance errors (initial mass released at the source minus plume mass at the specified time) for each of the four cases shown in Figures 3 and 4. All the errors are small (a few percent or less) except the error for Case d at time $N_t = 300$. In this case the plume spread so much that some of it (about 15% of the initial mass) crossed the outflow boundary and left the computational domain. Some of the other plumes shown in Figures 3 and 4 extended slightly outside the viewing windows defined in Figure 1 and so appear to be truncated. However, the mass balance results show that only the Case d $N_t = 300$ plume actually spread outside the computational domain.

It is interesting to note that Figure 9 indicates that the Case c solute concentrations are close to Case a but much higher than Case d at early times. As time progresses the Case c concentrations become smaller relative to Case a until they approach the values of Case d. This evolutionary process suggests that the two-scale plume is first influenced primarily by the smaller-scale velocity variations which also control the dispersion and dilution of the Case a plume. As larger-scale features take effect, they appear to contribute additional mixing which lowers concentrations well below the levels observed in the small-scale plume. The moments for Case c suggest that this large-scale mixing effect may have a different form than classical Fickian macrodispersion but it does, nevertheless, disperse and dilute the solute. It is possible that the large-scale mixing effect may be described as Fickian macrodispersion over larger time scales than considered here, but it should be noted that our two-scale plume has traveled well over 100 large correlation scales without reaching this state. Even if the plume eventually reaches a asymptotic Fickian (i.e. Gaussian) state, this will occur after the solute has become very dilute, with concentrations orders of magnitude below the source value. Asymptotic Fickian models which apply only at such large times are of limited practical usefulness.

3. Discussion and Conclusions

The numerical experiment described here, like a field tracer experiment, provides information about solute movement for particular conditions. If these conditions change we can expect the resulting plumes to change as well. In particular, plumes obtained from different velocity replicates can be expected to have features which differ qualitatively from those shown in Figures 3 and 4. Although some care must be exercised in generalizing from the results reported here, we have repeated our experiment with different replicates and have observed similar overall behavior. Furthermore, we believe that the plumes discussed in this paper have traveled sufficiently far to sample a wide range of velocity fluctuations, both at the small and large scales. So we feel that it is valid to draw some general conclusions from our results.

We have observed (for Case a) that the time derivatives of the spatial moments of an individual solute plume moving through a small-scale constant mean velocity field tend to asymptotic limits. Furthermore, this plume approaches the well-mixed Gaussian state associated with a relative dilution index of 1.0. Since Gaussian plumes are completely characterized by their first two spatial moments we might expect that a model which reproduces these asymptotic moments will provide good predictions in other respects, at least for large times.

We have investigated this hypothesis with a transport model which does not resolve small-scale velocity fluctuations but is able to reproduce the spatial moments of the small scale plume with an appropriately adjusted macrodispersion coefficient. Predictions from the model (Case b) match the overall shape of the small-scale plume and give reasonable approximations to the large-time dilution index, peak value, and cumulative areal distribution function. So the first part of our experiment confirms that the most of the effects of small-scale velocity variability can be accounted for with a macrodispersion coefficient, at least for large times, if there is no large-scale variability.

The situation appears to change when small and large-scale velocity fluctuations are both present (Case c). The resulting plume is qualitatively different from the classical Gaussian plume since it reflects the cumulative influence of all the large-scale features it has passed through. The longitudinal second spatial moment time derivative does not converge to an asymptotic value, even after the plume has traveled over one hundred large correlation lengths and the peak concentration has dropped to less than 1–2% of the initial source concentration. The corresponding transverse derivative appears to approach an asymptotic value, but this value is significantly larger than the one observed for the small-scale plume of Case a. The dilution index confirms that the two-scale plume remains far from Gaussian throughout the simulation period. Although it might be argued that the plume should eventually become Gaussian it appears that this state could only be obtained after the concentrations have been diluted far below the small levels observed at the end of our experiment.

The primary question motivating our investigation is whether the large-scale features of a two-scale solute plume can be reproduced by a model which perfectly resolves the large-scale component of the velocity field but uses macrodispersivities to account for small-scale variability (Case d). As might be expected, the plume predicted by such a model shares some qualitative features with the two-scale experimental plume, since they both depend on the same large-scale velocity field. But the predicted plume is more well-mixed and consistently underestimates peak solute concentrations at all times. So it appears that a model which uses conventional macrodispersivities to account for small-scale variability fails to reproduce some key features (especially peak concentrations) of the two-scale plume. It may be argued that since an ensemble mean concentration model based on effective parameters is intended to produce a smoothed version of the actual plume the results shown in Figure 4(d) should come as no surprise. The problem with the prediction of Figure 4(d) is not that it is smooth but that it fails to reproduce either the first two spatial moments of the two-scale plume (the weak ergodicity requirement) or its peak values, which may be more important than the moments in practical applications.

It should be pointed out that the two-scale plume considered in our experiment may be too narrow (with respect to the length of large-scale velocity variations) to display ergodic behavior over the time scales we have simulated. This is, in fact, one of our primary points. If large-scale velocity variations are larger than source sizes, as they are in many practical modeling applications (where advective velocities may be defined over grid spacings on the order of hundreds of meters and sources may be only tens of meters across), we cannot expect ensemble mean dispersion theories to accurately predict the behavior of individual plumes. The source sizes and large-scale velocity correlation lengths we have used are typical of many field situations. If the requirements needed for stochastic macrodispersion theories to hold are too confining these theories may be of limited practical use, even if they are correct for a particular set of assumptions. Non-Fickian and non-local theories of macrodispersion provide alternatives to the simple approach considered in Figure 4(d) (see, for example, Graham and McLaughlin, 1988). However, these theories generally replace traditional effective property concepts and associated transport equations with more complex alternatives. Also, they have not been extensively tested in field settings where large-scale velocity variations are important.

In many ways this experiment raises as many questions as it answers. How valid are conclusions drawn from a two-dimensional experiment which ignores the effects of mixing in the missing third dimension? How would the results differ if the velocity field were generated from a nonlinear flow simulator rather than a random field generator based on a linearizations of Darcy's law and the flow equation? Could a different macrodispersivity coefficient (e.g. one which accounts in some way for the presence of large-scale variability) do a better job in predicting dilution or concentration peaks? How about a non-local stochastic dispersion theory or one which treats the large-scale velocity component as a spatially variable ensemble mean? How would the situation change if the macrodispersivities were derived from statistics (e.g. velocity covariances) conditioned on the resolved large-scale velocity field, rather than from unconditional statistics? These are all questions which go beyond the scope of this paper but which deserve careful attention.

But the critical issue raised here is whether it is valid to assume that small and large-scale velocity fluctuations can be treated independently. We believe that the substantial difference in behavior observed for our two-scale case, even after travel over distances unprecedented in other field or numerical experiments, can best be explained by scale interactions resulting from the nonlinear transformation from velocity to concentration (Ruan, 1997). If this transformation were linear, our two-scale velocity field would be just one more statistically homogeneous (spatially stationary) input to classical macrodispersion theory. We would expect the spatial moment derivatives and dilution index to approach classical values, albeit more slowly than in the small-scale case. Since this does not appear to happen we suspect that scale interactions are important. This hypothesis is considered further in Ruan (1997), where a number of numerical sensitivity studies are described. However, the issue will probably be resolved only through more careful theoretical analysis of the nonlinear aspects of the transport process.

Acknowledgements

The authors wish to thank Prof. Wolfgang Kinzelbach, who offered many valuable suggestions and comments, and Prof. Lynn Gelhar, who provided helpful reviews of earlier versions of this work.

References

Adams, E. E. and Gelhar, L. W.: 1992, Field study of dispersion in a heterogeneous aquifer: spatial moments analysis, *Water Resour. Res.* **28**(12), 3293–3308.

Dagan, G.: 1984, Solute transport in heterogeneous porous formations, *J. Fluid Mech.* **145**, 813–833.

Dagan, G.: 1986: Statistical theory of groundwater flow and transport: pore to laboratory, laboratory to formation and formation to regional scale, *Water Resour. Res.* **22**(9), 120S–134S.

Dagan, G.: 1989, *Flow and Transport in Porous Formations*, Springer-Verlag, New York.

Dagan, G.: 1990, Transport in heterogeneous porous formations: spatial moments, ergodicity, and effective dispersion, *Water Resour. Res.* **26**(6), 1281–1290.

Freyberg, D. L.: 1986, A natural gradient experiment on solute transport in a sand aquifer, 2, Spatial moments and the advection and the dispersion of nonreactive tracers, *Water Resour. Res.* **22**(13), 2031–2046.

Gelhar, L. W. and Axness, G. L.: 1983, Three-dimensional stochastic analysis of macrodispersion in aquifers, *Water Resour. Res.* **19**(1), 161–180.

Gelhar, L. W.: 1986, Stochastic subsurface hydrology from theory to applications, *Water Resour. Res.* **22**(9), 135S–145S.

Gelhar, L. W.: 1993 *Stochastic Subsurface Hydrology*, Prentice Hall, Englewood Cliffs, New Jersey.

Hess, K. M., Wolf, S. H. and Celia, M. A.: 1992, Large-scale natural gradient tracer test in sand and gravel, Cape Cod, Massachusetts, 3, Hydraulic conductivity and calculated macrodispersivities, *Water Resour. Res.* **28**(8), 2011–2027.

Indelman, P. and Rubin, Y.: 1996, Solute transport in nonstationary velocity fields, *Water Resour. Res.* **32**(5), 1259–1267.

Kitanidis, P. K.: 1994, The concept of dilution index, *Water Resour. Res.* **30**(7), 2011–2026.

McKay, D. M., Freyberg, D. C., Roberts, P. V. and Cherry, J. A.: 1986, A natural gradient experiment on solute transport in a sand aquifer, 1, Approach an overview of plume movement, *Water Resour. Res.* **22**(13), 2017–2030.

Neuman, S. P., Winter, C. L. and Newman, C. M.: 1987, Stochastic theory of field-scale Fickian dispersion in anisotropic porous media, *Water Resour. Res.* **23**(3), 453–466.

Papoulis, A.: 1984, *Probability, Random Variables, and Stochastic Processes*, McGraw-Hill, New York, NY.

Rajaram, H. and McLaughlin, D.: 1990, Identification of large-scale spatial trends in hydrologic data, *Water Resour. Res.* **26**(10), 2411–2423.

Rajaram, H. and Gelhar, L. W.: 1991a, Three dimensional spatial moment analysis of the Borden tracer test, *Water Resour. Res.* **27**(6), 1239–1251.

Rehfeldt, K. R., Boggs, J. M. and Gelhar, L. W.: 1992, Field study of dispersion in heterogeneous aquifer, 3, Geostatistical analysis of hydraulic conductivity, *Water Resour. Res.* **28**(12), 3309–3324.

Ruan, F.: 1997, A numerical investigation of solute plumes in nested multi-scale porous media, PhD thesis, Mass. Instit. Tech., Cambridge, MA.

Ruan, F. and McLaughlin, D. B.: 1998, An efficient multivariate random field generator using the fast fourier transform, *Advances Water Resour.* **21**(5), 385–399.

Ruan, F. and McLaughlin, D. B.: 1999, An investigation of Eulerian–Lagrangian methods for solving heterogeneous advection-dominated transport problems, *Water Resour. Res.* (in press).

Rubin Y. and Seong, K.: 1986, Investigation of flow and transport in certain cases of nonstationary conductivity fields, *Water Resour. Res.* **30**(11), 2901–2911.

Sudicky, E. A.: 1986, A natural gradient experiment on solute transport in a sand aquifer: spatial variability of hydraulic conductivity and its role in the dispersion process, *Water Resour. Res.* **22**(13), 2069–2082.

Transport in Porous Media **42:** 155–179, 2001.

Role of Molecular Diffusion in Contaminant Migration and Recovery in an Alluvial Aquifer System

ERIC M. LABOLLE and GRAHAM E. FOGG
Hydrologic Sciences, University of California, Davis, 95616, U.S.A.

(Received: 10 August 1999; in final form: 30 December 1999)

Abstract. Highly-resolved simulations and flow and transport in an alluvial system at the Lawrence Livermore National Laboratory (LLNL) site explore the role of diffusion in the migration and recovery of a conservative solute. Heterogeneity is resolved to the hydrofacies scale with a discretization of 10.0, 5.0 and 0.5 m in the strike, dip and vertical directions of the alluvial-fan system. Transport simulations rely on recently developed random-walk techniques that accurately account for local dispersion processes at interfaces between materials with contrasting hydraulic and transport properties. Solute migration and recovery by pump and treat are shown to be highly sensitive to magnitude of effective diffusion coefficient. Further, transport appears significantly more sensitive to the diffusion coefficient than to local-scale dispersion processes represented by a dispersivity coefficient. Predicted hold back of solute mass near source locations during ambient migration and pump-and-treat remediation is consistent with observations at LLNL, and reminiscent of observations at the MADE site of Columbus Air Force Base, Mississippi. Results confirm the important role of diffusion in low-conductivity materials and, consequently, its impact on efficacy of pump-and-treat and other remedial technologies. In a typical alluvial system on a decadal time scale this process is, in part, fundamentally nonreversible because the average thickness of low-K hydrofacies is considerably greater than the mean-square length of penetration of the solute plume.

Key words: stochastic, groundwater, dispersion, contaminant transport, heterogeneity, diffusion, remediation.

1. Introduction

Alluvial aquifer systems commonly contain substantial volume fractions (0.2–0.8) of fine-grained aquitard materials encapsulating networks of ancient stream channel, levee and debris flow deposits (e.g. Galloway and Hobday, 1996). Hydraulic conductivity (K) often varies over five or more orders of magnitude (e.g. Fogg *et al.*, 1998). Recently developed geologic characterization methods (Carle *et al.*, 1998) can capture these salient features in detailed stratigraphic models for the study of physical, chemical and biological processes affecting solute migration and transformation in alluvial systems.

In this paper, detailed stratigraphic models (Carle *et al.*, 1998) facilitate a study of the role of diffusion in the migration and recovery of a conservative solute in

Figure 1. Location Lawrence Livermore National Laboratory.

a typical alluvial aquifer system. The study site (Figure 1) underlies Lawrence Livermore National Laboratory (LLNL). This work would not have been possible without recently developed random-walk simulation methods (LaBolle *et al.*, 1999) that accurately account for local dispersion processes at interfaces between materials with contrasting hydraulic and transport properties. To our knowledge, this paper presents the first use of detailed (on the order of meters) site-specific characterization of a complex heterogeneous system to explore the role of diffusion in solute migration on a decadal time scale.

2. Conceptual Framework

Much of the heterogeneity affecting flow and transport in geologic formations arises from hydrofacies-scale variability in K, which commonly ranges over 5 orders of magnitude or more. In contrast, reported values of effective pore-scale molecular diffusion coefficient D^* vary over about 1 order of magnitude (e.g., see Grathwohl, 1998, for measured values of D^* in clays). Consequently, many studies either give little regard to the magnitude and variability of D^*, or neglect it based on (1) the assumption that it is negligible compared to macroscopic dispersion arising from a heterogeneous local-scale material structure (e.g. Cvetkovic *et al.*, 1991), or on (2) its negligible role compared to mechanical dispersion in homogeneous media. The most significant effects of D^* on solute migration, however, tend to manifest only in the presence of heterogeneity in K (e.g. see Gillham *et al.*, 1984).

By and large, studies delineating the role of diffusion in transport have relied on simplistic models of heterogeneity, composites of flowing and non-flowing zones arranged in stratified or matrix/fracture systems (e.g. Gillham *et al.*, 1984; Feenstra *et al.*, 1984), as well as more general multi-regional geometric configurations (for a review, see Haggerty and Gorelick, 1995). These and similar studies have demonstrated that molecular diffusion in non-flowing zones can play an important role in the migration of solutes.

Studies based on more complex (stochastic Gaussian) models of heterogeneity (e.g. Gelhar and Axness, 1983; Gelhar, 1986; Dagan, 1989; Gelhar, 1993) suggest that macrodispersion is insensitive to local dispersion (which necessarily includes diffusion) and that the primary effect of local dispersion is the dilution of concentration (see Kapoor and Gelhar, 1994). Further, Cvetkovic, *et al.* (1991) suggest that "except for stratified formation(s) with flow parallel to stratifications and large travel time (Matheron and de Marsily, 1980) pore-scale dispersion has been found to play a minor role compared to heterogeneity ...". In their study testing the applicability of stochastic models to contaminant migration and recovery in the LLNL system, Tompson *et al.* (1998) emphasize the need for improved geologic characterization over the stochastic Gaussian representations of permeability structure. Thus, the present study represents a logical extension of the analysis of Tompson *et al.* (1998).

Our model of complex alluvial heterogeneity is based on detailed simulations of hydrofacies architecture that describe the character of the subsurface, and honor lithologic (borehole) data as well as geologic concepts (see Carle *et al.*, 1998). Instead of attempting to model spatial variability of K directly, we develop a spatial model of hydrofacies units (Carle and Fogg, 1997) that captures the extreme as well as the intermediate values of K (Fogg *et al.*, 1998) in a scientific-based manner that honors fundamental geologic principles (Carle *et al.*, 1998). In this fashion, we produce a plausible model of the heterogeneity which is consistent with field interference-well tests (Fogg *et al.*, in press) and incorporates nearly all available subsurface data.

Assigning values of K to each hydrofacies defines the dominant structures in the K field in detail. Dispersion is modeled directly on the resulting detailed computational grid by numerically solving the governing flow and transport equations. Therein, effects of heterogeneity at scales within hydrofacies are handled through a conventional advection–dispersion equation (ADE) and local-scale dispersivity coefficients. The main advantages of this direct numerical simulation (DNS) approach are that virtually any spatial pattern in K can be represented without restrictions on complex site-specific boundary conditions. Thus, the approach is especially suited for modeling actual systems in which the site complexity and investigative objectives (e.g. process elucidation, risk assessment, contaminant remediation) would preclude, or present difficulties, in spatial averaging, or more generic modeling approaches. Moreover, if the heterogeneity that causes dispersion is adequately represented, scale dependence of the dispersion process ensues naturally through use of an ADE, assumed to govern transport at the local scale. This eliminates the need to invoke macro-scale constitutive equations, and their associated constraints and restrictions. Disadvantages of DNS are that characterizing the heterogeneity in K at appropriate spatial scales in three dimensions can be difficult (especially with typically sparse data on K) and computational limitations may prevent inclusion of the relevant details in heterogeneity.

3. Study Area

Lawrence Livermore National Laboratory (LLNL) is located in Livermore valley of the Coast Range of California (Figure 1). The aquifer system consists of a network of ancient stream channel, levee and debris flow deposits embedded in fine-grained low-permeability floodplain deposits (Noyes, 1990; Fogg *et al.*, 1998). This study focuses on a portion of the upper aquifer, the detailed study area described in Carle (1996). Geologic characteristics of this system are typical of many alluvial aquifer systems found in sedimentary basins of the western US and elsewhere.

Groundwater contamination, consisting primarily of trichloroethylene (TCE) and perchloroethylene (PCE), is believed to have originated from multiple sources some 50 years ago (Tompson *et al.*, 1998). Dispersion and dilution appear to be the primary mechanisms for the natural attenuation of contaminant migration. Prior to initiating pump and treat (PAT), high concentrations were generally observed close to suspected or known contaminant source areas. During the course of more than 8 years of remediation, distal plumes have shown dramatic reductions in concentration whereas concentrations of source areas have generally persisted (Hoffman, 1997).

4. Simulation Procedure

The simulation procedure consists of (1) generating conditional simulations of hydrostratigraphic architecture, (2) specifying hydraulic and transport properties of hydrofacies, (3) simulating steady-state ambient and stressed (in response to pumping) groundwater flow fields, and (4) simulating scenarios of ambient contaminant migration followed by migration in response to PAT remediation. Although our simulation procedure relies on geostatistical simulation methods, we use these to investigate specific transport processes in geologically plausible conditional realizations of the heterogeneity rather than to study ensemble characteristics of plumes. A total of 10 realizations of hydrostratigraphic architecture were used (LaBolle, 1999), and, in the interest of brevity, we present results for five realizations that encompass a broad range of transport behaviors.

4.1. SUBSURFACE CHARACTERIZATION

Details of the subsurface characterization are described in Carle (1996) and Fogg *et al.* (1998). A brief description of the approach is given here.

Table I. Hydrofacies properties, including hydraulic conductivity (m/day), mean length (m), average Peclet number, and controlling transport process

Hydrofacies	Direction and mean length			K (m/d)	P_e^{a}	Dominant process
	Strike (m)	Dip (m)	Vertical (m)			
Debris flows	8.0	24.0	1.1	0.432	5.9×10^3	Advection
Floodplain	26.6	66.9	2.1	4.32×10^{-5}	5.1×10^{-2}	*Diffusion*
Levee	6.0	20.0	0.8	0.173	2.0×10^3	Advection
Channel	10.0	50.0	1.3	5.184	1.5×10^5	Advection

[a] $P_e = Kia/D^*\Theta$, where $\Theta = 0.35$, the regional gradient $i = 0.002$, and $D^* = 10^{-5}$ m^2/day. Due to the influence of vertical gradients, calculations use horizontal length scales for channel, levee and debris flow hydrofacies, and vertical length scales for floodplain hydrofacies.

Four hydrofacies, debris flow (7%), floodplain (56%), levee (19%), and channel (18%), were identified from detailed interpretations (Noyes, 1990; Fogg *et al.*, 1998) of 5,500 m of core and subsequently used in the classification of the remainder of the core through interpretation of geophysical logs and textural descriptions of site geologists. Vertical transition probability models were measured directly from the data whereas strike and dip direction models were established through geologic interpretation as described in Carle and Fogg (1997) and Carle *et al.* (1998). Mean lengths of the four hydrofacies in the dip, strike and vertical directions are summarized in Table I. Finally, the geostatistical techniques described in Carle (1996) and Carle *et al.* (1998) were used to conditionally simulate realizations of hydrostratigraphy on a rectilinear computational grid with discretization Δx (strike) = 5.0 m, Δy (dip) = 10.0, and Δz (vertical) = 0.5 m, and dimension 81 cubed, as shown in Figure 2 for realization 5. The top of the simulation roughly corresponds to the observed water table. On the average, approximately 80% of the simulated channel nodes form a connected network extending in all directions and connecting all six boundaries of the domain.

4.2. FLOW SIMULATIONS

Steady-state saturated flow in the system illustrated in Figure 2 is described by

$$\frac{\partial}{\partial x_i}\left(K(\mathbf{x})\frac{\partial h(\mathbf{x})}{\partial x_i}\right) = \sum_w Q_w \delta(\mathbf{x} - \mathbf{x}_w), \tag{1}$$

where $h(\mathbf{x})$ is hydraulic head [L], $K(\mathbf{x})$ is hydraulic conductivity [LT^{-1}] (assumed to be isotropic) and Q_w[L^{3}T^{-1}] is the rate withdrawal due to pumping at location $\mathbf{x}_w$. General head boundary conditions (GHBs) (Anderson and Woessner, 1992), formulated to avoid boundary effects (LaBolle, 1999) while allowing flow to and from the model on all sides of the domain (Figure 3), with the exception of the top. Recharge is distributed to nodes at the water table at a rate of 0.034 m/year.

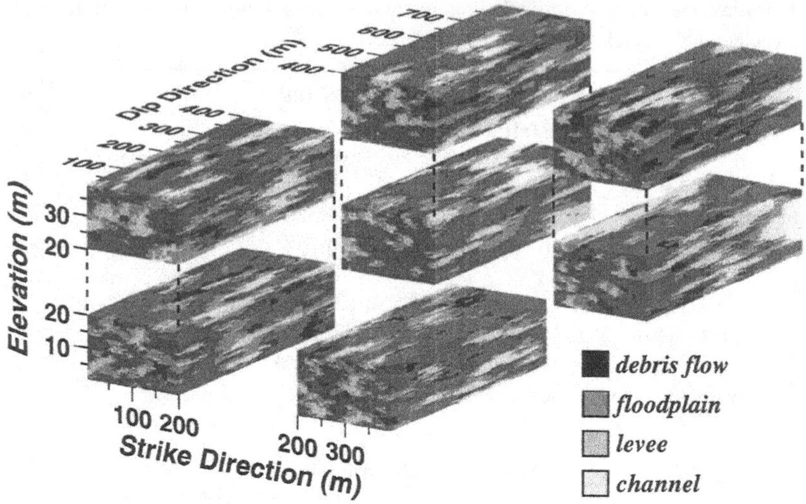

Figure 2. Conditional simulation of hydrostratigraphic architecture for realization 5.

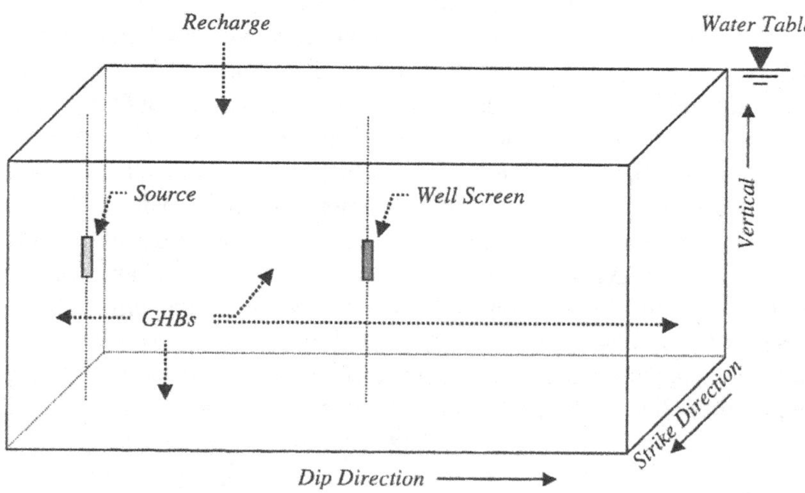

Figure 3. Schematic of flow system. The mean direction of flow is in the dip direction.

The GHBs maintain ambient a regional gradient of approximately 0.002 in the dip direction and a small vertical gradient consistent with the recharge boundary. Saturated flow is simulated with a block-centered finite difference method (McDonald and Harbaugh, 1988).

Each hydrofacies unit is assigned a different value of hydraulic conductivity (K) (Table I) based on extensive analysis of core data and field hydraulic tests as described in Carle (1996) and Fogg *et al.* (1998). These data strongly suggest that spatial variability of K within hydrofacies is small compared to variability between hydrofacies. Further, recent measurements of hydraulic conductivity on core samples representing floodplain hydrofacies from the site (Tompson *et al.*, 1998) are consistent with the assumed range.

Two steady-state hydraulic scenarios are considered, natural gradient (or ambient) and stressed. The latter corresponds to pumping from well MW-612 (Carle, 1996) (located at $x = 405.0$ m, $y = 202.5$ m in model coordinates, that is, the center of the system in plan view) on the interval $38.0 < z < 43.0$ (see Figure 3), at a rate of 50 m^3/day.

4.3. TRANSPORT SIMULATION

The solute is assumed to be conservative and nonsorbing. Given the assumption of limited variability in hydraulic properties within hydrostratigraphic units, compared to between units, we describe transport within units by a local-scale ADE of the form

$$\frac{\partial c}{\partial t} = -\nabla \cdot (\mathbf{v}c) + \nabla \cdot (\mathbf{D} \cdot \nabla c), \tag{2}$$

where c is aqueous-phase contaminant concentration, $\mathbf{v} = -K\nabla h/\Theta$ is pore-water velocity, where Θ is an effective porosity (approximated as a constant), and $\mathbf{D}$ is a local hydrodynamic dispersion tensor of the form

$$\mathbf{D} = (|\mathbf{v}|\alpha_{\mathrm{T}} + D^*)\mathbf{I} + (\alpha_{\mathrm{T}} - \alpha_{\mathrm{L}})\mathbf{v}\mathbf{v}/|\mathbf{v}|, \tag{3}$$

where α_{T} and α_{L} are transverse and longitudinal dispersivity [L], respectively, D^* is an effective molecular diffusivity [L^2T^{-1}], as described in the following section.

4.3.1. *Transport Parameters*

Parameters of Equation (2) include α_{L}, α_{T}, Θ and D^*. Simulations show that solute migration is sensitive to α_{T}, but insensitive to α_{L}, even when α_{L} is varied over the range from 10^{-3} to 10^{-1} m while maintaining α_{T} at 10^{-2} m (LaBolle, 1999). Therefore, to optimize performance of the RWPM algorithm (LaBolle *et al.*, 1999) simulations use an isotropic and constant local-scale dispersivity of $\alpha_{\mathrm{L}} = \alpha_{\mathrm{T}} = 10^{-2}$ m, based on the assumption of a moderate degree of local-scale heterogeneity.

Effective pore-scale aqueous-phase molecular diffusivity in a porous medium is assumed to be a function of the pore geometry of that medium and the aqueous-phase molecular diffusivity D_{aq} through definition of the mean-squared displacement:

$$\left\langle \sum_i \Delta X_i^2 \right\rangle \propto \tau^{-2} D_{\mathrm{aq}} \Delta t = D^* \Delta t, \tag{4}$$

where the tortuosity $\tau = l_e/l$ accounts for the tortuous paths of length l_e [L] molecules must travel on average to move a distance l in a straight line, and Δt is a time interval. Molecular diffusion may be the dominant transport process where

Table II. Transport parameters

Parameter	Value
Porosity, Θ	0.35
D^* (m^2/day)	$10^{-6} - 5.2 \times 10^{-5}$
$\alpha_L = \alpha_T$ (m)	0.01

time scales associated with advection $\langle \Delta t_v \rangle = O(a/|\mathbf{v}|)$ are smaller than those of molecular diffusion $\langle \Delta t_{D^*} \rangle = O(a^2/D^*)$ (where we neglect mechanical dispersion as we are primarily concerned with regions in which the velocity is relatively small), i.e., where $P_e = \langle \Delta t_{D^*} \rangle / \langle \Delta t_v \rangle \ll 1$, and a is a length scale associated with the hydrofacies unit of interest. Table I describes the relative influence of diffusion and advection in the four hydrofacies units. Calculations show that diffusion will tend to dominate transport in low-permeability floodplain hydrofacies. Thus, the magnitude of effective molecular diffusion coefficient in clays arises as a potentially important parameter.

Grathwohl (1998, p. 92) reported measured values of D^* for TCE in natural clays at 20°C (approximately the average groundwater temperature at LLNL) ranging from approximately 1×10^{-6} to 5.0×10^{-4} m^2/day. Here, measured values of D^* may include sorption and intra-particle diffusion effects (see Werth and Reinhard, 1999). We assume a range of uncertainty for D^* between 10^{-6} and a theoretical maximum value of 5.2×10^{-5} m^2/day computed via Archie's law, $\tau^2 = \Theta^{1-m}$ (Archie, 1942); where m is an experimentally determined coefficient that depends on the structure of the porous medium with an assumed minimum value of $m = 1.5$ (Rügner *et al.*, 1998), $\Theta = 0.35$, and $D_{\text{aq}} \sim 8.8 \times 10^{-5}$ m^2/day. Transport parameters used in the experiments that follow are given in Table II.

4.3.2. *Numerical Simulation*

Equation (2) is simulated with a new random-walk method, based on a generalization of stochastic differential equations (SDEs), in which particles change positions with time via (LaBolle *et al.*, 1999)

$$\mathbf{X}(t + \Delta t) = \mathbf{X}(t) + \mathbf{v}(\mathbf{X}, t)\Delta t + \mathbf{B}(\mathbf{X} + \Delta\mathbf{Y}, t) \cdot \Delta\mathbf{W}, \tag{5a}$$

$$\Delta\mathbf{Y}(t) = \mathbf{B}(\mathbf{X}, t) \cdot \Delta\mathbf{W}, \tag{5b}$$

where $\mathbf{X}(t)$ is a particle location at time t, $\mathbf{B} \cdot \mathbf{B}^{\mathrm{T}} = 2\mathbf{D}$, $\mathbf{D}$ is isotropic, and $\mathbf{W}(t)$ is a Brownian motion process. In (5a), $\mathbf{B}$ is evaluated at location $\mathbf{X}(t) + \Delta\mathbf{Y}(t)$, that is, the process of moving the particles requires two steps. More complicated algorithms are necessary when $\mathbf{D}$ is anisotropic (LaBolle *et al.*, 1999). The new generalized SDEs and associated random walk method in (5a) eliminate the need to smooth discontinuities in the dispersion tensor by interpolation to satisfy limitations of standard stochastic theory (LaBolle *et al.*, 1996, 1998, 1999). This property of (5a) is essential to this study because interpolation would alter the dispersion tensor field, particularly on the fringes of low-K floodplain hydrofacies units where diffusive mass exchange with non-floodplain hydrofacies occurs.

When applying (5a), the time step is dynamically controlled and semi-analytical advective paths lines (Pollock, 1988) are incorporated to increase accuracy in regions dominated by advective transport. Absorbing boundary conditions are prescribed on all extents of the model grid, and in grid blocks corresponding to pumping under stressed hydraulic conditions. Analyses of particle mass distributions to compute concentrations, spatial moments and macroscopic dispersivities are described in LaBolle *et al.* (1996), Tompson and Gelhr (1990) and Tompson *et al.* (1998).

5. Transport Experiments

Transport experiments explore the role of diffusion in solute migration and recovery in the detailed subsurface characterizations of the LLNL site. Scenarios 1 and 2 (Table III) simulate transport under ambient hydraulic conditions for 40 years followed by pump and treat (PAT) and bracket a reasonable range for effective molecular-diffusion coefficients D^*. Scenarios 3 and 4 consider use of an advection diffusion equation ($\alpha_L = \alpha_T = 0$), and ADE with no diffusion ($D^* = 0$), respectively, to examine effects of diffusion processes relative to dispersion processes represented with the dispersivity coefficient.

In all scenarios, the contaminant source is represented by 40 000 particles released in the same four grid blocks near the upstream edge of the model as illustrated in Figure 3. These grid blocks correspond to observed channel lithologies used to condition geostatistical simulations. Independent analyses (LaBolle, 1999) demonstrate that the source location is part of the primary channel network (i.e., the largest connected channel network, typically comprised of approximately 80% of the total volume of channel hydrofacies) in all realizations, but is poorly connected with this network in realization 2. Experiments (not shown) demonstrate that field-scale transport is not particularly sensitive to the initial lateral distribution of the contaminant source within these four grid blocks. Therefore, particles of the source are distributed uniformly along a vertical line at $x = 207.5$ m, $y = 25.0$ m,

Table III. Scenarios 1–4

Scenario	D^* (m^2/day)	Equation
1	5.2×10^{-5}	ADE
2	10^{-6}	ADE
3	5.2×10^{-5}	Advection diffusion
4	0	ADE, No diffusion

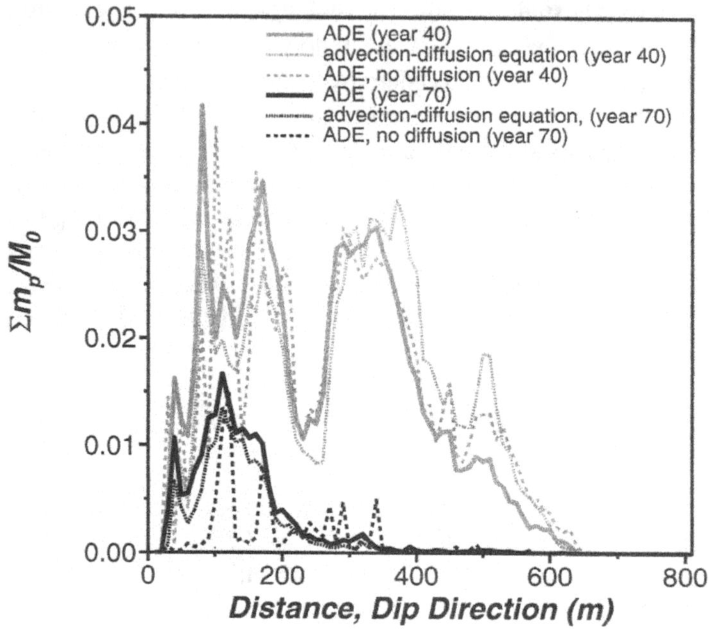

Figure 4. Normalized *x*-*z* average concentration for scenarios 1 (ADE), 3 (advection–diffusion equation, $\alpha_{\mathrm{L}} = \alpha_{\mathrm{T}} = 0$) and 4 (ADE, $D^* = 0$) of realization 3 at year 40 and year 70 (30 years after the start of PAT).

and $20.75\,\mathrm{m} \leqslant z \leqslant 22.25\,\mathrm{m}$ to simulate an instantaneous point source in plan view. In general, transport behavior is highly sensitive to initial conditions. For example, the release of contaminant mass in floodplain versus channel hydrofacies would be expected to produce dramatically different results than those which follow.

A brief summary of salient results that examine effects of diffusion processes relative to mechanical dispersion processes are presented first, followed by detailed results for ambient contaminant migration and PAT phases of scenarios 1 and 2.

5.1. RELATIVE ROLES OF DIFFUSION AND LOCAL-SCALE MECHANICAL DISPERSION

Knowledge of the relative sensitivity of contaminant migration and recovery to local-scale diffusion and dispersivity is essential in assessing impacts of diffusion to be presented in the sections that follow. Figure 4 plots normalized *x*-*z* average concentrations approximated by $(1/M_0)\sum_{p,\forall \mathbf{x}_p \in \Delta V_j^{x,z}} m_p(t)$ for scenarios 1 (ADE), 3 (advection–diffusion equation, $\alpha_{\mathrm{L}} = \alpha_{\mathrm{T}} = 0$) and 4 (ADE, $D^* = 0$) of realization 3 at year 40 and year 70 (30 years after the start of PAT). Here $\Delta V_j^{x,z}$ denotes the volume [L^3] of the *j*th *x*-*z* plane of grid blocks in the *y*-coordinate direction, m_p the mass of particle p, and M_0 the total mass at time 0. Advection diffusion equation results are quite close in character to ADE results, particularly at year 70, while results based on an ADE with $D^* = 0$ deviate much more noticeably from the straight ADE results, which include both mechanical dispersion and diffusion. The ADE $D^* = 0$ at year 40 results within 200 m of the source fluctuate erratically about the ADE results and miss most of the concentration peaks, while the advection–diffusion results track the ADE curve relatively closely. The ADE $D^* = 0$ at year 70 results grossly underestimate the mass within 200 m of the source compared to the ADE and advection–diffusion equation results. Virtually identical effects are observed in all the other realizations as well (LaBolle, 1999). In other words, neglecting pore-scale diffusion in this system leads to greater deviation from the ADE solution than does neglecting local-scale dispersivity in the ADE equation. These results indicate that the most significant local dispersion effects on field-scale transport are dominated by pore-scale diffusion in this system, and perhaps in most alluvial systems. Accordingly, the results that follow are aimed at elucidating the role of diffusion in contaminant migration and recovery in this rather typical alluvial system.

5.2. CONCENTRATION AND MASS DISTRIBUTIONS, AMBIENT MIGRATION

Three-dimensional concentration distributions after 40 years (14,600 days) of ambient contaminant migration (just prior to the start of PAT) are illustrated for 5 realizations in Figure 5 for scenario 1 ($D^* = 5.2 \times 10^{-5}$ m^2/day). Outer and inner isosurfaces represent the extent of the plume (i.e. grid blocks with $\geqslant 1$ particles) at the specified particle resolution and $c/c_0 = 2.5 \times 10^{-3}$, respectively.

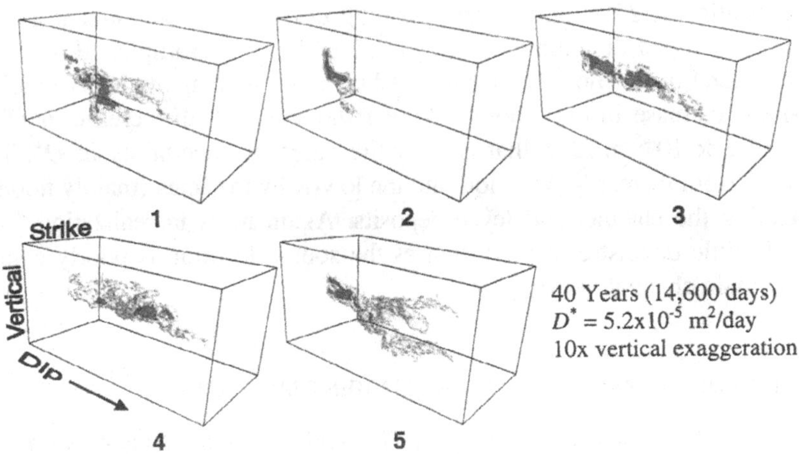

Figure 5. Three-dimensional isosurfaces of concentration for scenario 1 at year 40. The outer isosurface represents the extent of the plume at the current particle resolution. The Inner isosurface corresponds to $c/c_0 = 2.5 \times 10^{-3}$.

Results clearly illustrate the migration of mass within channels. Where primary flow-direction channel connectivity with the source location is achieved without significant branching of channels, plumes are linear in shape (realizations 3 and 4). Where connectivity is only achieved through branching, plumes tend to bifurcate in accordance with the channel network (realizations 1 and 5). Realization 2 shows relatively little downstream migration of mass because therein channel hydrofacies of the source location are poorly connected to the main channel network. Thus, transport behavior depends strongly on architecture of the channel network. Analysis of additional realizations (LaBolle, 1999) demonstrates that the distribution of transport behavior is bimodal. One mode corresponds to realizations in which channel hydrofacies of the source location are well-connected with the primary network of channels that percolates (connects) completely through the domain. The other mode corresponds to realizations in which channel hydrofacies of the source location are either not part of, or are poorly connected to the primary network of channels. Results are most consistent with observations of persistent high concentrations near source locations at the LLNL site (Hoffman, 1997), reminiscent of observed contaminant migration at the MADE site of Columbus Air Force Base, Mississippi (Adams and Gelhar, 1992; Boggs *et al.*, 1992).

Figure 6 plots the log of normalized 40-year x-z average concentrations for scenarios 1 and 2, realizations 1–5. Given that the channel (aquifer) facies in most of the realizations are extensively interconnected throughout the problem domain, one might be tempted to assume that transport would be controlled mainly by flow within the channels. Therefore, a plume generated for a latterally continuous aquifer with arbitrary $\alpha_L = 1.0$ m and $K = 5.184$ m/day, corresponding to the K of the channel hydrofacies, is shown for reference. The log scale tends to obscure some important differences between the two sets of results. Nevertheless, Figure 6 generally shows significant increase in holdback of mass, particularly close to the source location at 25 m, and a corresponding decrease in contaminant mass near the leading edge of the plume, with increase in D^*. For example, all realizations, except realization 2, show greater than a half order of magnitude (i.e. more than three-fold) increase in mass holdback near the source with increase in D^* from 10^{-6} to 5.2×10^{-5} m^2/day, that is, over the range of uncertainty in D^*. The effect arises from increased diffusion into the low-K hydrofacies (mainly floodplain) surrounding the channel and levee deposits. Again mass in realization 2 shows relatively little downstream migration as the source location is poorly connected with the main channel network.

5.3. MACRODISPERSIVE BEHAVIOR, AMBIENT MIGRATION

Longitudinal first $s(t)$ and second $\sum_{11}(t)$ spatial moments (LaBolle *et al.*, 1996; Tompson *et al.*, 1998) in the principle direction of migration as a function of time for scenarios 1 and 2, realizations 1–5, are shown in Figures 7 and 8. First moments show a noticeable increase in sequestration of mass with increase in D^*. In

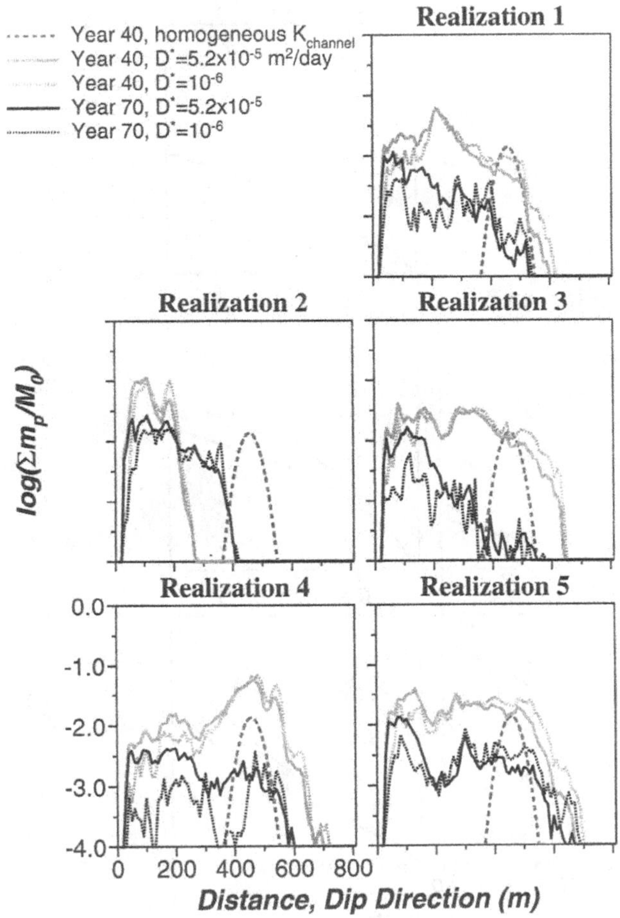

Figure 6. Normalized *x-z* average concentration at year 40 and year 70 (30 years after the start of PAT) for scenarios 1 and 2, realizations 1–5. A plume at year 40 generated for homogeneous $K = 5.184$ m/day and with $\alpha_L = 1.0$ m and is shown for reference.

realizations 3 and 5, for example, there is more than a 25% decrease in the extent of downstream migration of the center of mass at year 40 with increase in D^* from 10^{-6} to 5.2×10^{-5} m^2/day. Because source locations are in high-permeability channel hydrofacies, an increase in D^* generally results in an initial increase in the net flux of mass from high- to low-permeability hydrofacies, resulting in greater sequestration of contaminants. Simple calculations show that in the same time span, the leading edge of contamination may penetrate into a floodplain unit more than seven times further for $D^* = 5.2 \times 10^{-5}$ m^2/day than for $D^* = 10^{-6}$ m^2/day, that is, over the range of uncertainty in D^*.

With the exception of realization 4, second moments generally show little sensitivity to D^*, results consistent with stochastic theory as described in Kapoor and

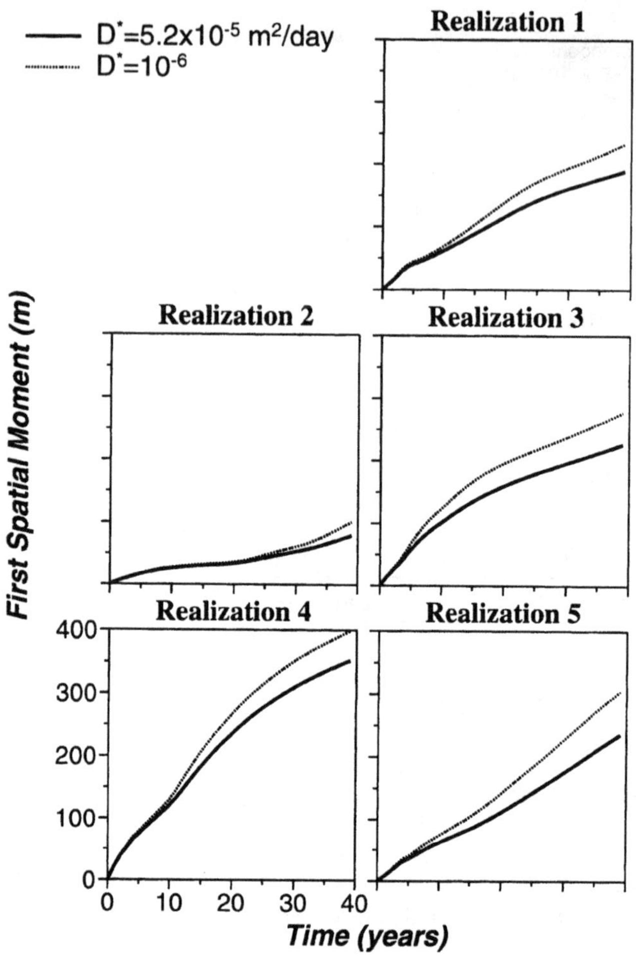

Figure 7. First spatial moments $s(t)$ for scenarios 1 and 2, realizations 1–5.

Gelhar (1994). An increase in D^* results in a consistent reduction of mass in the distal plume and a corresponding increase in mass near source locations. Apparently, despite its importance, this shift of mass commonly has little effect on second moment. In realization 4, however, increasing D^* flattens the peak of the mass distribution at 550 m enough to significantly increase the second moment. Thus, insensitivity of the second spatial moment to diffusion is a non-ubiquitous result. Furthermore, the second spatial moment fails to indicate important and consistent changes in the mass distribution.

The effect of diffusion on longitudinal macrodispersivity $A_{11} \approx \frac{1}{2} \mathrm{d} \sum_{11} / \mathrm{d}s$ for scenarios 1 and 2 is illustrated in Figure 9. At large times, results are particularly sensitive over the range of uncertainty in D^*, apparently due to the corresponding sensitivity of the time rate of change in the first moment. Increasing D^* consistently

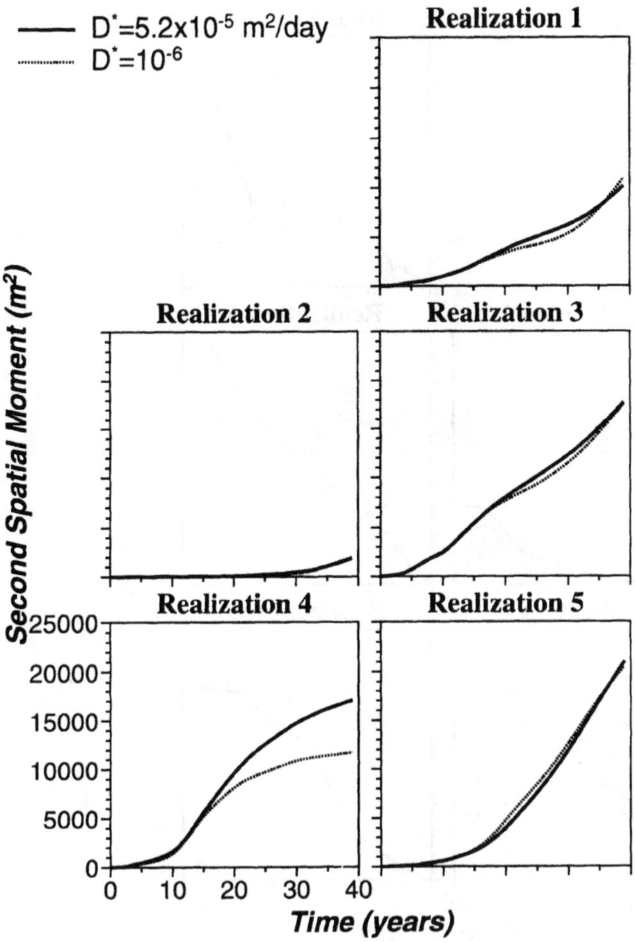

Figure 8. Second spatial moments $\Sigma_{11}(t)$ for scenarios 1 and 2, realizations 1–5.

increases A_{11}. Results presented in the following section show that the partitioning of mass among hydrofacies has not reached an 'asymptotic' state, suggesting that asymptotic behavior is not achieved within the 40-year time span. Further, at long times one should expect Taylor-dispersive behavior (Gelhar, 1993), that is, an inverse relationship between D^* and A_{11}, whereas the opposite is generally observed here. Indeed, these results are consistent with estimates for diffusion-rate limited residence times in floodplain hydrofacies on the order of centuries to millenia.

Results are contrary to the conclusions of Kapoor and Gelhar (1994) who, based on stochastic theory, suggested that diffusion has negligible effects on macrodispersion. Discrepancies with stochastic theory arise because direct numerical simulation does not rely on the simplifications commonly used to facilitate stochastic theoretical analysis. For example, our subsurface characterization is geologically-

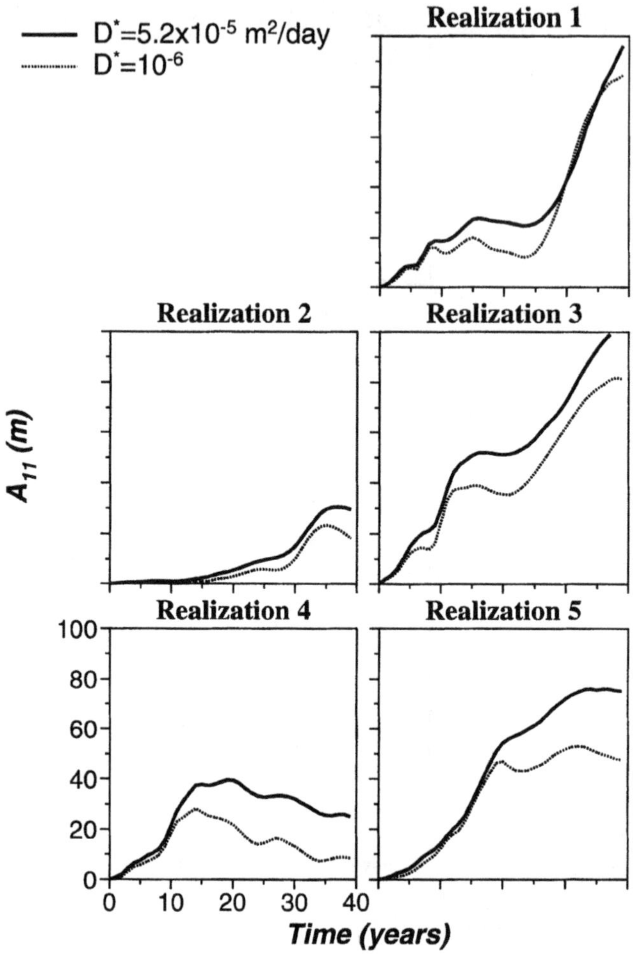

Figure 9. Macrodispersivity $A_{11}(t)$ for scenarios 1 and 2, realizations 1–5.

based, in contrast to the common Gaussian model with small ln K variance. Such a model applies more to local-scale transport entirely within an aquifer unit (e.g. our channel hydrofacies) than to aquifer-aquitard complexes such as this one. Moreover, our initial conditions are conditional (the source location is always located in channel hydrofacies) versus the usual unconditional initial conditions invoked in analytical stochastic analyses. Our results for macodispersivities, however, compare well with those of the Tompson *et al.* (1998) study of the LLNL site, based on a Gaussian geostatistcal model of heterogeneity. For example, Tompson *et al.* (1998) computed pre-asymptotic mean values of $A_{11} = 50\,\text{m}$ for $s \approx 350\,\text{m}$. Herein, s of realizations 1 and 3–5 ranges from $\sim$200 to 400 m at year 40 with A_{11} ranging from approximately 25 to more than 100 m.

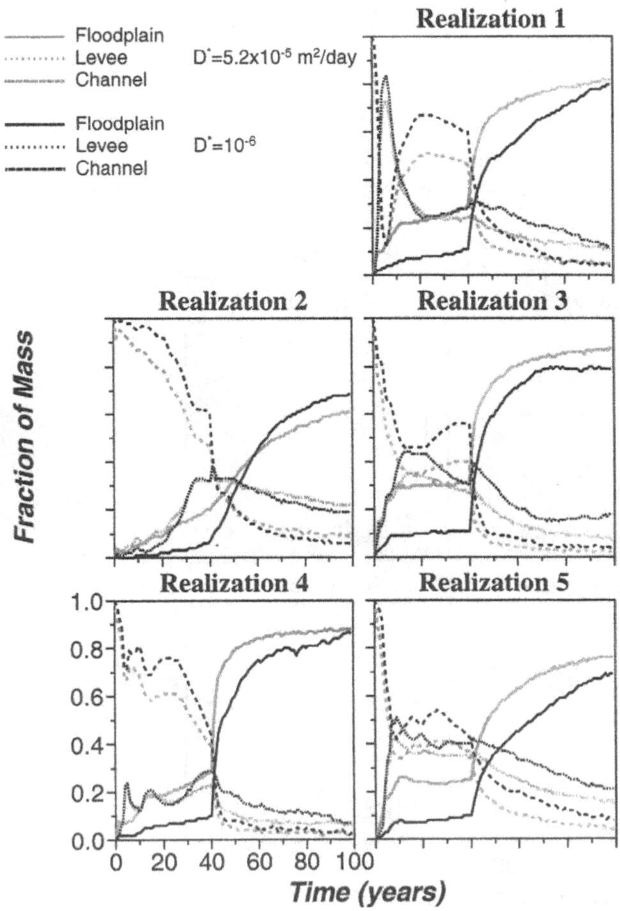

Figure 10. Fraction of total mass contained in each of the four hydrofacies as a function time for scenarios 1 and 2, realizations 1–5. PAT starts at year 40.

5.4. PARTITIONING OF MASS AMONG HYDROFACIES, AMBIENT MIGRATION

Figure 10 depicts the partitioning of mass among hydrofacies with time in scenarios 1 and 2. To simplify the presentation, fraction of mass in the less abundant debris flow hydrofacies is not shown. In all realizations, increasing D^* results in a significant increase in mass within floodplain hydrofacies and a corresponding decrease in mass within channel and levee hydrofacies. For example, in both realizations 3 and 5, increasing D^* from 10^{-6} to 5.2×10^{-5} m^2/day results in approximately three times as much mass in floodplain hydrofacies at year 40.

Figure 11 shows the partitioning of mass among hydrofacies at year 40 as a function of distance in the dip direction for scenario 1. Realizations 1 and 3–5, in which channel hydrofacies percolate and are well-connected to the source, tend to

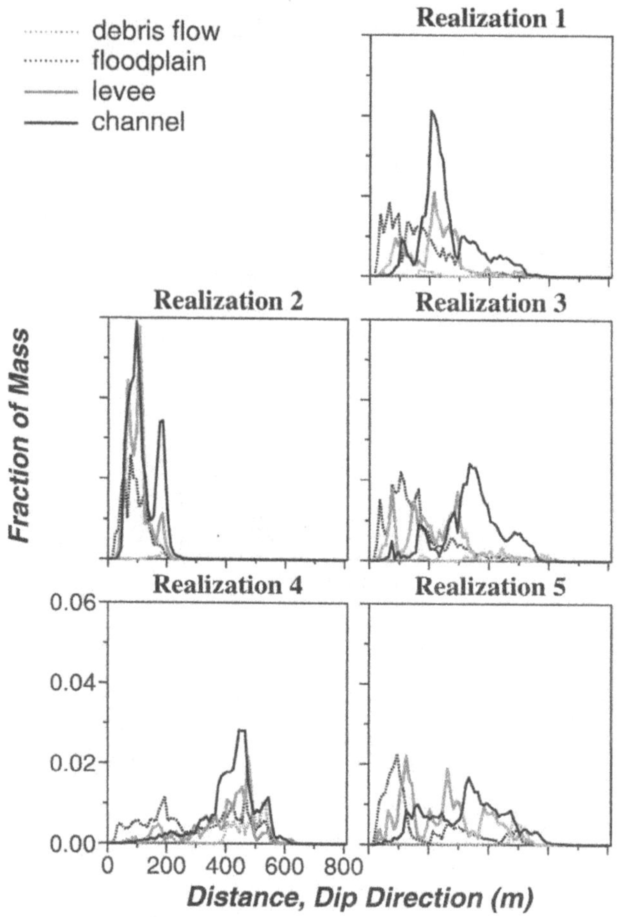

Figure 11. Fraction of total mass contained in each of the four hydrofacies as a function of distance in the dip direction for scenarios 1 and 2, realizations 1–5, year 40.

show two distinct plumes, one associated with floodplain hydrofacies and the other associated with channel hydrofacies. In all cases, mass in floodplain hydrofacies significantly lags behind mass in channel hydrofacies; separation between plumes is most pronounced in realizations 3 and 5. Results compare with those of stratified or matrix/fracture systems (e.g. Gillham *et al.*, 1984; Feenstra *et al.*, 1984). The following sections present results for the PAT remediation phase of scenarios 1 and 2.

5.5. MASS RECOVERY DURING PUMP AND TREAT

Figure 12 shows the normalized mass remaining $M(t)/M_0$ as a function of time from the start of pumping at year 40 for all five realizations of heterogeneity.

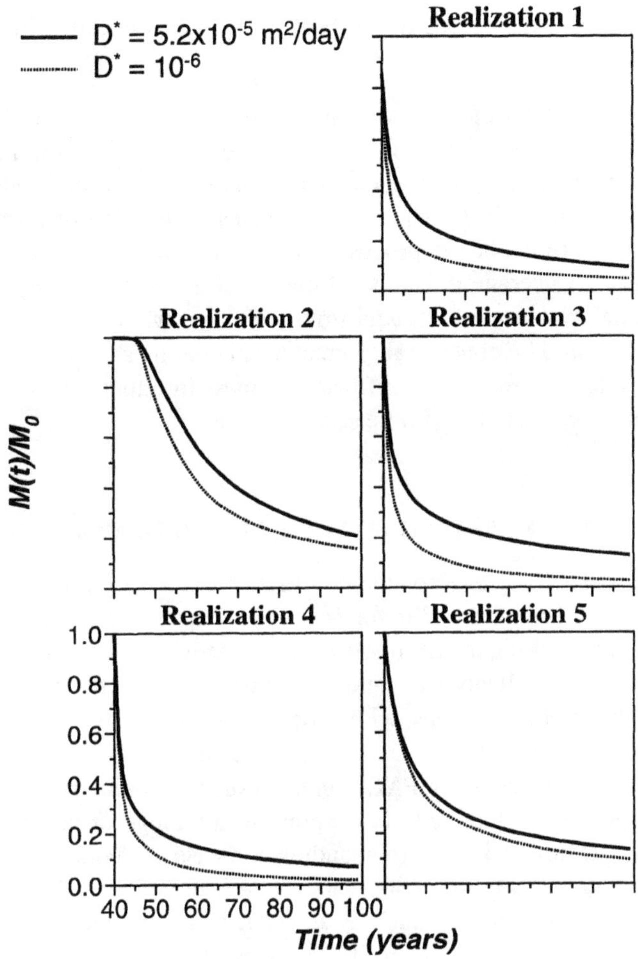

Figure 12. Mass remaining $M(t)/M_0$ as a function of time from year 40, the start of PAT, for scenarios 1 and 2, realizations 1–5.

Results show a decrease in mass remaining in the system with decrease in D^*. The long tails are consistent with observed performance of PAT technology (Mackay and Cherry, 1989; Mercer *et al.*, 1990; Nyer, 1993; NRC, 1994). In realization 2, where channel hydrofacies of the source location are poorly connected with the primary network of channels, the plume at the start of pumping in year 40 is more isolated near the source than for the other realizations. In this case, PAT with the pumping well located at the center of the system results in less recovery of contaminant mass.

Note, that an increase in D^* decreases residence time $\langle \Delta t_{D^*} \rangle$ of contaminant mass in floodplain hydrofacies which could, in some instances, result in more rapid remediation. For example, in the present case, residence times for $D^* = 10^{-6}$ m^2/day are approximately 50 times those for $D^* = 5.3 \times 10^{-5}$ m^2/day.

5.6. PARTITIONING OF MASS AMONG HYDROFACIES DURING PUMP AND TREAT

Figure 10 shows the partitioning of mass among hydrofacies as a function of time during the PAT process (starting from the year 40). During ambient migration, an increase in D^* results in corresponding increase of mass fraction in floodplain hydrofacies. Much of this mass in floodplain hydrofacies at the start of pumping is sequestered during the remediation process, as shown. This process depends largely on length scales, or average thickness (Table I), of the floodplain hydrofacies. In a typical alluvial system on a decadal time scale, the average thickness of floodplain hydrofacies would be considerably greater than the mean-square length of penetration of the solute plume. Thus, diffusion of mass into these materials is virtually a nonreversible process on this time scale.

5.7. CONCENTRATION AND MASS DISTRIBUTIONS DURING PUMP AND TREAT

Figure 6 plots normalized x-z average concentrations at the start of pumping (year 40) and after 30 years of PAT (year 70) for scenarios 1 and 2, realizations 1–5. The plots illustrate the significant decrease in mass recovery with increase in D^*. After 10 years of PAT, realizations 1 and 3–5 show significantly greater attenuation of mass near source locations with an increase in D^*. In realizations where channel hydrofacies percolate and are well-connected with the source, mass diminishes considerably as a result of PAT. Again, results are very sensitive to D^*, considering the specified values only correspond to the range of uncertainty, with an increase in D^* resulting in a corresponding increase in sequestration of contaminants during remediation. After 30 years of PAT, realizations 1 and 3–5 for $D^* = 5.2 \times 10^{-5}$ m^2/day show as much as an order of magnitude greater x-z average concentration than results for $D^* = 10^{-6}$ m^2/day.

In realizations where channel hydrofacies percolate and are well-connected with the source, Figure 6 shows the largest relative decreases in mass occur in the distal plume, downstream from the well located at 405 m. Therefore, the distal plume is more easily remediated than the plume near the source, even though the source is located upstream from the well whereas portions of the distal plume are not. This effect is primarily due to the greater extent of contaminant migration into floodplain hydrofacies near source locations and the difficulty in accessing this contamination by advection, and, to a much lessor extent, location of the pumping well in the distal plume. The effect of diffusion on sequestration of mass near the source is clearly illustrated in Figure 13 showing normalized z-average concentrations for realization 3 after 30 years of PAT for scenarios 1 ($D^* = 5.2 \times 10^{-5}$ m^2/day) and 2 ($D^* = 10^{-6}$ m^2/day). The increase in mass hold back near the source location with increase in D^* is remarkable. These results clearly illustrate that NAPLs need not be present for significant sequestration of mass near source locations.

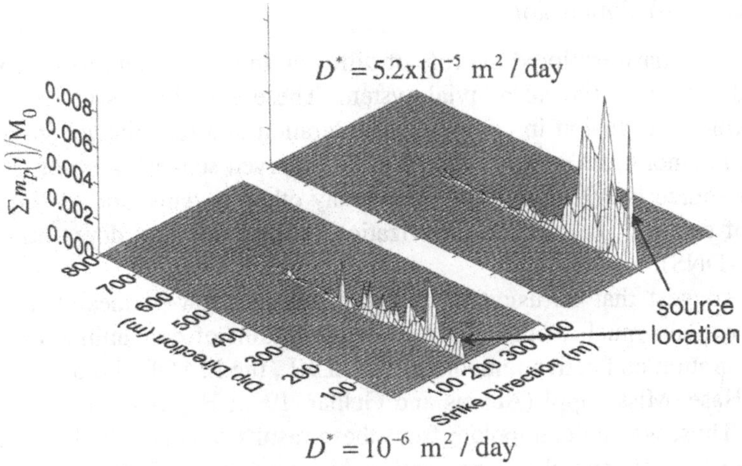

Figure 13. Normalized *z*-average concentrations at year 70, 30 years after the start of PAT, for scenarios 1 and 2, realization 3.

5.8. TRANSFERABILITY OF RESULTS

We anticipate that the hold back (sequestration) of contaminant mass near the source in these simulations would be less extreme in systems containing smaller fractions of low-*K* materials. Nevertheless, the volume fraction of silts and clays in the LLNL aquifer system is by no means anomalously high for an alluvial aquifer system (Fogg *et al.*, 1998).

One might wonder if the *K* assigned uniformly to floodplain hydrofacies in our LLNL geostatistical model is unrealistically low, thereby accentuating dispersion and the influence of diffusion. Heterogeneity in this hydrofacies would lead to higher effective *K*'s and greater advection through portions of the floodplain materials. Nonetheless, we believe $K_{\text{floodplain}} = 4.32 \times 10^{-5}$ m/day (Table I) is appropriate for two reasons. First, analysis of relatively abundant core from the site (Fogg *et al.*, 1998; Noyes, 1990) as well as recent observations made in extensive excavations into the deposit (Personal comm., Steven Carle, 1998) show that the fine-grained floodplain facies are not only the most voluminous of the hydrofacies, but also fairly 'massive' (i.e., generally lacking sedimentary structures and sand or gravel size particles). This is consistent with modes of deposition in floodplain environments. Second, the numerical pumping test simulations of Carle (1996), based on 20 realizations of the same geostatistical model used herein, indicate that $K_{\text{floodplain}} = 4.32 \times 10^{-5}$ is an excellent estimate. Nevertheless, if $K_{\text{floodplain}}$ were increased by more than an order of magnitude, transport within floodplain hydrofacies would still be diffusion dominated (see Table I). Thus, significant volumes of strata in which diffusion dominates over advection are obviously present. Clearly though, additional work is needed to determine how changes in site heterogeneity and *K* values affect plume behavior.

6. Summary and Conclusions

In this paper we have explored the role of diffusion in contaminant migration and PAT remediation in a typical alluvial system. These experiments (1) reveal the important role of diffusion in contaminant migration and remediation in alluvial systems, (2) demonstrate a mechanism for the observed sequestration of contaminants near source locations at LLNL and many other systems, and (3) highlight the value of detailed geologic characterization in the context of direct numerical simulation (DNS).

Results suggest that diffusion of contaminants into low-permeability hydrofacies can explain much of the observed sequestration of contamination during ambient migration and active remediation at LLNL, the MADE site of Columbus Air Force Base, Mississippi (Adams and Gelhar, 1992; Boggs *et al.*, 1992), and elsewhere. Thus, we can extrapolate from these results to conclude that effective molecular diffusivity, and abundance and architecture of low-K materials are key factors governing time to clean-up in many remediation projects.

The magnitude of effective molecular diffusion coefficient D^* generally governs the mass transfer between floodplain and non-floodplain hydrofacies. For example, calculations show that in the same time span, the leading edge of contamination may penetrate into a floodplain unit more than seven times further for $D^* = 5.3 \times 10^{-5}\,\mathrm{m}^2/\mathrm{day}$ than for $D^* = 10^{-6}\,\mathrm{m}^2/\mathrm{day}$, that is, over the range of uncertainty in D^*. Similarly, with a decrease in the rate of diffusion, the mean residence time $\langle \Delta t_{D^*} \rangle$ within diffusion-dominated hydrofacies increases; in the present case, residence times for $D^* = 10^{-6}\,\mathrm{m}^2/\mathrm{day}$ are approximately 50 times those for $D^* = 5.3 \times 10^{-5}\,\mathrm{m}^2/\mathrm{day}$. Additionally, contaminant migration and recovery showed significantly greater sensitivity to diffusion coefficient than to local-scale dispersion processes represented by the dispersivity coefficient (to the extent that one could make a case for ignoring the latter herein), emphasizing the important role of diffusion between hydrofacies in many alluvial aquifer systems.

Ambient contaminant migration and PAT remediation performance were found to be acutely sensitive to the magnitude of the diffusion coefficient. In ambient transport experiments, increasing diffusion coefficient (within the range of uncertainty) from 10^{-6} to $5.2 \times 10^{-5}\,\mathrm{m}^2/\mathrm{day}$ resulted in as much as a 300% increase in the total mass of contaminants residing within floodplain hydrofacies. This effect corresponded to as much as a 25% decrease in the extent of downstream ambient migration in the center of mass by the year 40, as exhibited in results for longitudinal first spatial moments. Results show that (1) insensitivity of the second spatial moment to diffusion is non-ubiquitous, (2) the second spatial moment fails to indicate important and consistent changes in the mass distribution (see Kapoor and Gelhar, 1994), and (3) longitudinal macrodispersivity can be sensitive to diffusion. Further, by the year 40 and several correlation lengths of migration, ambient transport was still pre-asymptotic with regard to macrodispersive behavior (Gelhar, 1993; Dagan, 1989). Indeed, this result is consistent with estimates for diffusion rate limited residence times in floodplain hydrofacies on the order of centuries to millenia.

The PAT phase of scenarios 1 and 2 showed an increase in the mass remaining in the system, with increase in diffusion coefficient, due primarily to a corresponding increase in the extent of migration of contaminants into floodplain hydrofacies during ambient migration. Remediation of distal contamination was more rapid than contamination near source locations, primarily due to the greater extent of diffusive contaminant migration into floodplain hydrofacies near source locations and the difficulty in accessing this contamination by pumping-controlled advection. On a decadal time scale, this diffusive transport process is virtually nonreversible because it depends largely on length scales of the floodplain hydrofacies, which in most alluvial systems are considerably greater than the mean-square length of penetration of the solute plume.

Experiments bring to light the merits of detailed subsurface characterization in a DNS approach, reminding us of the potentially important role of hydrofacies architecture in contaminant migration and remediation. In this regard, the connectivity of channel hydrofacies arises as an important factor controlling migration and remediation. On the average, approximately 80% of channel hydrofacies (18% by volume) at the LLNL site form a connected network extending in all directions (Fogg *et al.*, in press). Transport behavior depends on the details of this network. Results show that the distribution of transport outcomes or plume characteristics is bimodal: One mode corresponds to realizations in which channel hydrofacies of the source location are a part of the primary network of channels. The other mode corresponds to realizations in which channel hydrofacies of the source location are either not part of, or are poorly connected with, the primary network of channels.

Finally, the significant impact of diffusion on transport in alluvial systems suggests that future characterization efforts would benefit from studies elucidating the magnitude and variability of effective diffusion coefficient.

Acknowledgements

This work was supported by grants from the National Science Foundation (Award No. 9870342), U.S.E.P.A. (R819658) Center for Ecological Health Research at U.C. Davis, N.I.E.H.S. Superfund Grant (ES-04699), University of California Toxic Substances Teaching and Research Program and Lawrence Livermore National Laboratory. Although the information in this document has been funded in part by the United States Environmental Protection Agency, it may not necessarily reflect the views of the Agency and non official endorsement should be inferred.

References

Adams, E. E. and Gelhar, L. W.: 1992, Field study of dispersion in a heterogeneous aquifer. 2. Spatial moments analysis, *Water Resour. Res.* **28**(12), 3293–3307.

Anderson, M. P. and Woessner, W. W.: 1992, *Applied Groundwater Modeling, Simulation of Flow and Advective Transport*, Academic Press.

Archie, G. E.: 1942, The electrical resistivity log as an aid in determining some reservoir characteristics, *Trans. Am. Inst. Min. Metall. Pct. Eng.* **146**, 54–62.

Boggs, J. M., Young, S. C., Beard, L. M., Gelhar, L. W., Rehfeldt, K. R. and Adams, E. E.: 1992, Field study of dispersion in a heterogeneous aquifer. 1. Overview and site description, *Water Resour. Res.* **28**(12), 3281–3291.

Carle, S. F.: 1996, A transition probability-based approach to geostatistical characterization of hydrostratigraphic architecture, PhD Dissertation, University of California, Davis.

Carle, S. F. and Fogg, G. E.: 1997, Modeling spatial variability with one- and multidimensional Markov chains, *Math. Geol.* **28**, 453–476.

Carle, S. F., LaBolle, E. M., Weissmann, G. S., VanBrocklin, D. and Fogg, G. E.: 1998, Geostatistical simulation of hydrofacies architecture: A transition probability/Markov approach, In: G. S. Fraser and J. M. Davis (eds), *SEPM Concepts in Hydrogeology and Environmental Geology No. 1, Hydrogeologic Models of Sedimentary Aquifers*, SEPM (Society for Sedimentary Geology), Tulsa, Oklahoma.

Cvetkovic, V. D., Dagan, G. and Shapiro, A. M.: 1991, An exact solution of solute transport by one-dimensional random velocity fields, *Stoch. Hydrol. Hydraulics* **5**, 45–54.

Dagan, G.: 1989, *Flow and Transport in Porous Formations*, Springer-Verlag, Berlin, Heidelberg, 465 pp.

Feenstra, S., Cherry, J. A., Sudicky, E. A. and Haq, Z.: 1984, Matrix diffusion effects on contaminant migration from an injection well in fractured sandstone, *Ground Water* **22**(3), 307–316.

Fogg, G. E., Noyes, C. D. and Carle, S. F.: 1998, Geologically-based model of heterogeneous hydraulic conductivity in an alluvial setting, *Hydrologeol. J.* **6**, 131–143.

Fogg, G. E., Carle, S. F. and Green, C.: A connected-network paradigm for the alluvial aquifer system, In: D. Zhang (ed.), *Theory, Modeling and Field Investigation in Hydrogeology: A Special Volume in Honor of Shlomo P. Neuman's 60th Birthday*, Geological Society of America Special Publication, in press.

Galloway, W. E. and Hobday, D. K.: 1996, *Terrigenous Clastic Depositional Systems: Applications to Fossil Fuel and Groundwater Resources*, 2nd edn, Springer-Verlag, 491 pp.

Gelhar, L. W. and Axness, C. L.: 1983, Theree-dimensional stochastic analysis of macrodispersion in aquifers, *Water Resour. Res.* **19**(1), 161–180.

Gelhar, L. W.: 1986, Stochastic subsurface hydrology from theory to applications, *Water Resour. Res.* **22**(9), 135S–145S.

Gelhar, L.W.: 1993, *Stochastic Subsurface Hydrology*, Prentice Hall, New Jersey.

Gillham, R. W., Sudicky, E. A., Cherry, J. A. and Frind, E. O.: 1984, An advection diffusion concept for solute transport in heterogeneous unconsolidated geologic deposits, *Water Resour. Res.* **20**(3), 369–378.

Grathwohl, P.: 1998, *Diffusion in Natural Porous Media: Contaminant Transport, Sorption/Desorption and Dissolution Kinetics*, Kluwer Academic Publishers, Norwell, Massachusetts.

Haggerty R. and Gorelick, S. M.: 1995, Multiple-rate mass transfer for modeling diffusion and surface reactions in media with pore-scale heterogeneity, *Water Resour. Res.* **31**(10), 2383–2400.

Hoffman, F., Blake, R. G., Demir, Z., Gelinas, R. J., McKereghan, P. F. and Noyas, C. D.: 1997, A conceptual model and remediation strategy for VOCs in Low Organic Carbon Unconsolidated Sediments, Lawrence Livermore National Laboratory, UCRL-JC-125199 Rev. **1**.

Kapoor, V. and Gelhar, L. W.: 1994, Transport in three-dimensionally heterogeneous aquifers, 1, Dynamics of concentration fluctuations, *Water Resour. Res.* **30**(6), 1775–1788.

LaBolle, E. M., Fogg, G. E. and Tompson, A. F. B.: 1996, Random-walk simulation of transport in heterogeneous porous media: local mass-conservation problem and implementation methods, *Water Resour. Res.* **32**(3), 583–593.

LaBolle, E. M., Quastel, J. and Fogg, G. E.: 1998, Diffusion theory for transport in porous media: Transition-probability densities of diffusion processes corresponding to advection–dispersion equations, *Water Resour. Res.* **34**(7), 1685–1693.

LaBolle, E. M., Quastel, J., Fogg, G. E. and Gravner, J.: 1999, Diffusion processes in composite porous media: Generalized stochastic differential equations and their numerical integration by random walks, *Water Resour. Res.* in press.

LaBolle: 1999, Simulation of diffusion processes in porous media, PhD Dissertation, U.C. Davis.

Mackay, D. M. and Cherry, J. A.: 1989, Groundwater contaminantion: Pump-and-treat remediation, *Environ. Sci. Technol.* **23**, 630–636.

Matheron, G. and de Marsily G.: 1980, Is transport in porous media always diffusive? A counter-example, *Water Resour. Res.* **16**, 901–917.

McDonald, M. G. and Harbaugh, A. W: 1988, A modular three-dimensional finite-difference ground-water flow model. *US Geological Survey Techniques of Water-Resources Investigations Book 6*, Chapter A1, p. 586.

Mercer, J. W., Skipp, D. C. and Giffin, D.: 1990, *Basics of Pump-and-Treat Ground-Water Remediation Technology*, EPA/600/8-90/003, Ada, Okla., EPA, R.S. Kerr Environmental Research Laboratory, p. 31.

National Research Council (NRC): 1994, Alternative for Ground Water Cleanup, National Academy Press, Washington, D.C.

Noyes, C. N.: 1990, Hydrostratigraphic analysis of the Pilot Remediation Test Area, LLNL, Livermore, California, MS Thesis, University of California, Davis, 165 pp.

Nyer, E. K.: 1993, Aquifer restoration: pump and treat and the alternatives, *Groundwater Monitor. Rev.* Winter, pp. 89–92.

OSWER: 1997, Directive 9200.4-17, use of monitored natural attenuation at superfund, RCRA corrective action, and underground storage tank sites, U.S. EPA office of Solid Waste and Emergency Response Directive 9200.4-17, 29 pp.

Pollock, D. W.: 1988, Semianalytical computation of path lines for finite-difference models, *Ground Water* **26**(6), 743–760.

Rügner, J., Kleineidam, S. and Grathwohl, P.: 1998, Long-term sorption kinetics of phananthrene in aquifer materials, *Environ. Sci. Technol.* **33**(10), 1645–1651.

Tompson, A. F. B. and Gelhar, L. W.: 1990, Numerical simulation of solute transport in three-dimensional, randomly heterogeneous porous media, *Water Resour. Res.* **26**(10), 2541–2562.

Tompson, A. F. B., Falgout, R. D., Smith, S. G., Bosl, W. J. and Ashby, S. F.: 1998, Analysis of subsurface contaminant migration and remediation using high performance computing, *Adv. Water Resour.* **22**(3), 203–221.

Tompson A. F. B., Carle, S. F., Rosenberg, N. D. and Maxwell, R. M.: 1999, Analysis of groundwater migration from artificial recharge in a large urban aquifer: a simulation perspective, *Water Resour. Res.* **35**(10), p. 2981.

Werth, C. J. and Reinhard, M.: 1999, Counter-diffusion of isotopically labeled trichloroethylene in silica gel and geosorbent micropores: column results, *Eniron. Sci. Technol.* **33**, 730–736.

Transport in Porous Media **42:** 181–198, 2001.

Methods of Quantum Field Theory in the Physics of Subsurface Solute Transport

GARRISON SPOSITO
Department of Civil and Environmental Engineering, University of California, Berkeley, CA 94720-1710, U.S.A.

(Received: 16 November 1998; in final form: 19 November 1999)

Abstract. The stochastic theory of subsurface solute transport has received stimulus recently from modeling techniques originating in quantum field theory (QFT), resulting in new calculations of the solute macrodispersion tensor that derive from the solving Dyson equation with a subsequent renormalization group analysis. In this paper, we offer a critical evaluation of these techniques as they relate specifically to the derivation of a field-scale advection–dispersion equation. An approximate Dyson equation satisfied by the ensemble-average solute concentration for tracer movement in a heterogeneous porous medium is derived and shown to be equivalent to a truncated cumulant expansion of the standard stochastic partial differential equation which describes the same phenomenon. The full Dyson equation formalism, although exact, is of no importance to the derivation of an improved field-scale advection–dispersion equation. Similarly, renormalization group analysis of the macrodispersion tensor has not yet provided results that go beyond what is available currently from the cumulant expansion approach.

Key words: Advection–dispersion equation, cumulant expansion, Dyson equation, macrodispersion, renormalization group, solute transport.

Dedicated to the memory of Shu-Yuan Chu (1939–1998)

1. Introduction

The innovations currently in vogue among practitioners of theoretical subsurface hydrology include modeling based on methodologies from quantum field theory. Prototypical examples are in recent papers by Glimm *et al.* (1993), Dean *et al.* (1994, 1995, 1996), and Jaekel and Vereecken (1997), all of which join the quest for a tractable model to describe solute movement through spatially-heterogeneous porous media at field scales. A characteristic of these studies is their complacent reliance on mathematical techniques and terminology that are scarcely known to hydrology ('irreducible diagrams', 'Dyson equation', 'renormalization group', and the like). This curtain of unfamiliar argot may drop from a methodological heritage for physicists, but to hydrologists it screens an already opaque approach whose conceptual advances are not yet apparent (Dagan, 1998). One option is to consign the esoterica of quantum field theory (QFT) to its fans among the mathematically-inclined. However, to choose this option is also to lose access to the parallel QFT literature on subsurface fluid flows (King, 1987, 1989; Hristopulos and Christakos, 1997), which now has begun to offer concrete applications important to hydrologic practice (King *et al.*, 1993; King, 1996; Nötinger and Gautier, 1998).

The purpose of the present paper is to offer a brief but self-contained critical account of the methods of QFT as applied to the derivation of a field-scale advection–dispersion equation for subsurface solute transport. The emphasis is not on computational details (for example, no Feynman diagrams appear), but instead on evaluating the conceptual pathways in QFT that reflect standard practice (Peskin and Schroeder, 1995; Zagoskin, 1998), while staying close to the Eulerian picture of solute transport well known to hydrologists. Our approach will lead us naturally to consider one of the core methodologies in statistical physics, the cumulant expansion (van Kampen, 1974). In this respect, the present paper can be read as a complement to recent studies discussing the role of the cumulant expansion in the Lagrangian picture of subsurface solute transport (Sposito, 1999); in the relation between the Lagrangian and Eulerian pictures (Wood, 1998); and in the utility of the Corrsin conjecture to model solute mixing (Zhang, 1995).

2. Defining a Propagator

A useful model of solute migration in heterogeneous porous media can be developed by interpreting one or more of the coefficients in a local-scale partial differential equation for transport as a realization of a random field (Sposito *et al.*, 1991). Perhaps the most common example is found in the advection–dispersion equation for a tracer solute, in which the solute advection velocity is prescribed to be a random function parameterized by space and time coordinates (Sposito and Barry, 1987). More complicated examples include advection–dispersion equations that incorporate terms for solute adsorption by the porous matrix or solute disappearance through first-order decay (see Table I).

This approach yields a stochastic partial differential equation (in an infinite spatial domain) whose generic mathematical form is

$$\frac{\partial c}{\partial t} = [\mathcal{L}_0(\mathbf{x}, t) + \mathcal{L}_1(\mathbf{x}, t; \beta)]c(\mathbf{x}, t; \beta) \quad (t > t' \geqslant 0), \tag{2.1}$$

Table I. Examples of the deterministic operator $\mathcal{L}_0$ and the stochastic operator $\mathcal{L}_1$ in (2.1)

Transport mechanisms	$\mathcal{L}_0(\mathbf{x}, t)$	$\mathcal{L}_1(\mathbf{x}, t; \beta)$	Random fields
Dispersion and advection	$\nabla \cdot \overline{\overline{D}} \cdot \nabla - \mathbf{U} \cdot \nabla$	$-\mathbf{v}(\mathbf{x}, t; \beta) \cdot \nabla$	Advection velocity
Advection and linear adsorption	$-\mathbf{U}_\mathrm{R} \cdot \nabla$	$-\mathbf{u}(\mathbf{x}, t; \beta) \cdot \nabla$	Retarded advection velocity
Dispersion, advection and first-order decay	$\nabla \cdot \overline{\overline{D}} \cdot \nabla - \mathbf{U} \cdot \nabla - \mathrm{B}$	$-\mathbf{v}(\mathbf{x}, t; \beta) \cdot \nabla - b(\mathbf{x}; \beta)$	Advection velocity and decay parameter

where $c(\mathbf{x}, t; \beta)$ is a random solute concentration, $\mathcal{L}_0(\mathbf{x}, t)$ is a deterministic operator field, $\mathcal{L}_1(\mathbf{x}, t; \beta)$ is a stochastic operator field, and β indexes a realization of $\mathcal{L}_1$ in a probability space Ω (Sposito *et al.*, 1986). An initial condition on (2.1), necessary for the uniqueness of its solution, is specified by the equation,

$$c(\mathbf{x}, t') = f(\mathbf{x}, t') \quad (t' \geqslant 0) \tag{2.2}$$

where $f(\mathbf{x}, t')$ is a known deterministic function of the space coordinate $\mathbf{x}$ at an initial time t'. The operator $\mathcal{L}_0(\mathbf{x}, t)$ collects all deterministic terms in the partial differential equation that affect the rate of change of the solute concentration at some Eulerian point $\mathbf{x}$, whereas the stochastic operator $\mathcal{L}_1(\mathbf{x}, t; \beta)$ collects together the fluctuating terms whose behavior is prescribed probabilistically as a way of modeling the effects of spatial heterogeneity in the porous medium. Usually $\mathcal{L}_0$ represents an ensemble-average value of a random field and $\mathcal{L}_1$ represents its fluctuation about this average value. Taking as a very simple example the first-order decay of a non-migrating solute distributed within a heterogeneous porous medium, one would set $\mathcal{L}_0$ equal to minus the ensemble-average value B of the random decay parameter $\lambda(\mathbf{x}; \beta)$ over the probability space Ω, with $\mathcal{L}_1$ then taken to be minus the fluctuating difference between the random decay parameter and its ensemble-average value, $b(\mathbf{x}, \beta) \equiv \lambda(\mathbf{x}; \beta) - \mathrm{B}$ (Table I).

The exact, formal solution of the stochastic partial differential equation (2.1) subject to the condition in (2.2) is:

$$\begin{aligned} c(\mathbf{x}, t; \beta) &= \overleftarrow{T} \exp\left[\int_{t'}^{t} \mathcal{L}_0(\mathbf{x}, s)\, \mathrm{d}s\right] \overleftarrow{T} \exp\left[\int_{t'}^{t} \tilde{\mathcal{L}}_1(\mathbf{x}, s; \beta)\right] \times \\ &\quad \times f(\mathbf{x}, t') \quad (t > t') \end{aligned} \tag{2.3}$$

where, for example,

$$\begin{aligned} \overleftarrow{T} \exp\left[\int_{t'}^{t} \mathcal{L}_0(\mathbf{x}, s)\, \mathrm{d}s\right] &\equiv I + \sum_{m=1}^{\infty} \int_{t'}^{t} \mathrm{d}s_1 \int_{t'}^{s_1} \mathrm{d}s_2 \ldots \int_{t'}^{s_{m-1}} \times \\ &\quad \times \mathrm{d}s_m \mathcal{L}_0(\mathbf{x}, s_1) \ldots \mathcal{L}_0(\mathbf{x}, s_m) \end{aligned} \tag{2.4}$$

defines a time-ordered exponential operator involving repeated applications of $\mathcal{L}_0(\mathbf{x}, t)$, with I being the identity operator. The 'time-ordering', denoted symbolically by $\overleftarrow{T}$ in (2.4), requires always placing an operator $\mathcal{L}_0$ with the larger value of the time-variable, $s_i\,(i = l, \ldots, m)$, to the left of another operator $\mathcal{L}_0$ with a smaller value of s_i (hence the superscript arrow pointing left). Thus $\mathcal{L}_0(\mathbf{x}, s_1)\mathcal{L}_0(\mathbf{x}, s_2)$ is a proper time-ordering if $s_1 > s_2$ as in (2.4). The important point to note for the moment is how (2.3) neatly factorizes the deterministic and stochastic operator effects on the time-dependence of $c(\mathbf{x}, t; \beta)$. The exponential operator containing $\tilde{\mathcal{L}}_1$ acts on the initial concentration $f(\mathbf{x}, t')$, then the exponential operator containing $\mathcal{L}_0$ acts on the result. If there were no stochastic operator, (2.3) would be a formal way to express the solution of a typical partial differential equation in terms of its initial value, with the operator in (2.4) serving to 'propagate' the initial value forward in time by small increments along a trajectory generated by $\mathcal{L}_0(\mathbf{x}, t)$. This incremental process has been demonstrated in simple examples by Wood (1998, Appendix A) and Wood and Kavvas (1999).

The mathematical pathway from (2.1) to (2.3) begins with the definition of a random field $u(\mathbf{x}, t; \beta)$ which is to be propagated along a trajectory of $\mathcal{L}_0(\mathbf{x}, t)$ to create the solution of (2.1):

$$c(\mathbf{x}, t; \beta) \equiv \overleftarrow{T} \exp\left[\int_{t'}^{t} \mathcal{L}_0(\mathbf{x}, s)\,\mathrm{d}s\right] u(\mathbf{x}, t; \beta). \tag{2.5}$$

The time-ordering in (2.5) is required because the operators $\mathcal{L}_0(\mathbf{x}, s_m)$ in (2.4) do not necessarily commute at different times. Once time-ordering is prescribed, however, these operators can be manipulated mathematically just as if they did commute, so long as they are properly time-ordered in the final result. It is this fact that permits the compact exponential representation in (2.5), which strictly would otherwise apply only to functions, not to operators [see Chapter 1 in Zagoskin (1998)]. The inverse of the exponential operator in (2.5) is defined by

$$\left\{\overleftarrow{T} \exp\left[\int_{t'}^{t} \mathcal{L}_0(\mathbf{x}, s)\,\mathrm{d}s\right]\right\}^{-1} \equiv \overrightarrow{T} \exp\left[-\int_{t'}^{t} \mathcal{L}_0(\mathbf{x}, s)\,\mathrm{d}s\right], \tag{2.6}$$

where the reversed time-ordering on the right side, denoted by a superscript arrow pointing to the right, requires an operator $\mathcal{L}_0$ with the larger time-variable to stand to the right of one with a smaller time-variable. A geometric illustration of the effect of time-ordering is given by Peskin and Schroeder (1995, Chapter 4).

By direct calculation using (2.1) and (2.5), one finds that the random field $u(\mathbf{x}, t; \beta)$ satisfies the stochastic partial differential equation,

$$\begin{aligned}\frac{\partial u}{\partial t} &= \overrightarrow{T} \exp\left[-\int_{t'}^{t} \mathcal{L}_0(\mathbf{x}, s)\,\mathrm{d}s\right] \mathcal{L}_1(\mathbf{x}, t; \beta)\, \overleftarrow{T} \exp\left[\int_{t'}^{t} \mathcal{L}_0(\mathbf{x}, s)\,\mathrm{d}s\right] u(\mathbf{x}, t; \beta)\\ &\equiv \tilde{\mathcal{L}}_1(\mathbf{x}, t; \beta) u(\mathbf{x}, t; \beta)\end{aligned} \tag{2.7}$$

which defines a stochastic operator $\tilde{\mathcal{L}}_1(\mathbf{x}, t; \beta)$. The formal solution of (2.7) is then

$$u(\mathbf{x}, t; \beta) = \overleftarrow{T} \exp\left[\int_{t'}^{t} \tilde{\mathcal{L}}_1(\mathbf{x}, s; \beta)\,\mathrm{d}s\right] f(\mathbf{x}, t') \tag{2.8}$$

given (2.2) and (2.5). Therefore, by (2.5) and (2.8), the formal solution of (2.1) can be expressed as in (2.3). Evidently $u(\mathbf{x}, t; \beta)$ is created by propagating the initial concentration along a trajectory of the stochastic operator $\tilde{\mathcal{L}}_1(\mathbf{x}, t; \beta)$, then $c(\mathbf{x}, t; \beta)$ is created by propagating $u(\mathbf{x}, t; \beta)$ along a trajectory of the deterministic operator $\mathcal{L}_0(\mathbf{x}, t)$. If there were no deterministic terms in (2.1), (2.8) would be its formal solution as an initial-value problem.

A technical definition of the propagator $G(\mathbf{x}t; \mathbf{x}', t|\beta)$ corresponding to (2.3) now can be given as the space integral (Zagoskin, 1998),

$$c(\mathbf{x}, t; \beta) \equiv \int G(\mathbf{x}, t; \mathbf{x}', t'|\beta) f(\mathbf{x}', t')\,\mathrm{d}^3x' \tag{2.9}$$

such that

$$G(\mathbf{x}, t; \mathbf{x}', t'|\beta) \equiv \overleftarrow{T} \exp\left[\int_{t'}^{t} \mathcal{L}_0(\mathbf{x}, s)\, \mathrm{d}s\right] \times \times \overleftarrow{T} \exp\left[\int_{t'}^{t} \tilde{\mathcal{L}}_1(\mathbf{x}, s; \beta)\, \mathrm{d}s\right] \delta(\mathbf{x} - \mathbf{x}')\theta(t - t'), \quad (2.10)$$

where $\theta(t - t')$ is a unit step-function and $\delta(\mathbf{x} - \mathbf{x}')$ is a Dirac delta-'function'. The propagator defined in (2.10) is termed a retarded Green's function for the stochastic partial differential equation (2.1). It contains the physical information implicit in the partial differential equation, while obviating any influence of the arbitrary initial condition. As indicated in (2.9), it propagates the solute concentration field from its initial deterministic value at $(\mathbf{x}', t')$ to its current stochastic value at $(\mathbf{x}, t)$ under the 'nonanticipating constraint', that the past determines the future, hence the unit-step function in (2.10).

3. Calculating the Solute Concentration

The concentration field that solves (2.1) is a stochastic quantity whose relation to a measured concentration field, such as exists in a solute plume, remains to be prescribed. In most applications of (2.1) to tracer movement through groundwater, the ensemble-average concentration field (i.e., the average, denoted $\langle c(\mathbf{x}, t)\rangle$, of $c(\mathbf{x}, t; \beta)$ computed over the probability space Ω) is assumed to have direct physical relevance (Dagan, 1989). In practice, this amounts to associating the ensemble-average concentration field with a resident concentration field measured synoptically in an aquifer through which a solute plume is moving (Sposito and Barry, 1987). Whether this conceptual association is reasonable depends on knowing if the aquifer volume to which a resident concentration field pertains is large enough to encompass all of the variability among the realizations $c(\mathbf{x}, t; \beta)$ as the index β is varied for fixed $(\mathbf{x}, t)$ (Sposito *et al.*, 1986), an issue that remains controversial.

The commutativity of spatial integration with ensemble-averaging permits (2.9) to be written in terms of ensemble-average values:

$$\langle c(\mathbf{x}, t)\rangle = \int \langle G(\mathbf{x}, t; \mathbf{x}', t')\rangle f(\mathbf{x}', t')\, \mathrm{d}^3x', \quad (3.1)$$

where

$$\langle G(\mathbf{x}, t; \mathbf{x}, t)\rangle \equiv \overleftarrow{T} \exp\left[\int_{t'}^{t} \mathcal{L}_0(\mathbf{x}, s)\, \mathrm{d}s\right] \times \times \left\langle \overleftarrow{T} \exp\left[\int_{t'}^{t} \tilde{\mathcal{L}}_1(\mathbf{x}, s)\, \mathrm{d}s\right]\right\rangle \delta(\mathbf{x} - \mathbf{x}')\theta(t - t') \quad (3.2)$$

is the ensemble-average propagator, according to (2.10). The problem of calculating $\langle c(\mathbf{x}, t)\rangle$ is therefore that of calculating the ensemble average on the right side of (3.2), with the operator $\tilde{\mathcal{L}}_1(\mathbf{x}, t; \beta)$ defined as in (2.7). The ensemble average in (3.2) turns out to have an exact expansion in terms of the ordered cumulants of a suitable probability distribution defined on the space Ω (van Kampen, 1974). This approach, pioneered in subsurface solute transport theory by S.-Y. Chu (Chu and Sposito, 1980, 1981), has been exploited systematically in special cases of (2.1) by Barry and Sposito (1987), Kabala and Sposito (1991, 1998), Kavvas and Karakas (1996, 1998), Wood (1998), and Wood and Kavvas (1999). Technical mathematical issues pertaining to the use of the cumulant expansion for the solution of stochastic differential equations are discussed by Kubo (1962, 1963), van Kampen (1974, 1976), Terwiel (1974), Fox (1976), and Wood (1998).

The cumulant expansion of the ensemble mean on the right side of (3.2) can be expressed (van Kampen, 1974)

$$\left\langle \overleftarrow{T} \exp\left[\int_{t'}^{t} \tilde{\mathcal{L}}_1(\mathbf{x}, s)\, \mathrm{d}s\right]\right\rangle = \overleftarrow{T} \exp\left[\sum_{m=1}^{\infty} \frac{1}{m!} \times \right.$$
$$\left. \times \int_{t'}^{t} \cdots \int_{t'}^{t} \langle\langle \tilde{\mathcal{L}}_1(\mathbf{x}, \tau_1) \ldots \tilde{\mathcal{L}}_1(\mathbf{x}, \tau_m)\rangle\rangle\, \mathrm{d}\tau_1 \ldots \mathrm{d}\tau_m \right], \tag{3.3}$$

where $\langle\langle\,\rangle\rangle$ denotes a cumulant. The first two terms on the right side of (3.3) contain the cumulants,

$$\langle\langle \tilde{\mathcal{L}}_1(\mathbf{x}, \tau)\rangle\rangle = \vec{T} \exp\left[-\int_{t'}^{t} \mathcal{L}_0(\mathbf{x}, s)\, \mathrm{d}s\right] \times$$
$$\times \langle\langle \mathcal{L}_1(\mathbf{x}, \tau)\rangle\rangle\, \overleftarrow{T} \exp\left[\int_{t'}^{t} \mathcal{L}_0(\mathbf{x}, s)\, \mathrm{d}s\right] \tag{3.4}$$

and

$$\langle\langle \tilde{\mathcal{L}}_1(\mathbf{x}, \tau_1)\tilde{\mathcal{L}}_1(\mathbf{x}, \tau_2)\rangle\rangle = \vec{T} \exp\left[-\int_{t'}^{t} \mathcal{L}_0(\mathbf{x}, s)\, \mathrm{d}s\right] \times$$
$$\times \left\langle\!\!\left\langle \mathcal{L}_1(\mathbf{x}, \tau_1)\, \overleftarrow{T} \exp\left[\int_{t'}^{t} \mathcal{L}_0(\mathbf{x}, s)\, \mathrm{d}s\right] \vec{T} \exp\left[-\int_{t'}^{t_2} \mathcal{L}_0(\mathbf{x}, s)\, \mathrm{d}s\right] \times\right.\right.$$
$$\left.\left. \times \quad \mathcal{L}_1(\mathbf{x}, \tau_2)\right\rangle\!\!\right\rangle \overleftarrow{T} \exp\left[\int_{t'}^{t_2} \mathcal{L}_0(\mathbf{x}, s)\, \mathrm{d}s\right]. \tag{3.5}$$

The terms in the cumulant expansion (3.3) can be expressed further as sums over ensemble averages of the stochastic operators $\tilde{\mathcal{L}}_1$, following the 'van Kampen rules' (van Kampen, 1976; Fox, 1976). Suffice it to say in the present case that $\langle\langle\,\rangle\rangle$ in (3.4) can be replaced by a simple average $\langle\,\rangle$ as an expansion of the left side of (3.3) to first order readily demonstrates, and that $\langle\langle\,\rangle\rangle$ in (3.5) is the same as the covariance of the two stochastic operators involved, a result which reflects the fact that any cumulant vanishes whenever any one of the operators involved is statistically independent from the others (Kubo, 1962, 1963). This happens in particular when the time-variable on which the operator depends differs from all the others in the cumulant by an interval that greatly exceeds the correlation timescale for the second-order cumulant on the left side of (3.5) (Fox, 1976; Chu and Sposito, 1980; Wood, 1999). It is this 'cluster property' of the cumulants that makes the expansion in (3.3) effectively an expansion in powers of the correlation timescale (van Kampen, 1976).

To facilitate a connection between the cumulant expansion and the methods of QFT, it is useful to differentiate both sides of (3.2) with respect to time so as to derive an exact partial differential equation for the propagator,

$$\begin{aligned}\frac{\partial}{\partial t}\langle G(\mathbf{x},t;\mathbf{x}',t')\rangle &= \mathcal{L}_0(\mathbf{x},t)\langle G(\mathbf{x},t;\mathbf{x}',t')\rangle + \overleftarrow{T}\ \exp\left[\int_{t'}^{t}\mathcal{L}_0(\mathbf{x},s)\,\mathrm{d}s\right]\times\\ &\times\overleftarrow{T}\sum_{m=2}^{\infty}\frac{1}{m!}\int_{t'}^{t}\cdots\int_{t'}^{t}\langle\langle\tilde{\mathcal{L}}_1(\mathbf{x},t)\ldots\tilde{\mathcal{L}}_1(\mathbf{x},\tau_m)\rangle\rangle\mathrm{d}\tau_2\ldots\mathrm{d}\tau_m\times\\ &\times\overrightarrow{T}\ \exp\left[-\int_{t'}^{t}\mathcal{L}_0(\mathbf{x},s)\,\mathrm{d}s\right]\langle G(\mathbf{x},t;\mathbf{x}',t')\rangle+\delta(\mathbf{x}-\mathbf{x}')\delta(t-t'),\end{aligned}\tag{3.6}$$

where the cumulant expansion in (3.3) has been used along with the definition of $\langle G(\mathbf{x},t;\mathbf{x}',t')\rangle$ in (3.2) to obtain the second term on the right side. For the present purposes, we shall require only the second-order approximation to this exact partial differential equation; explicitly,

$$\begin{aligned}&\frac{\partial}{\partial t}\langle G(\mathbf{x},t;\mathbf{x}',t')\rangle\\ &\approx\mathcal{L}_0(\mathbf{x},t)\langle G(\mathbf{x},t;\mathbf{x}',t')\rangle+\langle\langle\tilde{\mathcal{L}}_1(\mathbf{x},t)\rangle\rangle\langle G(\mathbf{x},t;\mathbf{x}',t')\rangle+\\ &+\overleftarrow{T}\int_0^{t-t'}\left\langle\left\langle\mathcal{L}_1(\mathbf{x},t)\,\overleftarrow{T}\ \exp\left[\int_{t-\tau}^{t}\mathcal{L}_0(\mathbf{x},s)\,\mathrm{d}s\right]\mathcal{L}_1(\mathbf{x},t-\tau)\right\rangle\right\rangle\times\\ &\times\overrightarrow{T}\ \exp\left[-\int_{t-\tau}^{t}\mathcal{L}_0(\mathbf{x},s)\,\mathrm{d}s\right]\mathrm{d}\tau\langle G(\mathbf{x},t;\mathbf{x}',t')\rangle+\delta(\mathbf{x}-\mathbf{x}')\delta(t-t')\end{aligned}\tag{3.7}$$

which follows from (3.6) after the introduction of (3.4), (3.5), and the 'rule of composition',

$$\begin{aligned}\overleftarrow{T}\ \exp\left[\int_a^b\mathcal{L}_0(\mathbf{x},s)\,\mathrm{d}s\right]&=\overleftarrow{T}\ \exp\left[\int_{a'}^{b}\mathcal{L}_0(\mathbf{x},s)\,\mathrm{d}s\right]\overleftarrow{T}\ \times\\ &\times\exp\left[\int_a^{a'}\mathcal{L}_0(\mathbf{x},s)\,\mathrm{d}s\right]\end{aligned}\tag{3.8}$$

for any $a < a' < b$. Equation (3.7) is exact to second order in the correlation timescale for the covariance of the stochastic operator $\mathcal{L}_1(\mathbf{x},t;\beta)$ [cf. Wood and Kavvas (1999)]. Van Kampen (1976) has pointed out that the time-ordering operator $\overleftarrow{T}$ acting overall on the third term on the right side of (3.7) can be dropped safely, because the error committed in doing so is of higher order than the inaccuracy in (3.7) itself. Time-ordering in the two exponential operators containing $\mathcal{L}_0$, of course, is still required.

4. Connecting to QFT

The exact, formal solution of the stochastic partial differential equation (2.1) for the random solute concentration, given in (2.3), now has been put into the more familiar form of (3.1), which represents the solute concentration as a space integral of the retarded Green's function for the initial-value problem posed in (2.1) and (2.2), an approach that is well known in the theory of parabolic partial differential equations [see Chapter 3 in Cannon (1984)]. The relationship of this approach to QFT comes through comparison with the cumulant expansion of the ensemble-averaged Green's function, as given formally in (3.2) and, equivalently, in (3.6). For reasons that will become quite transparent, making the connection between the modeling of (3.6) as reported in the stochastic hydrology literature (Sposito and Barry, 1987; Kabala and Sposito, 1991; Kavvas and Karakas, 1996; Wood, 1998; Wood and Kavvas, 1999) and the methods of QFT (Glimm *et al.*, 1993; Dean *et al.*, 1994, 1995, 1996; Jaekel and Vereecken, 1997) requires only the approximate form in (3.7).

Perhaps the simplest physical context in which to do this is that of tracer dispersion by a random, solenoidal velocity field which is taken to be a wide-sense stationary and homogeneous stochastic process (Dagan, 1989). In this case (see Table I),

$$\mathcal{L}_0(\mathbf{x}, t) \equiv \nabla \cdot \overline{\overline{D}} \cdot \nabla - \mathbf{U} \cdot \nabla \tag{4.1}$$

$$\mathcal{L}_1(\mathbf{x}, t; \beta) \equiv -\mathbf{v}(\mathbf{x}, t; \beta) \cdot \nabla, \tag{4.2}$$

where $\overline{\overline{D}}$ is a constant, uniform, deterministic local-dispersion tensor whose principal axes are congruent with the Eulerian coordinate axes of $\mathbf{x}$; $\mathbf{U}$ is the constant, uniform ensemble-mean value of the random advection velocity; and $\mathbf{v}(\mathbf{x}, t; \beta)$ is its solenoidal fluctuation about the ensemble mean, with the elements of the covariance tensor $\overline{\overline{\Gamma}}$ then defined by

$$\langle v_i(\mathbf{x}, t) v_j(\mathbf{x}', t')\rangle = \Gamma_{ij}(\mathbf{x} - \mathbf{x}', t - t') \tag{4.3}$$

for $i, j = 1, 2, 3$. Equations (4.1) and (4.2) are based on a local-scale advection–dispersion equation for tracer movement in a heterogeneous medium whose porosity varies spatially over much longer length scales than does the solute concentration field (Kabala and Sposito, 1998).

The explicit form of (3.7) corresponding to the physical model in (4.1) and (4.2) is:

$$\begin{aligned}\frac{\partial}{\partial t}\langle G(\mathbf{x},t;\mathbf{x}',t')\rangle &\approx \left(\nabla\cdot\overline{\overline{D}}\cdot\nabla-\mathbf{U}\cdot\nabla\right)\langle G(\mathbf{x},t;\mathbf{x}',t')\rangle+\\&+\int_0^{t-t'}\left\langle\mathbf{v}(\mathbf{x},t)\cdot\nabla\exp\left[\left(\nabla\cdot\overline{\overline{D}}\cdot\nabla-\mathbf{U}\cdot\nabla\right)\tau\right]\mathbf{v}(\mathbf{x},t-\tau)\cdot\nabla\right\rangle\times\\&\times\exp\left[-\left(\nabla\cdot\overline{\overline{D}}\cdot\nabla-\mathbf{U}\cdot\nabla\right)\tau\right]\mathrm{d}\tau\langle G(\mathbf{x},t;x',t')\rangle+\\&+\delta(\mathbf{x}-\mathbf{x}')\delta(t-t'),\end{aligned}\tag{4.4}$$

where the covariance term on the right side is represented in dyadic form and account has been taken of the zero ensemble-mean value of $\mathcal{L}_1(\mathbf{x},t;\beta)$, noting the constant value of $\mathbf{U}$. Equation (4.4) is a special case of the ensemble-average solute transport equation presented by Kabala and Sposito (1991, 1998), who considered a nonsolenoidal advection velocity and a reactive solute susceptible to first-order decay processes and linear sorption by the porous matrix. The gradient operators in (4.4) act on all functions to their right within each term.

The next step is to express (4.4) in terms of Fourier space-time transforms, generalizing the approach of Sposito and Barry (1987) and Kabala and Sposito (1991):

$$\begin{aligned}\langle G(\mathbf{x},t;\mathbf{x}',t')\rangle &\equiv (2\pi)^{-4}\int \mathrm{d}^3p\int \mathrm{d}\omega g(\mathbf{p},\omega)\times\\&\times\exp\{i[\mathbf{p}\cdot(\mathbf{x}-\mathbf{x}')+\omega(t-t')]\},\end{aligned}\tag{4.5a}$$

$$\begin{aligned}\delta(\mathbf{x}-\mathbf{x}')\delta(t-t') &\equiv (2\pi)^{-4}\int \mathrm{d}^3p\int \mathrm{d}\omega\times\\&\times\exp\{i[\mathbf{p}\cdot(\mathbf{x}-\mathbf{x}')+\omega(t-t')]\},\end{aligned}\tag{4.5b}$$

$$\mathbf{v}(\mathbf{x},t;\beta)\equiv(2\pi)^{-4}\int \mathrm{d}^3p\int \mathrm{d}\omega\psi(\mathbf{p},\omega)\exp[i(\mathbf{p}\cdot\mathbf{x}+\omega t)],\tag{4.5c}$$

$$\begin{aligned}\overline{\overline{\Gamma}}(\mathbf{x}-\mathbf{x}',t-t') &\equiv (2\pi)^{-4}\int \mathrm{d}^3p\int \mathrm{d}\omega\overline{\overline{S}}(\mathbf{p},\omega)\times\\&\times\exp\{i[\mathbf{p}\cdot(\mathbf{x}-\mathbf{x}')+\omega(t-t')]\}.\end{aligned}\tag{4.5d}$$

Equations (4.3), (4.5c), and (4.5d) are mutually consistent only if the relation

$$(2\pi)^{-4}\langle\psi(\mathbf{p},\omega)\psi(\mathbf{p}',\omega')\rangle=\delta(\mathbf{p}+\mathbf{p}')\delta(\omega-\omega')\overline{\overline{S}}(\mathbf{p},\omega)\tag{4.5e}$$

is imposed to reflect the homogeneity and stationarity of the advection velocity field as a stochastic process. The solenoidal property of the advection velocity implies the additional constraint (Sposito *et al.*, 1991),

$$\mathbf{p}\cdot\overline{\overline{S}}(\mathbf{p},\omega)\cdot\mathbf{k}=\mathbf{k}\cdot\overline{\overline{S}}(\mathbf{p},\omega)\cdot\mathbf{p}=0\tag{4.5f}$$

Sposito and Barry (1987) noted that current models of subsurface tracer movement based on a stochastic advection velocity (Dagan, 1989; Wood, 1998) typically assume that velocity fluctuations remain 'frozen' over the timescale that solute migrates through a heterogeneous porous medium (see also Rehfeldt and Gelhar (1992)). This assumption is tantamount to neglecting time dependence in the covariance tensor, such that its Fourier transform, $\overline{\overline{S}}(\mathbf{p}, \omega)$, is replaced by the integral,

$$\overline{\overline{\Delta}}(\mathbf{p}) \equiv (2\pi)^{-1} \int d\omega \overline{\overline{S}}(\mathbf{p}, \omega) \tag{4.5g}$$

as a special case of (4.5d), with a corresponding integration performed on both sides of (4.5e). With (4.5a–g) introduced, (4.4) becomes the integral equation

$$\begin{aligned}(2\pi)^{-4} \int d^3\mathbf{p} \int d\omega \exp\{i[\mathbf{p} \cdot (\mathbf{x} - \mathbf{x}') + \omega(t - t')]\} \times \\ \times g(\mathbf{p}, \omega)\{i\omega + \mathbf{p} \cdot \overline{\overline{D}} \cdot \mathbf{p} + i\mathbf{U} \cdot \mathbf{p} - \\ - (2\pi)^{-3} \int d^3q \frac{1}{2}\mathbf{p} \cdot \left[\overline{\overline{\Delta}}(\mathbf{q}) + \overline{\overline{\Delta}}^T(\mathbf{q})\right] \cdot \mathbf{p} F(\mathbf{q}, \mathbf{p}, t - t')\} \approx 1\end{aligned} \tag{4.6}$$

where a superscript T refers to the transpose and

$$F(\mathbf{q}, \mathbf{p}, t - t') \equiv \int_0^{t-t'} ds \exp\left[-\left(2\mathbf{p} \cdot \overline{\overline{D}} \cdot \mathbf{q} + \mathbf{q} \cdot \overline{\overline{D}} \cdot \mathbf{q} + i\mathbf{U} \cdot \mathbf{q}\right) s\right], \tag{4.7}$$

describes transient effects resulting from local-scale dispersion and advection of the solute by the ensemble-average velocity, $\mathbf{U}$. Equation (4.6) is equivalent to an integral equation,

$$g(\mathbf{p}, \omega) \approx g_0(\mathbf{p}, \omega) + g_0(\mathbf{p}, \omega)(2\pi)^{-1} \int d\omega' \Sigma(\mathbf{p}, \omega - \omega') g(\mathbf{p}, \omega'), \tag{4.8}$$

where

$$\begin{aligned}\Sigma(\mathbf{p}, \omega) \equiv (2\pi)^{-3} \int d^3q \frac{1}{2}\mathbf{p} \cdot \left[\overline{\overline{\Delta}}(\mathbf{q}) + \overline{\overline{\Delta}}^T(\mathbf{q})\right] \cdot \mathbf{p} \times \\ \times 2 \int_0^{\infty} F(\mathbf{q}, \mathbf{p}, \tau) \exp(-i\omega\tau) d\tau\end{aligned} \tag{4.9}$$

and

$$g_0(\mathbf{p}, \omega) \equiv (i\omega + \mathbf{p} \cdot \overline{\overline{D}} \cdot \mathbf{p} + i\mathbf{U} \cdot \mathbf{p})^{-1}. \tag{4.10}$$

Equation (4.8) is the celebrated Dyson equation of QFT (Zagoskin, 1998), derived here for the first time in the context of subsurface solute transport theory. It is, as put aptly by Reichl (1988, p. 436), 'the beautiful result we have been after'. The 'unperturbed ensemble-average propagator' $g_0(\mathbf{p}, \omega)$ in (4.10) is the Fourier space-time transform of the solution of (4.4) without a stochastic term containing the velocity fluctuations. Alternatively, $g_0(\mathbf{p}, \omega)$ is the Fourier space-time transform of the ensemble-average propagator in (3.2) after (4.1) is introduced and $\mathcal{L}_1$ has been set equal to zero on the right side. Given (4.1), it is an exact result. This is also its physical meaning in QFT (Zagoskin, 1998).

The second term on the right side of (4.8) reflects the influence of velocity fluctuations on the ensemble-average propagator. The function $\Sigma(\mathbf{p}, \omega)$ in this term, defined in (4.9), corresponds to the 'irreducible self-energy' in QFT (Zagoskin, 1998). Reichl (1988) and Zagoskin (1998) describe in a careful but approachable manner the calculation of the 'irreducible self-energy' in terms of a series expansion in Feynman diagrams. The salient point in this calculation with respect to the present study is that only the 'connected' (or 'irreducible') Feynman diagrams are involved – those diagrams that cannot be partitioned into sets of smaller, connected diagrams. But this 'connected-graph expansion' is precisely equivalent mathematically to an ordered cumulant expansion. Indeed, the second term on the right in (4.8) can be put into direct correspondence with one of the first-order 'irreducible' Feynman diagrams in the perturbation expansion of the self-energy (Dean *et al.*, 1994; Jaekel and Vereecken, 1997), where it is part of the lowest-order 'Hartree-Fock approximation'. Thus, (4.8) indeed has the same physical and mathematical meaning as does the Dyson equation, and the cumulant expansion is the conceptual key to knowing this fact.

5. Connecting to Tracer Dispersion

The physical relation between the Dyson equation and well-known models in the stochastic theory of subsurface tracer transport is exposed transparently by a consideration of (4.7) in the limit of solute travel-time $(t - t')$ much longer than the local dispersion timescale (Sposito *et al.*, 1991); i.e.,

$$\tau_c \gg (t - t') \gg \left(\mathbf{q} \cdot \overline{\overline{D}} \cdot \mathbf{q}\right)^{-1}, \tag{5.1}$$

where τ_c is a correlation time for the velocity covariance $\overline{\overline{\Gamma}}(\mathbf{x} - \mathbf{x}', t - t')$. This case, which corresponds to the Gelhar–Axness (1983) model of subsurface tracer movement, allows the upper limit on the time integral in (4.7) to approach $+\infty$, with the result that (4.9) reduces to its zero-frequency limit [see also Rehfeldt and Gelhar (1992)],

$$\Sigma(\mathbf{p}, \omega) = -(2\pi)^{-2} \int \mathrm{d}^3 q \frac{\frac{1}{2}\mathbf{p} \cdot \left[\overline{\overline{\Delta}}(\mathbf{q}) + \overline{\overline{\Delta}}^T(\mathbf{q})\right] \cdot \mathbf{p}}{\left[2\mathbf{p} \cdot \overline{\overline{D}} \cdot \mathbf{q} + \mathbf{q} \cdot \overline{\overline{D}} \cdot \mathbf{q} + i\mathbf{U} \cdot \mathbf{q}\right]} \delta(\omega) \tag{5.2}$$

and (4.8) becomes

$$g(\mathbf{p}, \omega) \approx g_0(\mathbf{p}, \omega) - g_0(\mathbf{p}, \omega)\mathbf{p} \cdot \overline{\overline{D}}_M \cdot \mathbf{p} g(\mathbf{p}, \omega) \tag{5.3}$$

where the tensor $\overline{\overline{D}}_M$, defined by the integral,

$$\overline{\overline{D}}_M(\mathbf{p}) \equiv (2\pi)^{-3} \int \mathrm{d}^3 q \frac{\frac{1}{2}\left[\overline{\overline{\Delta}}(\mathbf{q}) + \overline{\overline{D}}^T(\mathbf{q})\right]}{\left[2\mathbf{p}\cdot\overline{\overline{D}}\cdot\mathbf{q} + \mathbf{q}\cdot\overline{\overline{D}}\cdot\mathbf{q} + i\mathbf{U}\cdot\mathbf{q}\right]} \tag{5.4}$$

now replaces the 'irreducible self-energy' in (4.8).

In the context of QFT, (5.3) yields a 'spectral representation' of the Fourier-transformed propagator, $g(\mathbf{p}, \omega)$ [see, also Reichl (1988, p. 436)]:

$$g(\mathbf{p}, \omega) \approx \left[i\omega + \mathbf{p}\cdot\overline{\overline{D}}\cdot\mathbf{p} + \mathbf{p}\cdot\overline{\overline{D}}_M\cdot\mathbf{p} + i\mathbf{U}\cdot\mathbf{p}\right]^{-1} \tag{5.5}$$

after the incorporation of (4.10). In QFT, the singularities ('poles') of $g(\mathbf{p}, \omega)$, when it is considered as a function of ω for fixed p, determine the frequency and lifetime of the 'excitations', or 'quasiparticles', in a physical system (Reichl, 1988; Zagoskin, 1998). In (5.5), the 'excitation energy' is analogous to the term in **U** plus the imaginary part of the term in $\overline{\overline{D}}_M$, whereas 'excitation damping' is analogous to $\overline{\overline{D}}$ plus the real part of $\overline{\overline{D}}_M$ (see Table II).

In the context of the Gehlar–Axness (1983) model, (5.5) is equivalent to a Fourier-transformed advection–dispersion equation for the retarded ensemble-average propagator [see Rehfeldt and Gelhar (1992)]. This fact emerges after re-writing (5.5) in the form,

$$\left[i\omega + \mathbf{p}\cdot\left(\overline{\overline{D}} + \overline{\overline{D}}_M\right)\cdot\mathbf{p} + i\mathbf{U}\cdot\mathbf{p}\right] g(\mathbf{p}, \omega) \approx 1, \tag{5.6}$$

with the first term representing the time-derivative, and the second and third terms in the bracket representing local-scale dispersion plus macrodispersion and solute

Table II. Analogy between the Dyson Equation and the Gelhar–Axness model

Quantity	QFT interpretation	Gelhar–Axness model
$g(\mathbf{p}, \omega)$	Propagator for a system of interacting particles	Green's function for a field -scale advection–dispersion equation
$\left[\mathbf{U} + \mathbf{p}\cdot\mathrm{Im}\left(\overline{\overline{D}}_M\right)\right]\cdot\mathbf{p}$	Excitation energy for a 'quasiparticle'	Advection term for tracer solute concentration
$\mathbf{p}\cdot\left[\overline{\overline{D}} + \mathrm{Re}\left(\overline{\overline{D}}_M\right)\right]\cdot\mathbf{p}$	Excitation damping constant for a 'quasiparticle'	Dispersion term for tracer solute concentration

advection, together with spatial derivatives. The real part of $\overline{\overline{D}}_M$,

$$\overline{\overline{D}}_M^{GA}(\mathbf{p}) \equiv (2\pi)^{-3} \int d^3q \frac{\frac{1}{2}\left[\overline{\overline{\Delta}}(\mathbf{q}) + \overline{\overline{\Delta}}^T(\mathbf{q})\right]\left(2\mathbf{p}\cdot\overline{\overline{D}}\cdot\mathbf{q} + \mathbf{q}\cdot\overline{\overline{D}}\cdot\mathbf{q}\right)}{\left[\left(2\mathbf{p}\cdot\overline{\overline{D}}\cdot\mathbf{q} + \mathbf{q}\cdot\overline{\overline{D}}\cdot\mathbf{q}\right)^2 + (\mathbf{U}\cdot\mathbf{q})^2\right]}, \quad (5.7)$$

is an integral expression for the macrodispersion tensor as calculated in the Gelhar–Axness model (Sposito and Barry, 1987). Further development in order to relate the tensor $\overline{\overline{\Delta}}(\mathbf{q})$ to spatial variability in the hydraulic conductivity is straightforward after making well-known simplifying physical assumptions about the spatial fluctuations in the hydraulic conductivity and those in the solute advection velocity (Gutjahr, 1984; Sposito *et al.*, 1991; Rehfeldt and Gelhar, 1992). However, there is still a notable difference between (5.7) and the model expression for the macrodispersion tensor presented originally by Gelhar and Axness (1983), because the term in $\mathbf{p}\cdot\overline{\overline{D}}\cdot\mathbf{q}$ does not appear in the latter.

6. Renormalizing the Macrodispersion Tensor

Smith and Woodruff (1998) review the renormalization group method as applied to time-evolution equations, such as (2.1) (see also Chapter 12 in Ma (1976) and Chapter 9 in McComb (1992) for critical reviews). Interestingly, the prototypical physical example explicated by Smith and Woodruff (1998) is the advection–dispersion equation for a passive scalar moving under the influence of local diffusion in a random two-dimensional velocity field whose statistics are specified to be gaussian with zero ensemble mean, and whose spatial dependence is constrained to be along a direction perpendicular to the flow direction (Avellaneda and Majda, 1992). In this example, which is exactly analogous to the movement of a tracer solute in a perfectly-stratified aquifer (Matheron and de Marsily, 1980), the renormalization group analysis involves just two steps: (1) iterative removal of high-wavenumber, high-frequency Fourier components of $\overline{\overline{S}}(\mathbf{q}, \omega)$ from any contribution to the macrodispersion tensor $\overline{\overline{D}}_M$ in (5.4), and (2) removal of finite-wavenumber, finite-frequency Fourier modes in $g_0(\mathbf{p}, \omega)$ from any influence on the macrodispersion tensor. The first step ensures that rapid variations in $\mathbf{v}(\mathbf{x}, t; \beta)$ over very small length scales do not affect macrodispersion, in keeping with the physical interpretation of $\overline{\overline{D}}_M$ as arising from field-scale behavior (Dagan, 1989), whereas the second step explicitly recognizes that the inverse Fourier space-time transform of (5.6) applies on length and time scales that are long compared to those over which $\mathbf{v}(\mathbf{x}, t; \beta)$ varies (Avellaneda and Majda, 1992). A careful discussion of technical issues in the renormalization group method as applied to time-evolution equations has been given by Eyink (1994). Applications of the method to tracer movement under a random velocity field are outlined by Dean *et al.* (1994) and Jaekel and Vereecken (1997).

Step (2) above is most effective if taken at the outset, for then the terms that ultimately would be eliminated after the iterations in step (1) are completed do not get carried along gratuitously (Smith and Woodruff, 1998). Accordingly, (5.4) is considered in the 'long wavelength limit', $\mathbf{p} \downarrow \mathbf{0}$:

$$\overline{\overline{D}}_M = (2\pi)^{-3} \int \mathrm{d}^3 q \frac{\frac{1}{2}\left[\overline{\overline{\Delta}}(\mathbf{q}) + \overline{\overline{\Delta}}^T(\mathbf{q})\right]}{\left(\mathbf{q} \cdot \overline{\overline{D}} \cdot \mathbf{q} + i\mathbf{U} \cdot \mathbf{q}\right)}. \tag{6.1}$$

This step, which is tantamount to the neglect of $\mathbf{p}$-dependence in (4.7), was also invoked by Sposito and Barry (1987), who were motivated by the large-scale nature of macrodispersion. In mathematical terms, little contribution to $\langle G(\mathbf{x}, t; \mathbf{x}', t')\rangle$ is expected from $g(\mathbf{p}, \omega)$ at p-values that are comparable to the inverse of length scales over which local-scale dispersion occurs, because of the resultant rapid oscillation of the accompanying Fourier kernel, $\exp[-i\mathbf{p}' \cdot (\mathbf{x} - \mathbf{x}')]$ in the inverse Fourier transform of $g(\mathbf{p}, \omega)$, as $(\mathbf{x} - \mathbf{x}')$ is varied over the long length scales at which macrodispersion applies. The real part of $\overline{\overline{D}}_M$ in (5.8), the Gelhar–Axness macrodispersion coefficient, is thereby rendered equivalent to an integral expression for $\overline{\overline{D}}_M^{GA}$ derived by Sposito and Barry (1987):

$$\overline{\overline{D}}_M^{GA} = (2\pi)^{-3} \int \mathrm{d}^3 q \frac{\frac{1}{2}\left[\overline{\overline{\Delta}}(\mathbf{q}) + \overline{\overline{\Delta}}^T(\mathbf{q})\right]\left(\mathbf{q} \cdot \overline{\overline{D}} \cdot \mathbf{q}\right)}{\left[\left(\mathbf{q} \cdot \overline{\overline{D}} \cdot \mathbf{q}\right) + (\mathbf{U} \cdot \mathbf{q})^2\right]}. \tag{6.2}$$

Step (1) above amounts to imposing a restriction on the range of integration over q in (6.2), since there is no frequency dependence to consider. Therefore,

$$\overline{\overline{D}}_M^{GA}(\Lambda_0) \equiv (2\pi)^{-3} \int_{q > \Lambda_0} \mathrm{d}^3 q \frac{\frac{1}{2}\left[\overline{\overline{\Delta}}(\mathbf{q}) + \overline{\overline{\Delta}}^T(\mathbf{q})\right]\left(\mathbf{q} \cdot \overline{\overline{D}} \cdot \mathbf{q}\right)}{\left[\left(\mathbf{q} \cdot \overline{\overline{D}} \cdot \mathbf{q}\right)^2 + (\mathbf{U} \cdot \mathbf{q})^2\right]}, \tag{6.3}$$

where Λ_0 is a wavenumber that is comparable in magnitude to the inverse of the length scale over which local-scale dispersion occurs, described by the tensor $\overline{\overline{D}}$ (Smith and Woodruff, 1998). The iteration of (6.3) involves a difference equation in the variable $\Lambda < \Lambda_0$,

$$\begin{aligned} &\overline{\overline{D}}_M^{GA}(\Lambda) - D_M^{GA}(\Lambda - \delta\Lambda) \\ &\quad = -(2\pi)^{-3} \int_{\Lambda - \delta\Lambda < q < \Lambda} \mathrm{d}^3 q \frac{\frac{1}{2}\left[\overline{\overline{\Delta}}(\mathbf{q}) + \overline{\overline{\Delta}}^T(\mathbf{q})\right]\left(\mathbf{q} \cdot \overline{\overline{D}}_M^{GA}(\Lambda) \cdot \mathbf{q}\right)}{\left[\left(\mathbf{q} \cdot \overline{\overline{D}}_M^{GA}(\Lambda) \cdot \mathbf{q}\right)^2 + (\mathbf{U} \cdot \mathbf{q})^2\right]} \end{aligned} \tag{6.4}$$

where $\overline{\overline{D}}_M^{GA}(\Lambda)$ now replaces the local dispersion tensor $\overline{\overline{D}}$ in the integrand as Λ decreases below Λ_0. In the limit $\delta\Lambda \downarrow 0$, (6.4) becomes an integrodifferential equation,

$$\frac{\mathrm{d}\overline{\overline{D}}_M^{GA}}{\mathrm{d}\Lambda} = -\frac{-\Lambda^2}{(2\pi)^3}\int_{\Omega_\Lambda} \mathrm{d}\Omega_\Lambda \frac{\frac{1}{2}\left[\overline{\overline{\Delta}}(\Lambda) + \overline{\overline{\Delta}}^T(\Lambda)\right]\left(\Lambda \cdot \overline{\overline{D}}_M^{GA}(\Lambda)\cdot \Lambda\right)}{\left[\left(\Lambda \cdot \overline{\overline{D}}_M^{GA}(\Lambda)\cdot\Lambda\right)^2 + (\mathbf{U}\cdot\Lambda)^2\right]}, \tag{6.5}$$

where $\mathrm{d}\Omega_\Lambda$ is an element of solid angle on the surface of a sphere of radius Λ centered at $\mathbf{q} = \mathbf{0}$. The boundary condition on (6.5) is $\overline{\overline{D}}_M^{GA}(\Lambda_0) = \overline{\overline{D}}$, whereas the value of $\overline{\overline{D}}_M^{GA}(\Lambda)$ inferred from solving (6.5) as $\Lambda \downarrow 0$ is a renormalized macrodispersion tensor that is uninfluenced by small-scale fluctuations in $\mathbf{v}(\mathbf{x}, t; \beta)$. Jaekel and Vereecken (1997) found little difference between $\overline{\overline{D}}_M^{GA}$ in (6.2) and this renormalized macrodispersion tensor with respect to its 'longitudinal' element, but they did observe a significant difference in respect to the 'transverse' elements (i.e., those representing directions perpendicular to that of **U**. The physical interpretation of this result, however, is not resolved and is a matter of continuing research (Jaekel *et al.*, 1998).

7. Applying QFT to Solute Transport

The principal results of this study are contained in (4.8), the Dyson equation, and in (6.5), the renormalization group model for the macrodispersion coefficient. Under an assumption of separation of timescales, as in (5.1), the Dyson equation can be written in an approximate, factorized form,

$$g(\mathbf{p}, \omega) \approx g_0(\mathbf{p}, \omega) + g_0(\mathbf{p}, \omega)\Sigma(\mathbf{p}, 0)g(\mathbf{p}, \omega) \tag{7.1}$$

leading to the approximate spectral representation

$$g(\mathbf{p}, \omega) \approx \left[i\omega + i\mathbf{U}\cdot\mathbf{p} + \mathbf{p}\cdot\overline{\overline{D}}\cdot\mathbf{p} - \Sigma(\mathbf{p}, 0)\right]^{-1}, \tag{7.2}$$

for the Fourier transform of the propagator. This representation illustrates the key point that any finite-order approximation for $\Sigma(\mathbf{p}, 0)$ leads to an infinite-order approximation for the propagator, as obtained from summing an infinite subseries of the conventional perturbation series that would be found directly from (3.2) after Fourier transformation [see Zagoskin (1998)]. The Dyson equation also can be derived as an exact analog of (3.6) after Fourier transformation and the introduction of (4.1) and (4.2), thus permitting replacement of the approximate equation (7.1) by an exact equality. This possibility, which also follows from the correspondence between the cumulant expansion and the connected-graph expansion of QFT, would indeed improve on the accuracy of (4.9). But it would also yield terms of higher order than quadratic in **p** – which, in turn, means that inverse Fourier transformation of (5.6) (with an improved model of $\Sigma(\mathbf{p}, \omega)$ replacing $\mathbf{p}\cdot\overline{\overline{D}}_M\cdot\mathbf{p}$) would yield spatial derivatives of order higher than two in the governing partial differential equation.

In other words, as pointed out by Kabala and Sposito (1991), utilizing the cumulant expansion beyond its first non-vanishing terms indeed provides improved accuracy, but at a significant physical cost. There is no longer a governing partial differential equation having the form of an advection–dispersion equation for the ensemble-average solute concentration. Therefore, the exact result that the Dyson equation ultimately represents holds no more physical promise for solving (3.6) accurately than does keeping the entire cumulant expansion. Given (4.1) and (4.2), the approximate (4.4) is the partial differential equation of choice, and the approximate (4.8) is the Dyson equation of choice, if the physical concept of advection with macrodispersion is to be preserved as a scale-invariant notion in the governing partial differential equation. Thus, QFT offers no more physically than we already have in hand from the cumulant expansion approach.

Acknowledgements

The research leading to this paper was supported in part by NSF grant EAR-9704630. Thanks to Angela Zabel for beautiful preparation of the typescript.

References

Avellaneda, M. and Majda, A. G.: 1992, Approximate and exact renormalization theories for a model for turbulent transport, *Phys. Fluids A* **4**, 41–57.

Cannon, J. R.: 1984, *The One-Dimensional Heat Equation*, Addison-Wesley, Menlo Park, CA.

Chu, S. -Y. and Sposito, G.: 1980, 1981, A derivation of the macroscopic solute transport equation for homogeneous, saturated, porous media, *Water Resour. Res.* **16**, 542–546, **17**, 1238.

Dagan, G.: 1989, *Flow and Transport in Porous Formations*, Springer, New York.

Dagan, G.: 1998, Comment on "Renormalization group analysis of macrodispersion in a directed random flow" by U. Jaekel and H. Vereecken, *Water Resour. Res.* **34**, 3197–3198.

Dean, D. S., Drummond, I. T. and Horgan, R. R.: 1994, Perturbation schemes for flow in random media, *J. Phys. A* **27**, 5135–5144.

Dean, D. S., Drummond, I. T. and Horgan, R. R.: 1995, Perturbation theory for effective diffusivity in random gradient flows, *J. Phys. A* **28**, 1235–1242.

Dean, D. S., Drummond, I. T. and Horgan, R. R.: 1996, Renormalization of drift and diffusivity in random gradient flows, *J. Phys. A* **29**, 7867–7879.

Eyink, G. L.: 1994, The renormalization group method in statistical hydrodynamics, *Phys. Fluids* **6**, 3063–3078.

Fox, R. F.: 1976, Critique of the generalized cumulant expansion method, *J. Math. Phys.* **17**, 1148–1153.

Gelhar, L. W. and Axness, C. L.: 1983, Three-dimensional stochastic analysis of macrodispersion in aquifers, *Water Resour. Res.* **19**, 161–180.

Glimm, J., Lindquist, W. B., Pereira, F. and Zhang, Q.: 1993, A theory of macrodispersion for the scale-up problem, *Transport in Porous Media* **13**, 97–122.

Gutjahr, A.: 1984, Stochastic models of subsurface flow: Log linearized Gaussian models are 'exact' for covariances, *Water Resour. Res.* **20**, 1909–1912.

Hristopulos, D. T. and Christakos, G.: 1997, Diagrammatic theory of effective hydraulic conductivity, *Stoch. Hydrol. Hydraul.* **11**, 369–395.

Jaekel, U. and Vereecken, H.: 1997, Renormalization group analysis of macrodispersion in a directed random flow, *Water Resour. Res.* **33**, 2287–2299.

Jaekel, U., Schwarze, H. and Vereecken, H.: 1998, Reply, *Water Resour. Res.* **34**, 3199–3200.

Kabala, Z. J. and Sposito, G.: 1991, A stochastic model of reactive solute transport with time-varying velocity in a heterogeneous aquifer, *Water Resour. Res.* **27**, 341–350.

Kabala, Z. J. and Sposito, G.: 1998, Technical comment on: "On the stochastic theory of solute transport by unsteady and steady groundwater flow in heterogeneous aquifers," by M.L. Kavvas and A. Karakas, *J. Hydrol.* **207**, 136–138.

Kavvas, M. L. and Karakas, A.: 1996, On the stochastic theory of solute transport by unsteady and steady groundwater flow in heterogeneous aquifers, *J. Hydrol.* **179**, 321–325.

Kavvas, M. L. and Karakas, A.: 1998, Response to comments of Z. J. Kabala and G. Sposito, *J. Hydrol.* **207**, 139–143.

King, P. R.: 1987, The use of field theoretic methods for the study of flow in a heterogeneous porous medium, *J. Phys. A* **20**, 3935–3947.

King, P. R.: 1989, The use of renormalization for calculating effective permeability, *Transport in Porous Media* **4**, 37–58.

King, P. R., Muggeridge, A. H. and Price, W. G.: 1993, Renormalization calculations of immiscible flow, *Transport in Porous Media* **12**, 237–260.

King, P. R.: 1996, Upscaling permeability: Error analysis for renormalization, *Transport in Porous Media* **23**, 337–354.

Kubo, R.: 1962, Generalized cumulant expansion method, *J. Phys. Soc. Japan*, **17**, 1100–1120.

Kubo, R.: 1963, Stochastic Liouville equations, *J. Math. Phys.* **4**, 174–183.

Ma, S. -K.: 1976, *Modern Theory of Critical Phenomena*, Benjamin, Reading, MA.

Matheron, G. and de Marsily, G.: 1980, Is transport in porous media always diffusive? A counter-example, *Water Resour. Res.* **16**, 901–917.

McComb, W. D.: 1992, *The Physics of Fluid Turbulence*, Clarendon Press, Oxford, UK.

Nötinger, B. and Gautier, Y.: 1998, Use of the Fourier–Laplace transform and of diagrammatical methods to interpret pumping tests in heterogeneous reservoirs, *Advan. Water Resour.* **21**, 581–590.

Peskin, M. E. and Schroeder, D. V.:1995, *An Introduction to Quantum Field Theory*, Addison-Wesley, Reading, MA.

Rehfeldt, K. R. and Gelhar, L. W.: 1992, Stochastic analysis of dispersion in unsteady flow in heterogeneous aquifers, *Water Resour. Res.* **28**, 2085–2099.

Reichl, L. E.: 1988, *A Modern Course in Statistical Physics*, Univ. of Texas Press, Austin, TX, pp. 435–437.

Smith, L. M. and Woodruff, S. L.: 1998, Renormalization-group analysis of turbulence, *Annu. Rev. Fluid Mech.* **30**, 275–310.

Sposito, G., Jury, W. A. and Gupta, V. K.: 1986, Fundamental problems in the stochastic convection-dispersion model of solute transport in aquifers and field soils, *Water Resour. Res.* **22**, 77–88.

Sposito, G. and Barry, D. A.: 1987, On the Dagan model of solute transport in groundwater: Foundational aspects, *Water Resour. Res.* **23**, 1867–1875.

Sposito, G., Barry, D. A. and Kabala, Z. J.: 1991, Stochastic differential equations in the theory of solute transport through inhomogeneous porous media, *Advan. Porous Media* **1**, 295–309.

Sposito, G.: 1999, The statistical physics of subsurface solute transport, Chapter 3 in: M. Parlange and J.W. Hopmans (eds.), *Vadose Zone Hydrology: Cutting Across Disciplines*, Oxford University Press, New York.

Terwiel, R. J.: 1974, Projection operator method applied to stochastic linear differential equations, *Physica* **74**, 248–265.

van Kampen, N. G.: 1974, A cumulant expansion for stochastic linear differential equations, *Physica* **74**, 215–247.

van Kampen, N. G.: 1976, Stochastic differential equations, *Phys. Rept.* **24**, 171–228.

Wood, B. D.: 1998, A connection between the Lagrangian stochastic-convective and cumulant expansion approaches for describing solute transport in heterogeneous porous media, *Advan. Water Resour.* **22**, 319–332.

Wood, B. D. and Kavvas, M. L.: 1999, Ensemble-averaged equations for reactive transport in porous media under unsteady flow conditions, *Water Resour. Res.* **35**, 2053–2068.

Zagoskin, A. M.: 1998, *Quantum Theory of Many-Body Systems,* Springer, New York.

Zhang, Q.:1995, The asymptotic scaling behavior of mixing induced by a random velocity field, *Advan. Appl. Maths.* **16**, 23–58.

Transport in Porous Media **42:** 199–209, 2001.

The Tracer Transit-Time Tail in Multipole Reservoir Flows

JOEL KOPLIK
Benjamin Levich Institute and Department of Physics, City College of the City University of New York, New York, NY 10031, U.S.A.

(Received: 16 October 1998; in final form: 1 October 1999)

Abstract. Some general aspects of the transit-time distribution for a passive tracer released in a reservoir flow between localized sources and sinks are discussed. Because of hydrodynamic dispersion, tracer particles which sample different regions of the reservoir tend to arrive at different times, and the resulting distribution encodes information about the flow and the large-scale geometry of the reservoir in a simple way. In particular, the transit-time distribution has a region of power-law decay, whose exponent depends only on the multipole order of the distribution of fluid sources and sinks, an exponential decay region, whose characteristics are related to stagnation points in the flow field, and a shoulder whose location is related to the system size. Tracer measurements may thus be used as a non-invasive surface diagnostic tool to characterize subsurface reservoirs.

Key words: hydrodynamic dispersion, long-time tail, multipole flow, passive tracer, power laws, reservoir heterogeneity.

1. Introduction

This paper concerns some little known but very generic qualitative features of the behavior of a passive tracer in the reservoir flows that arise in realistic hydrological or oil-recovery applications. Suppose that a pulse of passive tracer is released at a source well in a steady fluid flow, and collected at one or more producing wells in a reservoir, and consider the tracer transit time distribution as a function of time. This quantity, denoted $p(t)$ in this paper, can be thought of equivalently as the fraction of released tracer arriving at some collection well at time t, or as the first-passage probability distribution function for a biased random walker moving in the reservoir's velocity field to travel between a source and a sink. Because different packets of tracer leave the sink in different directions and sample different regions of the reservoir, $p(t)$ will be a non-trivial function of t. A typical result is shown in Figure 1. We shall argue that independently of the detailed fine structure of the porous medium and the nature of the fluids, $p(t)$ exhibits a region of power-law decay, a hump or plateau, and a region of exponential decay. Furthermore, the numerical values of the power-law exponent, the location in time of the hump, and the exponential decay constant are all related to geometric properties of the reservoir, as encoded in the flow field.

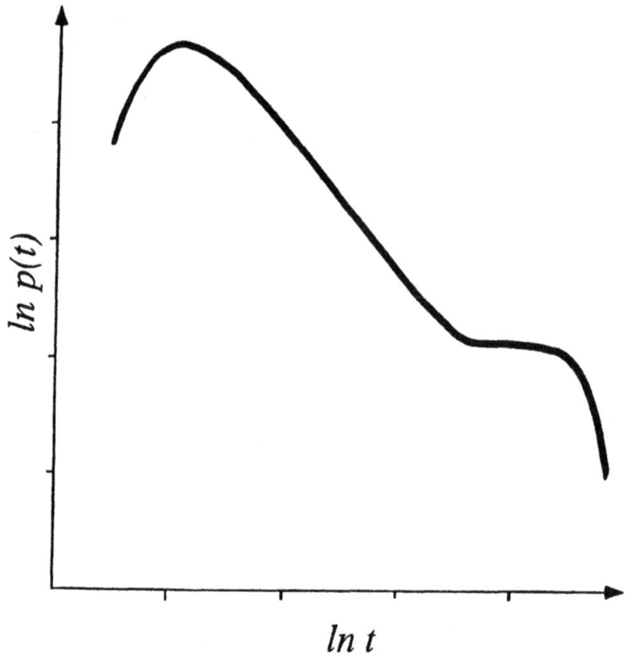

Figure 1. Typical form of the transit-time distribution for a tracer pulse in reservoir flows.

The reason that the generic behavior of the transit time distribution to be discussed here is not widely known is that it arises only in flows with a two or three-dimensional spatial variation, whereas most work is based on flows which are one dimensional on average. Field studies of hydrodynamic dispersion are straightforward but require considerable patience, and involve a subsurface reservoir which is at best partially characterized. Experimentalists quite naturally prefer to work with a 'core' of reservoir rock extracted during drilling which can be examined in detail, and for which laboratory studies require hours to days rather than months to years. Likewise, most theoretical work on the problem [1–5], including that of this author, is based on a quasi-one-dimensional configuration because the facility of comparison to core measurements, the convenience of fewer coordinates to keep track of, and in particular the availability of analytic solutions of the convection–diffusion equation (CDE). The analysis below is in fact based on using approximate and numerical solutions of the governing equations, without the advantage of a clean analytic solution exhibiting the relevant behavior. The detailed arguments for the results below may be found in [6–10], and here we will be somewhat general and heuristic.

Aside from the intrinsic physics interest in finding some new unifying behavior, and serving as a check on other calculations, the results have a number of potentially useful practical implications. Observations of $p(t)$ represent relatively easy measurements at the surface which can provide inexpensive, non-invasive and nontrivial information about the interior of a subsurface reservoir, such as estimates of its size, detection of unknown fractures or fluid sources and sinks, and testing for large-scale heterogeneities. Furthermore, in many field cases, surface tracer measurement data already exists.

2. Dipole Flow

The simplest relevant case is that of a single injection well and a single producing well in a porous layer, in the limit of very high Péclet number. The mathematical idealization is a particle which convects with the local velocity field produced by a single point source and a single point sink in an infinite plane. If the source and sink are at $\mathbf{r}_{1,2} = (0, \mp a)$, respectively, and the fluid flux is Q, then the velocity is the gradient of the dipole potential,

$$\mathbf{u} = \nabla\phi = \nabla \frac{Q}{2\pi} \ln \left| \frac{\mathbf{r} - \mathbf{r}_1}{\mathbf{r} - \mathbf{r}_2} \right| \tag{1}$$

and the motion of a tracer particle is given by

$$\frac{d\mathbf{r}(t)}{dt} = \mathbf{u}(\mathbf{r}(t)) \, . \tag{2}$$

Integrating, the transit time t for the particle to go from $\mathbf{r}_1$ to $\mathbf{r}_2$ can be shown [6] to be

$$t(\theta) = \frac{4\pi a^2}{Q}(1 - \theta \cot\theta) \csc^2\theta \, . \tag{3}$$

Here, θ is the polar angle with which the particle leaves the source, with the convention that $\theta = 0(\pi)$ is directly towards (away) from the sink; see Figure 2. As θ increases, t increases monotonically, as the trajectories vary from a straight line joining source to sink, to circular arcs, to large loops in the plane which start out almost backwards and eventually circle around to the sink. For $\theta \to \pi$ the transit time diverges as the loops extend further and further away from the source and sink, and the particles emitted at $\theta = \pi$ travel directly away from the sink and never arrive. Although each trajectory is deterministic, the tracer particles are emitted at random angles, with typically a uniform probability distribution $p(\theta)$, giving rise to a probability distribution for t

$$p(t) = p(\theta) \left| \frac{d\theta}{dt} \right| = \frac{1}{2\pi t'(\theta)} \underset{\theta\to\pi}{\sim} t^{-4/3} . \tag{4}$$

We have explicitly indicated the long-time limit of p, corresponding to tracer emitted 'backwards' away from the sink; this is the power-law decay regime mentioned earlier. Although strictly speaking this argument is valid for the asymptotic limit, the result holds when t is large compared to the only intrinsic time scale in the

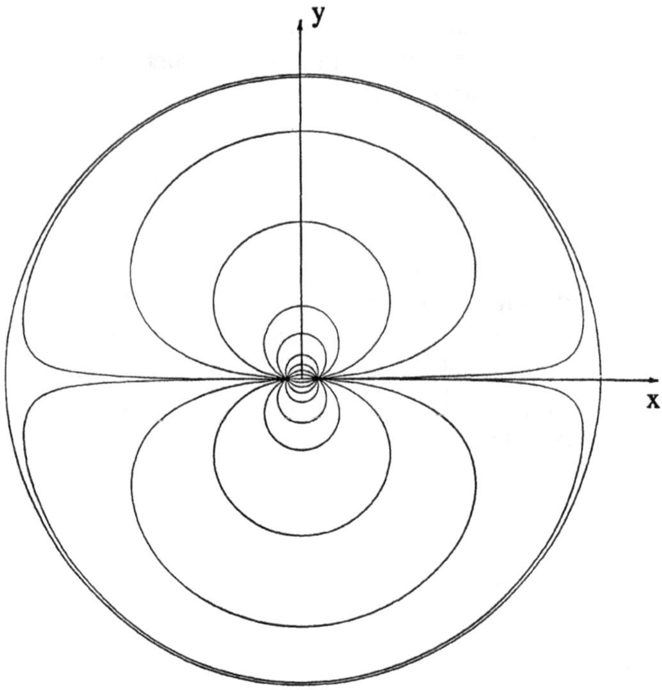

Figure 2. Streamlines for a dipole flow field in a circular domain.

problem, a^2/Q, corresponding to trajectories whose typical extent is much greater than a.

Next, we consider how this result changes when the simplifications in the preceding analysis are relaxed. First, the source and sink are not points but rather have finite size. In numerical simulations, this is a useful feature to incorporate because the dipole velocity field diverges at a point source or sink, but the effect on the transit time is negligible for the same reason. More importantly at the other extreme, all reservoirs have a finite spatial extent, and this will affect the most distant trajectories. Suppose that we have a large circular reservoir of radius R. If the outer boundary is impermeable, the velocity field changes because the distant trajectories must bend to remain inside the circle. In this case, the velocity can still be found analytically using the method of images [1, 7], but it is not possible to find the transit time in closed form. The numerical computation is easy, however, involving a pair of coupled ordinary differential equations, and one finds a power-law region in $p(t)$ over several decades, followed by an ultimate exponential decay. The distinction between the two regimes is that the first corresponds to trajectories which do not approach the outer boundary and are not significantly altered from the previous case, while the second is associated with trajectories starting sufficiently close to $\theta = \pi$ that they would have passed outside the boundary and are now forced to bend.

The reason that the tail of $p(t)$ is exponential can again be seen from an asymptotic analysis. Note that trajectories whose initial angle θ is just below (above) π move off into the upper (lower) half-plane, and there is a stagnation point in the flow on the boundary at $(r = R, \theta = \pi)$. In coordinates centered there, the velocity has the local form $\mathbf{u} = \mathrm{d}\mathbf{r}/\mathrm{d}t = (-Gx, Gy)$. If the stagnation point is approached along the x-axis, then trajectories which pass near the origin and then escape have the form $\mathbf{r}(t) = (x_0 e^{-Gt}, y_0 e^{Gt})$ with $y_0 \ll x_0$. The time T spent near the stagnation point can be defined by requiring that $\dot{y} \sim U_0$, where U_0 is some characteristic O(1) velocity. Since $\dot{y} \sim y_0 G e^{GT}$, this gives $T \sim G^{-1} \ln (U_0/y_0 G)$. Typically, for trajectories which pass near a stagnation point, the transit time between source and sink is dominated by the traversal time for this region. The corresponding distribution of transit times may now be obtained by accounting for the distribution of initial positions y_0. Consequently,

$$p(T) = p_0(y_0) \left| \frac{\mathrm{d}y_0}{\mathrm{d}T} \right| = p_0 \left(\frac{U_0}{G} e^{-GT} \right) \left| -U_0 e^{-GT} \right| \underset{T\to\infty}{\sim} p_0(0) U_0 \, e^{-GT} . \quad (5)$$

Thus the transit time distribution depends on the local shear rate G near the stagnation point, but not on the details of the initial spatial distribution of the tracer p_0.

Next we consider the effects of molecular diffusion on the tracer motion, or in other words, finite Péclet number. A nominal definition of the latter may be taken to be $\mathrm{Pe} = Q/2\pi D$, where D is the molecular diffusivity, but there is in effect a spatially-varying local Péclet number, corresponding to the roughly r^{-2} velocity variation in a dipole flow field. If the nominal Pe is sufficiently large, the velocity will be large everywhere in the reservoir, aside from the boundary stagnation points, and the preceding analysis should remain at least approximately valid. To address the question quantitatively, one may average over tracer trajectories given by

$$\Delta\mathbf{r} = \mathbf{u}(\mathbf{r})\Delta t + \hat{n} \left(\frac{4\Delta t}{Pe} \right)^{1/2} , \quad (6)$$

where $\hat{n}$ is a unit vector of random orientation. The first term is just a brute force method for numerically integrating (2), and the second yields upon averaging the (nondimensionalized) CDE for the average tracer concentration $c(\mathbf{r}, t)$

$$\frac{\partial c(\mathbf{r}, t)}{\partial t} + \mathbf{u}(\mathbf{r}) \cdot \nabla c(\mathbf{r}, t) = \frac{1}{\mathrm{Pe}} \nabla^2 c(\mathbf{r}, t) . \quad (7)$$

This method is numerically efficient at high Pe, where the second (random) term in (6) is a small correction, and gives a noisy version of Figure 1. At lower Pe the fluctuations are too severe, and the signal is masked by noise. For this limit a direct numerical solution of (7) is preferred and again yields results of the general form shown above. This method is only practical for relatively low Pe however, because otherwise too fine numerical grid is required, so unfortunately there is no single calculation which handles all regimes of Péclet number. There is however a new effect appearing – if the global Péclet number is not large, the tracer motion in the distant parts of the region will be dominated by molecular diffusion, so the stagnation point is of secondary importance. There is *still* an exponential cutoff in $p(t)$, from the fact that diffusion in a finite region gives a long-time behavior $c \sim e^{-t/\tau}$, where $\tau \sim r^2/D$ is the lowest nontrivial eigenvalue of the operator $-D\nabla^2$. A further consequence of diffusion is the hump or plateau appearing in Figure 1, whose origin is the rebound of diffusing tracer from the boundary. The relevant approximate analysis appears in Appendix D of [7].

Lastly, one may consider the modifications to the molecular diffusivity resulting from convection, In (7) we should in principle replace D by an effective diffusivity tensor [1–5] with longitudinal component $D_{\|} = D[1 + O(pe \ln pe)]$ and so on, where pe is a *local* Péclet number. This complication may be incorporated numerically, and turns out to have no observable effect on the power laws or the exponential decay rates in $p(t)$. The reason is that the modification to $D_{\|}$ is only significant when pe is large, but then the tracer motion is dominated by convection anyway, whereas diffusion is only important at low pe, but then the modification is negligible. This situation is distinctly different than the familiar quasi-one-dimensional flow, where all of the tracer convects on average at the same rate, and diffusion controls the spreading of a packet.

In order to test these theoretical arguments, a laboratory experiment was conducted [6] in a circular Hele Shaw cell, with sources and sinks near the center and a conducting solution used for measurements. The power-law decay was clearly seen, and the details of the time-dependence of the tracer concentration were in good agreement with simulations using (6) and the parameters of the experiment, but the measurement did not have adequate precision at low concentrations to test the exponential tail region quantitatively.

3. General Multipole Flows

If multiple sources and sinks are present, the analysis above may be repeated with the appropriate additional effort [7, 9, 10]. For circular and spherical reservoirs in two or three dimensions the velocity field may be found in closed form by superposition and the use of image sources to fit the boundary conditions. For regions of different shapes, generally a numerical solution of the Laplace equation is needed. Given the velocity field it is an easy computation to find $p(t)$ in the convective limit by integrating along trajectories, a marginally more difficult one at high Pe using (6), and a considerably more elaborate one based on (7) at lower values. One finds a curve of the same general shape – an early-time peak, followed by a power law decay, and then an exponential decay region perhaps with a rebound hump. The effects of diffusion and stagnation points are changed only in their numerical details, and the most significant new feature is that the value of the power in (4) is generally not 4/3. There remains some universality in its value however, in that it depends only on the *multipole order* of the distribution of sources and sinks. It is not possible to obtain an exact analytic result analogous to (3) in general, but the power may be predicted by various methods. In two dimensions, one can study the flow field in detail and estimate the distribution of convective times along streamlines [7, 8], but this method is based on the stream function and so does not extend to non-axisymmetric flows in three dimensions.

A very general argument for the shape of $p(t)$ in power-law region follows from scale invariance however [9]. Consider a set of sources and sinks in a finite region near the center of a much larger reservoir. The tail of the transit time distribution will result from tracer convecting on flow streamlines which make large excursions from the central region and then circle back. (Trajectories which approach the boundary, or tracer entering into regions where diffusion dominates will contribute to the exponential tail at still longer times.) At large distance from the source-sink region, it is reasonable to work with the multipole expansion of the velocity potential and retain the leading term only. For an nth order multipole in d-dimensions, this takes the form

$$\phi(\mathbf{r}) = Af(\Omega)r^{2-n-d}\,, \tag{8}$$

where Ω refers to the d-1 angular variables. (The power of r follows from the fact that a d-dimensional monopole has $\phi \sim r^{2-d}$, which in turn follows from the fact that its potential satisfies $\nabla^2\phi = \delta^d(\mathbf{r})$). Since the velocity is $\mathbf{u} = \nabla\phi$, dimensional analysis implies that the dimension of ϕ is $[\phi] = L^2/T$ where L and T are suitable length and time scales, respectively, and therefore $[A] = L^{n+d}/T$. Now consider the flux of tracer particles $F(t)$ which arrive at the source at times greater than t, for times in the power-law decay regime. This quantity depends on the tracer particles which are on the far field trajectories controlled by the asymptotic form of the potential (7) and depends only on the time t and whatever constants appear in ϕ itself, namely A. The velocity, and hence the transit time, for this region are ignorant of the outer size of the reservoir and any of the fine details of the source-sink distribution. Since the flux F has dimensions of velocity times (d-1 dimensional) area, $[F] = L^d/T$, so necessarily $F(t) = \alpha A^{d/(n+d)}t^{-n/(n+d)}$, where α is a pure number. The rate at which tracer particles arrive at the sink is then

$$p(t) \sim -F'(t) \sim t^{-\frac{2n+d}{n+d}}\,. \tag{9}$$

For the two-dimensional dipole, $n = 1$ and $d = 2$, leading to $p(t) \sim t^{-4/3}$, and more generally (8) agrees with all of the numerical simulations we have made.

Note that the exponent in (8) always lies between 1 and 2, which is a general constraint. The form given there for $p(t)$ would apply for all sufficiently large times in an infinite medium, and since all of the tracer arrives eventually at the sink $\int_0^\infty \mathrm{d}t\; p(t) = 1$, and the exponent must be greater than one. Furthermore, a general theorem [12] states that the mean transit time for a passive tracer in a steady flow equals the system volume divided by the flux, infinity in this case, so $\int_0^\infty \mathrm{d}t\; tp(t)$ must diverge, and the exponent cannot be larger than 2. A further remark concerns the approach to the asymptotic limit. The argument just given depends on use of the leading far-field term only, but a general source-sink distribution has subdominant terms corresponding to subsets of trajectories which explore a finite region. These can produce a regime of t where that exponent controls $p(t)$, and one can see crossover behavior from one power to another; examples are given in [7]. Likewise the configuration may be such as to produce an 'accidental' stagnation point in the interior of the flow domain, which gives an exponential component to $p(t)$ for those tracer trajectories which pass nearby. This subset of the emitted tracer manifests itself as an exponentially-decaying branch in $p(t)$, which may or may not be masked by other components of the tracer pulse.

Although we have emphasized two and three-dimensional reservoir situations, (8) is still valid for a one-dimensional flow, as it must be since only dimensional analysis was used in its derivation. In this case there is no divergence of streamlines, so the only interesting situation is a single source and a single sink. $p(t)$ may then be found explicitly using the method of images, and (9) verified directly. The tail of the 1-d transit time distribution is a very relevant quantity for the characterization of large-scale heterogeneities, but its usual analysis [5] is in terms of phenomenological models such as that of Coats and Smith for the entire distribution, rather than exponents.

4. Heterogeneities

The preceding arguments considered somewhat idealized situations, involving wells in the central region of a large, roundish, statistically homogeneous and isotropic reservoir, with a constant permeability. Real reservoirs are far more complicated [12], and we now consider the effects of relaxing these assumptions.

If the reservoir properties are approximately constant, that is to say the permeability has either small fluctuations everywhere about a well-defined mean or only small localized regions where its properties differ drastically, the results do not change at all. The reason is that the general behavior of $p(t)$ depends only on the large-scale geometry of the streamlines and the global reservoir shape. A zero-mean permeability fluctuation with a small correlation length generates some small scale wiggling in the shape of the streamlines, which leads in general to a small numerical change in the transit time on any particular trajectory. However, the power-law exponent for the tail of the distribution depends only on the global shape of the streamlines, which is not significantly changed in this case. The numerical verification of this argument consists of adding a local random fluctuation to the velocity, taking care to preserve incompressibility [9]

$$\mathbf{u}(\mathbf{r}) = \mathbf{u}_0(\mathbf{r}) + \mathbf{u}_r(\mathbf{r}) + \mathbf{u}_c(\mathbf{r}) . \tag{10}$$

In this expression, u_0 is the unperturbed velocity, and u_r the random perturbation, taken to be $\lambda\xi(\mathbf{r})\mathbf{u}_0(\mathbf{r})$ where ξ is uniformly distributed in $[-1, 1]$ independently for each $\mathbf{r}$ and $\lambda > 0$ is an adjustable strength. The term u_c is the correction term to insure incompressibility, and is chosen so that (10) is divergence free (obtained numerically using a fast Poisson solver). Given the fluctuating velocity field, one finds $p(t)$ by either averaging over individual trajectories or explicitly solving the CDE. For values $\lambda \lesssim 0.2$, one finds that the transit time distribution fluctuates with more or less noise about the power laws cited previously [9].

In the second situation, a velocity fluctuation which could be strong but which is confined to a localized region perturbs any tracer particle's transit time by an amount that is roughly the size of the region divided by the strength of the perturbation. Here, local permeability perturbations produce only local changes in streamline shape and is a negligible effect in this case as well. Another way of thinking about these effects is that local permeability changes produce shifts in tracer motion between nearby unperturbed streamlines, and if they have zero mean the net shift will be small. Although the numerical values of $p(t)$ can be somewhat altered by such weak fluctuations, power laws and decay rates are not.

Note that if the heterogeneities are sufficiently strong, corresponding to a rapid spatial variation in the physical properties of the reservoir rock, vorticity will be generated and there can be departures from Darcy's law [1]. However, as long as the rapid transition regions are small in size, any vorticity will decay away rapidly on a length scale whose order of magnitude is the square root of the permeability. (The physics of this effect can be captured conceptually, if not completely quantitatively, by means of the Brinkman equation, whose solutions for flow near a sharp permeability transition exhibit an exponential decay in vorticity*.) Furthermore, the relative change in velocity is at most O(1). As long as the non-Darcy deviations of the velocity occur over a small region, their effect on transit times will be correspondingly small.

Large-scale perturbations can of course alter the global shape of the streamlines and change the results. In cases where the flow field is known a modified calculation is straightforward. For example, in [7] we considered a point source in a wedge-shaped reservoir with absorbing sides in two dimensions, where the potential can be found by conformal mapping methods to behave as $\phi \sim r^{-\alpha/\pi}$ at large distances from the source, where α is the opening angle of the wedge. A numerical simulation gave $p(t) \sim t^{2\pi+2\alpha/\pi+2\alpha}$, consistent with (9). Another example studied concerned a large impermeable barrier separating a source and sink. In two dimensions, again a conformal mapping solution for the velocity potential could be found which gave an analytic result for the tail of $p(t)$ in agreement with numerical calculations [7] whereas in three dimensions [9] a numerical solution for the potential was needed, and gave a different exponent. Another example, suggested by a referee, is a high-permeability streak in a low-permeability background. There would be an early-time pulse of arriving tracer (perhaps with an observable

*Taking the curl of the Brinkman equation $\nabla p = -(\mu/k)\mathbf{u} + \mu\nabla^2\mathbf{u}$ gives the Helmholtz equation $(\nabla^2 + 1/k)\boldsymbol{\omega} = 0$ for the vorticity $\boldsymbol{\omega} = \nabla \times \mathbf{u}$. The fundamental Green's function for the Helmholtz equation in three dimensions is proportional to $\exp(-\sqrt{k}\,r)/r$, so vorticity decays exponentially away from any localized source.

exponential tail), corresponding to quasi-one-dimensional motion along the streak, in addition to the power-law regimes at later times associated with flow through the bulk of the reservoir. Although any given geometry can be analyzed directly, often by elementary reasoning, it is difficult to make any very general statements about large scale heterogeneities beacuse the possibilities are simply endless.

So far we have considered only the total amount of tracer arriving at time t, irrespective of its well of origin or the direction form which it enters the sink, since this is by far the simplest measurement available. If directional measurements or focussed emission of tracer are allowed, then considerably more information may be obtained, and in the presence of heterogeneities, further determination of the system geometry is possible. We mentioned above that crossover effects between different powers are associated with tracer emitted in different directions. As another example, if there is a single source, one can show [7] that the time-integrated tracer flux arriving at a particular point on a sink is proportional to the fluid velocity there, independent of Péclet number.

5. Conclusions

We have argued that the tail of the transit time distribution for a pulse of tracer released into a reservoir flow exhibits universal properties which are determined purely by the large scale features of the flow. There is a power law decay regime with an exponent given directly in terms of the spatial dimension of the reservoir (2 or 3) and the leading multipole order of the source-sink distribution in the simplest cases. Here 'simple' means that the sources and sinks are in the interior of a non-tortuous compact region and that there are no large-scale heterogeneities such as impermeable barriers, so that the asymptotic behavior of the velocity at large distances from the sources and sinks is known with little effort. More generally, some additional measurement or calculation is needed to determine the velocity, but then the power law exponent is readily determined from (8, 9). In addition to this behavior, there is a hump as well as an exponential component of $p(t)$, whose location and decay constant can likewise be related to certain features of the flow and reservoir geometry.

The tracer transit time distribution thus incorporates a considerable amount of information about the geometry if the reservoir and wells into its tail and may provide a valuable non-invasive diagnostic tool for the surface characterization of underground formations. We have focused here on what a particular geometry implies for $p(t)$, thereby identifying the significance of various features of a tracer calculations or experiment in a particular system, but equally one may attempt the inverse problem of inferring reservoir properties from the tracer measurements. One immediate application is to directly test whether a reservoir is planar or three dimensional in extent, by measuring the power-law exponent. Another obvious and relatively straightforward application is to test a subsurface region for 'leaks', unexpected paths for fluid and perhaps contaminants to leave the system. In this case there is an additional sink, which could manifest itself as a value of the exponent in the power law regime which differs than that obtained from the known set of wells, or produce some unexpected crossover behavior. Obviously, source-sink measurements involving different choices of specific wells in a reservoir will provide an even more informed picture. Such measurements are highly productive for the modest effort involved in making them, and we hope that these results prove useful in applications.

Acknowledgements

The work reported here was done in collaboration with my co-authors cited in references [6–10], and supported by the Geosciences Research Program of the Office of Basic Energy Sciences, U.S. Department of Energy.

References

1. Bear, J.: 1971, *Dynamics of Fluids in Porous Media*, Elsevier, Amsterdam.
2. Scheidegger, A. E.: 1974, *The Physics of Flow in Porous Media*, 3rd edition, University of Toronto Press.
3. Dullien, F. A. L.: 1991, *Porous Media: Structure and Fluid Transport*, 2nd edition, Academic, London.
4. Sahimi, M.: 1993, *Rev. Mod. Phys.* **65**, 1393.
5. Guyon, E., Nadal, J. -P. and Pomeau, Y. (eds): 1998, *Disorder and Mixing*, Kluwer, Dordrecht.
6. Kurowski, P., Ippolito, I., Hulin, J. P., Koplik, J. and Hinch, E. J.: 1994, *Phys. Fluids* **6**, 108.
7. Koplik, J., Redner, S. and Hinch, E. J.: 1994, *Phys. Rev. E* **50**, 4650.
8. Koplik, J., Redner, S. and Hinch, E. J.: 1995, *Phys. Rev. Lett.* **74**, 82. Some of the results of this paper were obtained independently by Mittag, L. and Stephen, M. J.: 1995, *J. Stat. Phys.* **78**, 377.
9. Koplik, J. and Zhang, M.: 1997, *Phys. Rev. E* **56**, 4244.
10. Zhang, M.: 1996, The First-passage properties of passive tracer in three-dimensional multipole flows, Ph. D. thesis, City University of New York.
11. Naumann, E. B. and Buffham, B. A.: 1983, *Mixing in Continuous Flow Systems*, Wiley, New York, 1983. [See also Appendix C of Koplik, J., Redner, S. and Wilkinson, D.: 1987, *Phys. Rev. A* **37**, 2619].
12. Lake, L. W. and Carroll, H. B. (eds.): 1986, *Reservoir Characterization*, Academic, New York.

Transport in Porous Media **42:** 211–240, 2001.

Fractional Dispersion, Lévy Motion, and the MADE Tracer Tests

DAVID A. BENSON[1], RINA SCHUMER[1], MARK M. MEERSCHAERT[2] and STEPHEN W. WHEATCRAFT[3]
[1]*Desert Research Institute, Division of Hydrologic Science, 2215 Raggio Parkway, Reno, NV 89512, U.S.A.*
[2]*Department of Mathematics, University of Nevada, Reno, NV 89557, U.S.A.*
[3]*Department of Geologic Sciences, University of Nevada, Reno, NV 89557, U.S.A.*

(Received: 21 January 1999; in final form: 20 December 1999)

Abstract. The macrodispersion experiments (MADE) at the Columbus Air Force Base in Mississippi were conducted in a highly heterogeneous aquifer that violates the basic assumptions of local second-order theories. A governing equation that describes particles that undergo Lévy motion, rather than Brownian motion, readily describes the highly skewed and heavy-tailed plume development at the MADE site. The new governing equation is based on a fractional, rather than integer, order of differentiation. This order (α), based on MADE plume measurements, is approximately 1.1. The hydraulic conductivity (K) increments also follow a power law of order $\alpha = 1.1$. We conjecture that the heavy-tailed K distribution gives rise to a heavy-tailed velocity field that directly implies the fractional-order governing equation derived herein. Simple arguments lead to accurate estimates of the velocity and dispersion constants based only on the aquifer hydraulic properties. This supports the idea that the correct governing equation can be accurately determined before, or after, a contamination event. While the traditional ADE fails to model a conservative tracer in the MADE aquifer, the fractional equation predicts tritium concentration profiles with remarkable accuracy over all spatial and temporal scales.

Key words: fractional derivative, fractional Laplacian, anomalous dispersion, Lévy motion, α-stable, heavy tails, Fokker–Planck equation, MADE site.

1. Introduction

The idea of a fractional derivative is nearly as old as the definition of an integer-order derivative (see the history compiled by Oldham and Spanier, 1974). Fractional derivatives effectively interpolate between successive integer-order derivatives when applied to many kinds of functions. For example, the qth fractional derivative with respect to x (as defined in Appendix A) of e^{ax} is $a^q e^{ax}$. Early in this century, a direct link was made between some stochastic processes and an integer-order differential equation. Einstein (1908) showed that a particle undergoing random Gaussian displacements fulfills the second-order diffusion equation. Kolmogorov extended the link between the second-order diffusion equation and generalized finite-variance processes (see Gnedenko and Kolmogorov, 1954; Feller, 1971). This model is extremely robust, primarily because it is a restatement of the central limit theorem (CLT). As a result, many different processes behave, in the long-term limit, in accordance with the classical diffusion equation. Einstein formulated his famous treatise on Brownian motion and the diffusion equation prior to Lévy's (1937) discovery of the stable densities, which are a superset of the Gaussian. Had Lévy's discoveries come several decades sooner, Einstein might have chosen a more general approach that would have led to a fractional diffusion equation 90 years earlier. Instead, the definitive link between all Lévy motions and a fractional differential equation has been developed only recently (Sheshadri and West, 1982; Gorenflo and Mainardi, 1998; Meerschaert *et al.*, 1999).

Lévy's stable laws are important because they are the limit distributions of sums of random variables. If a parcel of tracer is characterized as a particle that undergoes independent displacements as it moves through heterogeneous material, then the total displacement tends to converge to a stable variable. The motion is called a Lévy motion; Brownian motion is a subset. A large number of particles, i.e., the plume as a whole, fulfills a large portion of the probability distribution for a single particle. Then the ensemble concentration replaces the probability distribution for a single particle. Benson (1998) showed that the second-order advection–dispersion equation (ADE), even in a nonlinear form with the space-dependent dispersion tensor, cannot generally predict the stable Lévy motions. This is due to the Gaussian-density fundamental solution (Green function) of the ADE. An infinitude of nonlinear superpositions of the Gaussian must still scale proportional to the square root of time (cf., Crank, 1975). Further, combinations of the exponentially decaying tail of the Gaussian cannot model the 'heavy' power-law tails that characterize Lévy motion.

Recent work on random motions with power-law excursion probabilities (Fogedby, 1994; Compte, 1996; Saichev and Zaslavsky, 1997; Benson, 1998; Gorenflo and Mainardi, 1998; Meerschaert *et al.*, 1999; Compte and Caceres, 1998; Metzler *et al.*, 1998, 1999; Rocco and West, 1999; Grigolini *et al.*, 1999) has shown that the governing equations are generalizations, using fractional derivatives, of the classical ADE. In this paper, we review the derivation of the governing equation of stable Lévy motions in 1D. We then analyze data from the macrodispersion experiment (MADE) site in *a posteriori* and *a priori* modes to see (1) whether a fractional ADE is an appropriate model; and (2) to see whether the correct order of the fractional ADE, and the values of parameters, can be discerned before any contamination takes place.

We use the MADE site because it defies the traditional approaches used to model contaminant transport based on the classical integer-order ADE. The degree of K variability is very high (Rehfeldt *et al.*, 1992), and the plumes of conservative tracer are highly skewed. Moreover, the relative mass of tracer in the leading edge is very high, or in the vernacular of this study, 'heavy-tailed.' The first modeling approach used small-perturbation methods to give the local dispersion tensor within the classical ADE (Adams and Gelhar, 1992). Failure of this method led to more complex approaches, such as finely discretized numerical implementations of the classical ADE (Zheng and Jiao, 1998). This method still does not adequately reproduce the extreme tailing and skewness of the plume. Some have advocated a model based on diffusion-limited transport at this (C. Zheng, pers. comm., 1998) and other sites (Brusseau, 1992; Haggerty and Gorelick, 1995). The diffusion-limited, or two-site model, is a bimodal method that assumes a certain portion of the plume transfers in and out of a zero-velocity mode, and that Gaussian transport still occurs in each mode. Each of these successive approaches requires an added level of complexity, or required information, in the governing equation. The first needs a different dispersion parameter for each scale. The second adds spatial variability. The third adds transfer coefficients. Berkowitz and Scher (1998) proposed that the underlying transport process at the MADE site is not correctly modeled by a Gaussian process. We agree, and derive a simple (and lowest information-containing) equation that is based on the very large velocity contrasts experienced by the plume.

2. Stable Laws and Motions

Before developing the fractional advection–dispersion equation, it is useful to first investigate the generalized central limit theorem (GCLT) that leads to Lévy's stable laws. Complete discussions are given by Gnedenko and Kolmogorov (1954) and Feller (1971). In short, if $X_1, X_2, ..., X_n$ are independent, identically distributed (iid) random variables, then the sum of these converges in distribution to an α-stable random variable (Y). The distribution of Y scales predictably with the number of summands (n) according to

$$\frac{X_1 + X_2 + \cdots + X_n - n\mu}{\sigma n^{1/\alpha}} \Rightarrow Y, \tag{1}$$

where σ is a positive real constant and $\Rightarrow$ denotes convergence in probability. An example is the sum of finite-variance variables, which converge to (are in the domain of attraction of) a Gaussian random variable. In this case, the scaling rule dictates that $\alpha = 2$, σ is the standard deviation, and (1) is the traditional CLT. All finite-variance random variables are in the Gaussian domain of attraction. Many infinite variance random variables (e.g., Pareto, or power-law) are in the domain of attraction of an α-stable random variable, for which the rate of scaling in (1) follows $0 < \alpha < 2$. A special case of (1) arises when the summands (X) are themselves stable random variables: the sum of any number of iid stable variables is also a stable variable. The GCLT can also be generalized to vectors in any number of dimensions (Meerschaert, 1986).

Liu and Molz (1997a, b), Molz *et al.* (1997), and Painter (1996a, b, 1997) analyze the underlying heavy-tailed distribution of hydraulic conductivity (K) at several sites. In particular, Painter (1996b) shows clear evidence of these α-stable distributions in vertical borehole measurements of K. The stable variables are notably different from the normal or lognormal because the tails of the density functions decay according to a power-law. If Y is a stable variable, then for large arguments

$$P[|Y| > y] \approx W y^{-\alpha}, \tag{2}$$

where $P[\cdot]$ denotes probability and W is a positive constant. This leads to an infinite variance, since $\int y^2(y^{-\alpha-1})\mathrm{d}y$ diverges for all $\alpha < 2$. The power-law (Pareto) tail allows a higher probability of extreme values than a typical finite-variance distribution (even the lognormal) can predict. Any random variable that has a power-law tail (only the tail is important, virtually any function for small arguments [y] is permissible) belongs to the domain of attraction of an α-stable variable.

Densities of α-stable variables are represented by their Fourier transforms (Feller, 1971; Samorodnitsky and Taqqu, 1994). If Y is an α-stable random variable with a probability density function $f(y)$, then the Fourier transform of the density (denoted by the overhat and the change to the Fourier variable k) has the form

$$\hat{f}(k)=\int_{-\infty}^{\infty} \mathrm{e}^{-\mathrm{i}ky} f(y)\,\mathrm{d}y=\exp[-|k|^{\alpha}\sigma^{\alpha}\left(1+\mathrm{i}\beta\,\mathrm{sign}(k)\tan\left(\frac{\pi\alpha}{2}\right)\right)-\mu\mathrm{i}k], \tag{3}$$

where the parameters $0 < \alpha \leqslant 2$, $\sigma > 0$, $-1 \leqslant \beta \leqslant 1$, and μ describe the index of stability, the spread, the skewness, and the location of the density, respectively. The sign (k) function is -1 for $k < 0$ and 1 otherwise. The characteristic function for $\alpha = 1$ (the Cauchy density) is slightly different from (3) and will not be discussed here for the sake of brevity. An important feature of these densities is that they are invariant after shifting by μ and dividing by σ. All versions of the density are scaled and shifted versions of the *standard* density with $\sigma = 1$, $\mu = 0$. We also use an equivalent form

$$\hat{f}(k) = \exp[q\mathbb{C}(-\mathrm{i}k)^{\alpha} + p\mathbb{C}(\mathrm{i}k)^{\alpha} - \mu\mathrm{i}k], \quad \alpha \neq 1, \tag{4}$$

where p and q range between 0 and 1 with $p+q = 1$, and $\mathbb{C}$ is a positive constant. The two forms (3) and (4) are related by $\sigma^{\alpha} = \mathbb{C}|\cos(\pi\alpha/2)|$ and $\beta = p-q$. When the density is *symmetric*, the skewness parameter β is zero (so $p = q = 1/2$), and the symmetric characteristic function is

$$\hat{f}(k) = \exp(-\sigma^{\alpha}|k|^{\alpha} - \mu\mathrm{i}k)\,. \tag{5}$$

In this form, it is easy to see that the Gaussian (normal) density is α-stable with $\alpha = 2$. It is a simple matter to show that for $\alpha > 1$, the expectation $\mathbf{E}(Y) = \mu$. The mean is undefined for $\alpha \leqslant 1$. A standard, symmetric, α-stable distribution is characterized by the compact formula $\hat{f}(k) = \exp(-|k|^{\alpha})$.

The most important feature of the α-stable distributions is the characteristic exponent (also called the index of stability) α. The value of α determines how 'non-Gaussian' a particular density becomes (Figure 1). As the value of α decreases from a maximum of 2, more of the probability density shifts toward the tails. The densities appear near-Gaussian in untransformed coordinates; the difference lies in

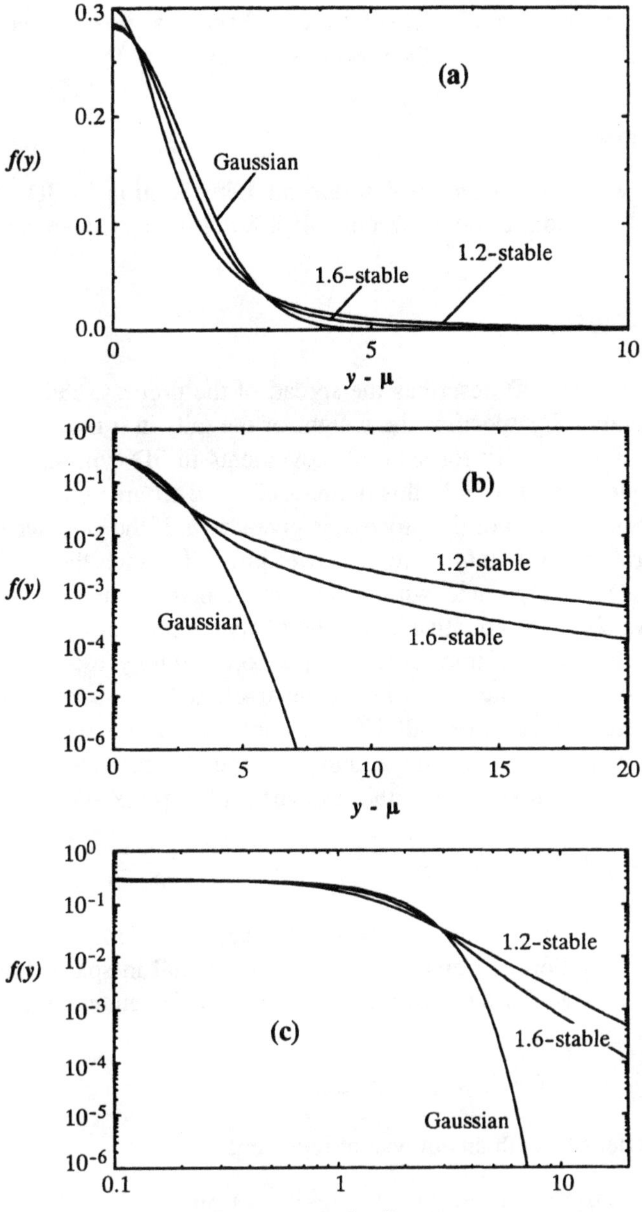

Figure 1. Plots of the positive density functions of several standard symmetric α-stable distributions using (a) linear axes, (b) semilog axes, and (c) log–log axes. For large arguments, the stable densities plot as straight lines with slope $-1-\alpha$ on log–log axes. For contaminant plumes, $\mu = vt$ is the mean travel distance, and $f(y)$ is the relative concentration (see Section 3).

the relative weight present in the tails. This difference becomes apparent using log axes (Figures 1b and 1c), which show the power-law character of the tails.

3. Governing Equation

Recently, the ADE was generalized to describe all Lévy motion in 3D after a general definition of multidimensional fractional differentiation (Meerschaert *et al.*, 1999):

$$\frac{\partial C}{\partial t} = -v\frac{\partial C}{\partial x} + \mathcal{D}\nabla_M^\alpha C\,, \tag{6}$$

where v is the drift velocity, $\mathcal{D}$ describes the spread of the process, and the fractional Laplacian operator ∇_M^α specifies the weight of the tails in a moving plume (via α), as well as the propensity for skewed movements in 3D (via the random measure M on the unit sphere). Within this definition, v and $\mathcal{D}$ are constants. The mean advective velocity, or drift of the process, is given by v. If the variance of the particle transition probability is infinite (e.g., power-law tails), then the nonlinear growth of the plume is incorporated within the fractional derivative, not within the leading parameter $\mathcal{D}$. The derivative is defined so that it correctly captures the nonlinear scaling of the particle transition density. To accurately model particle transport, one need only estimate the order of the fractional derivatives and the constant $\mathcal{D}$. This equation describes all Lévy motions including the subset of Brownian motion when $\alpha = 2$. In 1D, the random measure M reduces to $p+q = 1$, where the relative weight of p and q describes skewness (Appendix A):

$$\frac{\partial C}{\partial t} = -v\frac{\partial C}{\partial x} + p\mathcal{D}\frac{\partial^\alpha C}{\partial x^\alpha} + q\mathcal{D}\frac{\partial^\alpha C}{\partial(-x)^\alpha}\,. \tag{7}$$

Solutions to common solute transport boundary value problems (BVPs) are gained through Laplace or Fourier transforms. The 1D fractional-in-space Equation (7) is solved via Fourier transform (Appendix A) to give the fundamental (Green's function) solution:

$$\hat{C}(k,t) = \exp[q\mathcal{D}t(-\mathrm{i}k)^\alpha + p\mathcal{D}t(\mathrm{i}k)^\alpha - \mathrm{i}kvt] \tag{8}$$

which is an α-stable density with an equivalent representation:

$$\hat{C}(k,t) = \exp[-\mathcal{D}t|k|^\alpha|\cos(\pi\alpha/2)|(1+\mathrm{i}\beta\,\mathrm{sign}(k)\tan(\pi\alpha/2)) - \mathrm{i}kvt]\,. \tag{9}$$

This is the general form of a 1D Lévy motion (Samorodnitsky and Taqqu, 1994). The positive constant $\sigma = (|\cos(\pi\alpha/2)|\mathcal{D}t)^{1/\alpha}$ indicates that the stable density is invariant upon shifting by the mean (vt) and scaling by $t^{1/\alpha}$ (Figure 2). The nonlinear scaling (growth rate) results from the ability to replace the variable k with $t^{1/\alpha}k$ in (9). The distance between two points with constant relative concentration in a plume will grow proportional to $t^{1/\alpha}$, leading to a measured plume variance that is proportional to $t^{2/\alpha}$. Only if $\alpha = 2$ does this reduce to a Gaussian density

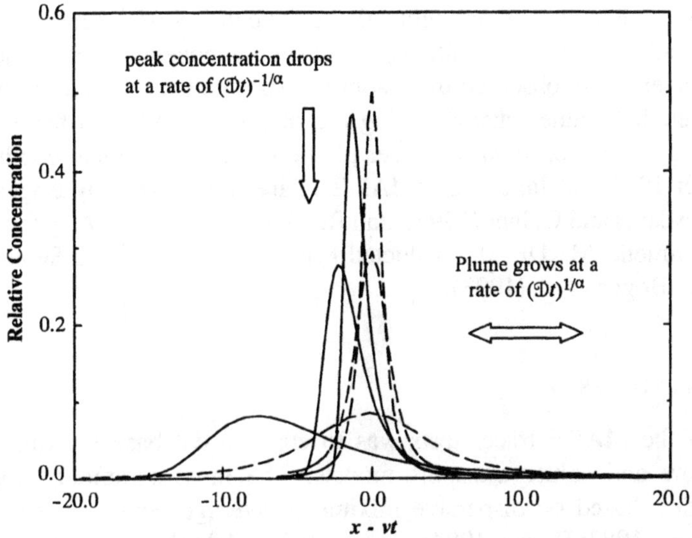

Figure 2. Comparison of the development of spatially symmetric (dashed lines) and positively skewed ($\beta = 1$, solid lines) plumes from a pulse source. Three dimensionless elapsed times (0.5, 1.0, and 5) are shown. All curves use $\alpha = 1.3$ and $\mathcal{D} = 1$. The point $x - vt = 0$ is the plume center of mass for all curves shown. Solutions were gained by numerical integration of (A.15).

with the characteristic Boltzmann scaling (growth) of the plume proportional to $\sqrt{t}$ and a measured variance growing linearly with time.

In addition to growing faster than the Boltzmann scaling predicts, the stable plumes have two significant characteristics. First, the skewness can be positive or negative, depending on whether the bulk of the contaminant is initially placed in lower- or higher-than-average velocity material. This leads to preferential sampling of portions of the velocity distribution. Second, one or both of the tails are 'heavier' than the second-order equation predicts. One or both of the tails asymptotically follow a power law (Figure 1) and can result merely from a heavy-tailed velocity distribution. The peak concentration may also be separated from the mean travel distance (vt) by large distances (Figure 2). The skewness and tailing can arise from the hydraulic characteristics of the aquifer, even in the absence of nonlinear sorption or completely stagnant, diffusion-limited zones.

4. Evidence for Fractional Dispersion in the MADE Tracer Tests

The MADE site is located on the Columbus Air Force Base in northeastern Mississippi. The two large-scale, natural-gradient tracer tests performed there differ from other large-scale tracer experiments such as the Borden site in Ontario, Canada (e.g., Sudicky, 1986) or the Cape Cod site in Massachusetts (e.g., LeBlanc *et al.*, 1991) because of the strong heterogeneity of the aquifer (Rehfeldt *et al.*, 1992;

Boggs *et al.*, 1993). The unconfined, alluvial aquifer consists of generally unconsolidated sands and gravels with smaller clay and silt components. Irregular lenses and horizontal layers were observed in an aquifer exposure near the site (Rehfeldt *et al.*, 1992). Detailed studies characterizing the spatial variability of the aquifer and the spreading of the conservative tracer plume for the experiment conducted between October 1986 and June 1988 (MADE-1) are summarized by Boggs and Adams (1992), Adams and Gelhar (1992), and Rehfeldt *et al.* (1992). A synopsis of the second experiment (MADE-2), conducted between June 1990 and September 1991, is given by Boggs *et al.* (1993).

4.1. PREVIOUS STUDIES

One purpose of the MADE tracer tests was to create a database for validating groundwater flow and solute transport models. The most commonly applied stochastic theories, based on dispersive mixing in heterogeneous aquifers (e.g., Gelhar and Axness, 1983; Dagan, 1984), were tested by Adams and Gelhar (1992). They found that classical Gaussian models were not able to reproduce the highly skewed plume evolution found at the MADE site. They presented no data concerning the heavy plume tails.

Berkowitz and Scher (1998) show how a very general, non-Gaussian, model of particle transport can be applied to fracture networks. Their work is based on particle transitions described by a joint space-time distribution. If one is able to identify this distribution, then a plume can be generated. Based on numerical generation of fracture networks and an analogy to the preferential flowpaths in the highly heterogeneous MADE site, the authors present several figures that show an improved fit over those presented by Adams and Gelhar. It is not clear whether the density of Berkowitz and Scher (1998) reproduces the extremely heavy tails of the plume or scales at the correct rate, since the authors predict that the ratio of the sample mean and standard deviation is a constant. These quantities measured in the MADE tests do not have the same slope on a log–log plot. Their methodology also depends on measuring an empirical joint space-time particle transition density.

Zheng and Jiao (1998) produced 3D numerical simulations for the MADE-2 experiment using K distributions derived from borehole flowmeter data and various geostatistical methods. They found that the simulated plume was highly sensitive to the distribution of the K field. Additionally, they were not able to concurrently reproduce the movement of the plume's center of mass and the significant spreading of the tracer at the plume's leading edge. Finally, they suggest that inconsistencies in modeling may be due to temporal flow field variations and preferential flowpaths not consistent with the assumption of a Gaussian or finite-variance pore velocity distribution. The large-scale preferential flowpaths lead to a higher probability of sustained high velocity events that the normal or lognormal cannot predict. Further, the K increments (Appendix B) measured at the smallest scale (a vertical lag of 15 cm) suggest that heavy-tailed velocity distribution precludes the use of the traditional ADE even at the smallest scale within finely discretized numerical models.

The fractional ADE can describe both the heavy-tailed transport and the skewed transitions. We develop a relationship between aquifer properties and the parameters within the fractional ADE, and compare the estimates to those gained from sampling a contaminant plume at early time. The former leads to estimation of parameters *a priori* a contamination event, while the latter leads to *a posteriori* parameter estimation.

4.2. *A PRIORI* ESTIMATION OF PARAMETERS

Recent studies show that permeability variations can be described with stable or other heavy-tailed distributions (e.g., Painter, 1996b). Liu and Molz (1997b) have suggested that the ln(K) distribution at the MADE site is heavy-tailed with a power-law index that is a function of sample separation. We re-examine the statistical distribution of the K field at the MADE site, since the ability to estimate the order of the fractional dispersion term *a priori* depends on the ability to accurately estimate the Lévy index (α) of aquifer K data. The values of α and $\mathcal{D}$ would follow immediately from a direct measurement of the natural velocity profiles; however, this is a difficult task. The basis of the following estimates is the assumption (as yet unverified) that the head gradient fluctuations within the soil are negligible compared to the magnitude of the K fluctuations. In this case, the small-scale velocity field is dominated by the many order-of-magnitude differences in K, rather than the smaller differences in head gradient. Therefore, if the K distribution is found to have a power-law tail with exponent α, we assume that the velocity distribution would have a similar exponent. Only the scale factor of the distribution (and possibly the low-velocity shape of the distribution) would change according to the mean gradient. An outline of the theory and application of four popular estimators is contained in Appendix B. One method for determining the Lévy index from a sample K or ΔK distribution appears to be most robust and stable; we present it here.

The Pareto is the most simple of the heavy-tailed distribution functions for a random variable X in an α-stable domain of attraction. The distribution function is defined by $P[X < x] = 1 - Wx^{-\alpha}$, where $\alpha > 0$, $W > 0$, and $x \geqslant W^{1/\alpha}$. Since one or both of the tails of a non-Gaussian stable distribution are asymptotically Pareto, it is also possible to model a stable distribution by evaluating its tails with a shifted Pareto distribution (Anderson and Meerschaert, 1998). Meerschaert and Scheffler (1998) develop a simple method for estimating the thickness of heavy tails for Pareto-like data with $W = 1$. When $0 < \alpha < 2$, $1/\alpha$ can be approximated by

$$\frac{1}{\alpha} \approx \frac{\gamma + \ln \sum_{i=1}^{n} (X_i - \overline{X})^2}{2(\gamma + \ln n)}, \tag{10}$$

where $\gamma \approx 0.5572$ is Euler's constant, $X_1, \ldots, X_n$ are the sample data, and $\overline{X}$ is the sample mean.

Details of borehole flowmeter logging, slug tests, and laboratory permeameter methods for measuring K at the MADE site are described by Rehfeldt *et al.* (1992). Application of Equation (10) to all normalized (each data point divided by the sample median) vertical K increment data yields an estimate of $\alpha = 1.1$. Another popular estimator is given by Hill (1975). Hill's estimator is similar to Meerschaert and Scheffler's except that does not allow the shift of the Pareto. However, Hill's estimator is able to provide the scale (width) parameter W, which is similar to the standard deviation of a Gaussian. It is this scale of the ΔK distribution that ultimately determines the magnitude of the dispersion coefficient.

The Hill's estimator fits the equation $P[\Delta K > x] = Wx^{-\alpha}$. Sums of Pareto data converge to an α-stable with an equality of the scale parameters of $\sigma^\alpha = W\Gamma(1-\alpha)\cos(\pi\alpha/2)$ (e.g., Feller, 1971; Janicki and Weron, 1994), where $\Gamma(\cdot)$ is the gamma function. From (9) we have $\sigma^\alpha = \mathcal{D}t|\cos(\pi\alpha/2)|$, which has units L^α. The Pareto scale factor W from the ΔK data has units (cm/s)$^\alpha$. Therefore, the ΔK increments must be translated to relative particle separation size over a short period of time. These separation distances are the heavy-tailed random variables in (1) that converge to an α-stable variable governed by (7). The gradient divided by the porosity (J/η) is assumed to be relatively uniform giving the velocity increments $\Delta v = (J/\eta)\Delta K$. The difference in the solute particle travel distances (Δd) is this random variable multiplied by some time interval τ, which integrates the temporal correlation, giving $P[\Delta d > (J\tau/\eta)x] = Wx^{-\alpha}$. With the change of variable $y = (J\tau/\eta)x$, the distribution becomes $P[\Delta d > y] = W(J\tau/\eta)^\alpha y^{-\alpha}$. Now equating the dimensionally correct scale factors of the α-stable solution and this Pareto Δd distribution, $\mathcal{D}\tau = W(J\tau/\eta)^\alpha|\Gamma(1-\alpha)|$. Therefore, the dispersion coefficient is given by the K statistics, the mean gradient and porosity, and a characteristic transition time by the relationship

$$\mathcal{D} = W(J/\eta)^\alpha \tau^{\alpha-1}|\Gamma(1-\alpha)|. \tag{11}$$

Note that as α approaches unity, the estimate becomes very insensitive to the choice of transition time (τ). For example, when $\alpha = 1.1$, a ten-fold decrease in τ only changes the estimate of $\mathcal{D}$ by 20%. For any value of $\alpha > 0.5$, the estimate of $\mathcal{D}$ is most strongly sensitive to the value of J/η. The assumed gradient also directly changes the mean velocity, so the estimated values of v and $\mathcal{D}$ must be tracked simultaneously.

Once again assuming that the gradient is relatively constant throughout the soil, Darcy's Law gives $v = \overline{K}(J/\eta)$. The arithmetic mean K value from the flowmeter and core tests is 24 m/d, which compares favorably to a pumping test that indicated a horizontal K of 17 m/d (Boggs and Adams, 1992). Hill's estimator for the K increments (Appendix B) gives $\alpha = 1.1$ and $W = 0.0028\,(\text{cm/s})^{1.1}$. Using $\tau = 1$ day in (11) gives $\mathcal{D} \approx 50\,\text{m}^{1.1}/\text{d}(J/\eta)^{1.1}$. Table I summarizes the coefficients v and $\mathcal{D}$ for a range of values (J/η). The gradient experienced by most of the plume

Table I. Relationship of the constants v and $\mathcal{D}$ to the assumed gradient for a range of values reasonable to the MADE site

J/η	v (m/d)	$\mathcal{D}$ ($m^{1.1}$/d)
0.005	0.12	0.14
0.010	0.24	0.32
0.015	0.36	0.50

(after a distance from the release point of approximately 10 m) is typically lower than 0.003 (Boggs and Adams, 1992), so we estimate a value of approximately 0.005 for J/η. The validity of this entire procedure can only be addressed through a detailed examination of the relationship between head gradient and the heavy-tailed K distribution, as well as the decay of particle time correlation on the order of 1 day.

4.3. *A POSTERIORI* ESTIMATION OF PARAMETERS

Two *a posteriori* methods for estimating the Lévy index are presented using information from the MADE database to validate the a priori estimate of $\alpha = 1.1$. These methods require far less data collection than extensive K measurements and are more practical methods for real plumes. We use the analyses of the MADE-I test presented by Adams and Gelhar (1992).

Approximately 10^4 L of water containing 2,500 mg/l of bromide were injected into the MADE aquifer for the first test. Over the next 20 months, seven sampling events using an extensive array of multi-level samplers (MLS) were performed. Using the spatial moments of plume growth for the MADE-1 experiment, Adams and Gelhar (1992) compare a normalized distribution of mass with a Gaussian distribution with similar 0th, 1st, and 2nd longitudinal moments. They noted that the longitudinal variance of the plume grew at an accelerated rate. A plot of the apparent plume variance along the travel direction on log–log axes confirms that plume growth is indeed nonlinear (Figure 3a). Because the density of an α-stable plume is scale invariant with $t^{1/\alpha}$, the variance of the discretely-sampled plume should grow proportionally with $t^{2/\alpha}$ (Benson, 1998). Thus, the slope of the graph (Figure 3a) provides a second estimate of the Lévy index ($\alpha = 1.2$), that agrees reasonably closely with the *a priori* estimate. The plume does not appear to follow the early- (slope = 2) or late- (slope = 1) time behavior predicted by local stochastic theories (e.g., Gelhar and Axness, 1983). Though the data show some scatter due to the difficulties of measurement, there is no systematic deviation from the general trend of $\alpha = 1.2$ (Figure 3a).

A simpler method for obtaining parameters for the fractional ADE arises from the fact that the plume is a 3D density that grows at a rate of $(\mathcal{D}t)^{1/\alpha}$ (Meerschaert *et al.*, 1999). The peak concentration should thus follow:

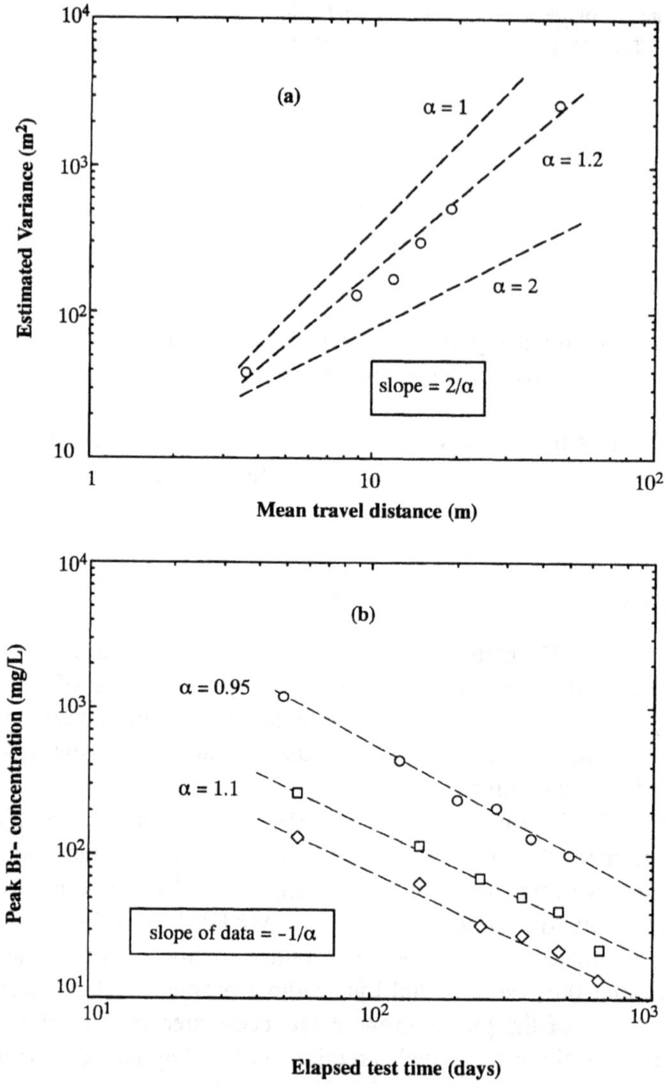

Figure 3. Estimation of Lévy index (α) from plume data: (a) calculated plume variance (Adams and Gelhar, 1992) along the mean direction of travel for the MADE-1 experiment and (b) maximum bromide concentration measured during each snapshot. Circles represent unfiltered data, squares and diamonds are spatially filtered (averaged, see Adams and Gelhar, 1992).

$$C(t) = \frac{C_0}{(\mathcal{D}t)^{1/\alpha}}. \tag{12}$$

The slope of a log–log plot of the peak concentration versus time should be $(-1/\alpha)$. Adams and Gelhar (1992) list the maximum bromide concentrations before and after spatial averaging (filtering). The unfiltered data indicate that $\alpha = 0.95$, while both sets of filtered data conform to a line with $\alpha = 1.1$ (Figure 3b).

The *a posteriori* estimates of the order of the fractional ADE (α) show some variability, from 0.95 to 1.2. It is not known which are the best estimates. The unfiltered (raw data) peak concentrations should have the highest degree of variability. The variance estimations suffer bias, especially at late time, from truncation of distant, low concentrations by the fixed wellfield and finite detection limits. The method based on maximum spatially-averaged concentrations is probably the most useful, since this represents the simplest form of data collection in real-world applications: monitoring wells are typically screened across large vertical distances, providing data similar to Adams and Gelhar's (1992) filtered data. Only the maximum concentration from a vertically averaged (filtered) wellfield at any time is identified and used as the peak concentration. Numerical integration is not needed, and the value of α should be apparent after the first few measurements. This method gives a value equal to the a priori estimate of $\alpha = 1.1$. Solutions (e.g., Figure 1) suggest that the recognition of stable transport is more important in predicting concentrations in the leading edge than exact determination of α. A sensitivity analysis is needed to assess the importance of accurate estimation of α and $\mathcal{D}$.

4.4. ANALYTIC SOLUTIONS

We use the parameters estimated from the aquifer *K* statistics (supported by Adams and Gelhar's analysis of the MADE-1 bromide plume) to predict the MADE-2 tritium plume. We have only recently received the MADE-1 bromide data and have not yet re-analyzed that test. During the MADE-2 experiment, 9.7 m^3 of tritiated water and four organic compounds were injected into the shallow alluvial aquifer at the test site. Over the next 15 months, five snapshots of the tracer concentration distributions were collected. Several sets of analytic solutions have been generated using the aquifer parameters estimated in previous sections (see Appendix B for a discussion). We compare the solutions of the traditional ADE ($\alpha = 2$) and the fractional ADE ($\alpha = 1.1$):

$$\frac{\partial C}{\partial t} + v\frac{\partial C}{\partial x} - \mathcal{D}\frac{\partial^2 C}{\partial x^2} = C_0 x_0 \delta(t, x), \tag{13}$$

$$\frac{\partial C}{\partial t} + v\frac{\partial C}{\partial x} - \mathcal{D}\frac{\partial^{1.1} C}{\partial x^{1.1}} = C_0 x_0 \delta(t, x), \tag{14}$$

where $C_0 x_0 \delta(t, x)$ is the initial solute concentration (C_0) spread over some injection distance x_0, which is mathematically concentrated into a Dirac delta function 'spike'. A value of 55,610 picoCuries (pCi) per $cm^3 \times 400$ cm (the approximate length of the injection area) $= 2.78 \times 10^7$ pCi/cm^2 was used in all solutions for the MADE-2 tritium plume. The values of v and $\mathcal{D}$ in the fractional ADE (14)

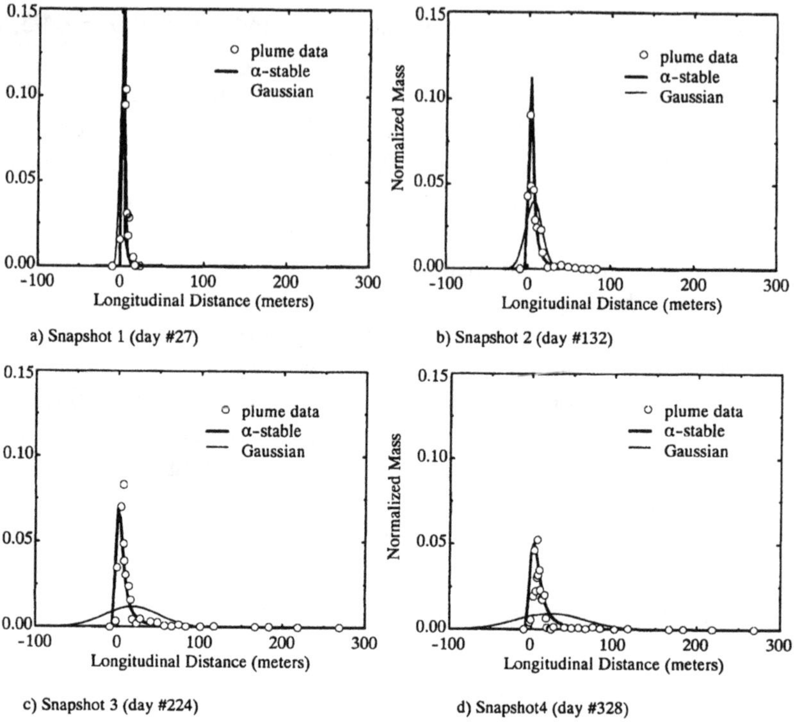

Figure 4. Linear plots of the MADE-2 normalized longitudinal tritium mass distribution at four intervals. Analytic solutions of the ADE and the fractional ADE were gained by numerical integration of (A.15).

are constants for all times of interest (0.12 m/d and 0.14 $m^{1.1}$/d, respectively). Similar to Adams and Gelhar (1992), we used 'best fit' values of v and $\mathcal{D}$ in the traditional ADE (13) for each time period by calculating the mean and variance of the measured plume.

Figure 4 shows the longitudinal distribution of total mass for MADE-2 snapshots 1–4. The data points represent the maximum concentration measured in vertical slices perpendicular to the direction of plume travel. These maxima were then integrated versus travel distance to get the 1D 'mass recovery'. The α-stable solutions (14) are superimposed on the normalized mass distributions. Also shown, in the manner of Adams and Gelhar (1992), is a Gaussian distribution with first and second moments equal to those of the normalized sample data. Note that this 1D analysis assumes that the mass lost to transverse dispersion is relatively small. The maximum concentrations were then normalized by the total mass recovered during each snapshot so that each plot has the same 1D mass. We use this method for ease of comparison with the data presented for the MADE-1 test (Adams and Gelhar, 1992). Since the concentration maxima along the plume core represent conditional densities of the entire 3D plume, their evolution will be similar to marginal (integrated) densities (Benson, 1998).

The initial 1D mass was calculated using a constant value of porosity of 0.35 (Boggs *et al.*, 1993). The normalizations represent mass recovery of 145%, 82%, 62%, and 37% for snapshots 1–4, respectively. Mass recovery in excess of 100% in early sampling rounds, followed by decreases in mass recovery to levels below 50% were also observed during the MADE-1 test (Adams and Gelhar, 1992) and the Borden tracer experiment (Freyberg, 1986; Rajaram and Gelhar, 1991). Adams and Gelhar (1992) discuss potential sources for these phenomena. In particular, they show that significant amounts of bromide mass passed beyond the MADE-1 sampling array later in the test. They also show that sampling bias toward higher local K material can seriously underestimate late-time mass. The heavy leading edges predicted by (14) suggest that some of the tritium may be downgradient of the sampling array. No snapshot was bounded in the downstream direction by nondetectable levels of tritium. We have not estimated this mass loss. Radioactive decay of tritium accounts for only about 5% of the lost mass after 328 days. Data for snapshot 5 are not presented because only a small portion of the plume was sampled.

Both the fractional ADE and Gaussian models reproduce the movement of the center of mass in the early time snapshots. In addition, the fractional ADE is able to capture the separation of the positions of the plume mean and the mode, or peak, concentration. The two late-time snapshots clearly show that the fractional ADE is a better model than a local (Gaussian) ADE in predicting the movement of the tritium plume. The fractional ADE was developed with heavy (power-law) tails in mind, so we also plot the data using logarithmic axes for the tritium mass (Figure 5). Here the predictive accuracy of the fractional ADE is dramatically evident, as is the inadequacy of thin-tailed second-order theories at later times. The traditional ADE, even using a best-fit value of $\mathcal{D}$, underpredicts concentrations throughout most of the downstream portion of the plume by many orders of magnitude. Using the ADE in a predictive mode with smaller values of $\mathcal{D}$ from classical stochastic theories would result in even worse estimates. Adams and Gelhar (1992) and Rehfeldt *et al.* (1992) discuss the fact that classical stochastic theory predicts a dispersion coefficient from 4 to 50 times smaller than the measured plume moments suggest. However, the traditional ADE seems to work very well for the earliest time period. The unsampled portion of the plume beyond 30 m might have measurable concentrations of tritium that would extend the tail of the real plume somewhat. It is also possible that the initial plume placement was far from the high velocity areas, and that some time was needed before the GCLT (1) was reasonably valid.

Since the fractional ADE predicts a leading edge of the plume that is a power function of distance, the decline in normalized mass (concentration) should appear linear on a log–log plot (Figure 1c). This is shown in the last three snapshots (for brevity we show the last two in Figure 6), supporting the notion that Lévy motion is a good model of transport at this site. We mention that unnormalized data concentrations would plot slightly lower (one to two diameters of the circles) on these plots.

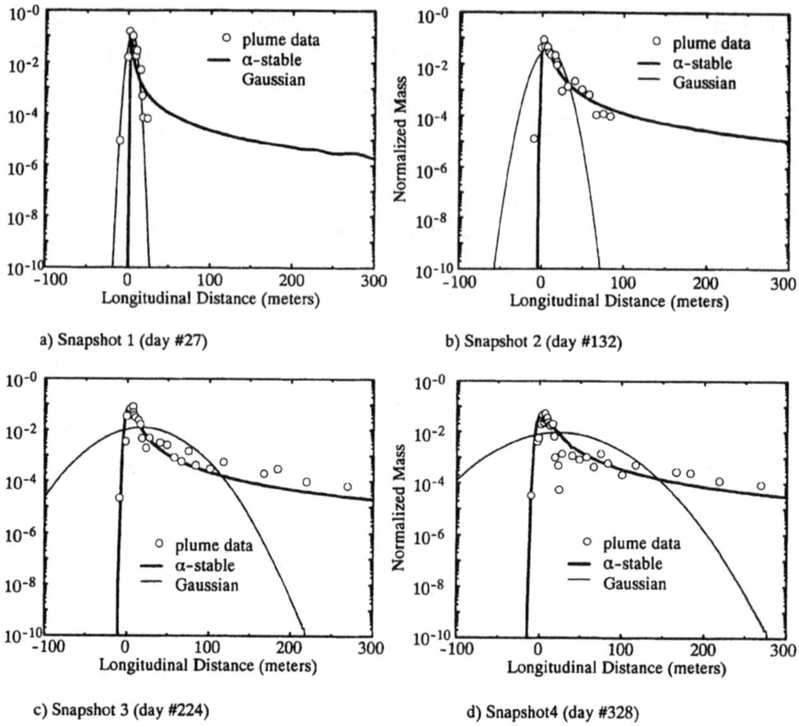

Figure 5. Semi-log plots of the MADE-2 tritium plume. Analytic solutions of the ADE and the fractional ADE were gained by numerical integration of (A.15).

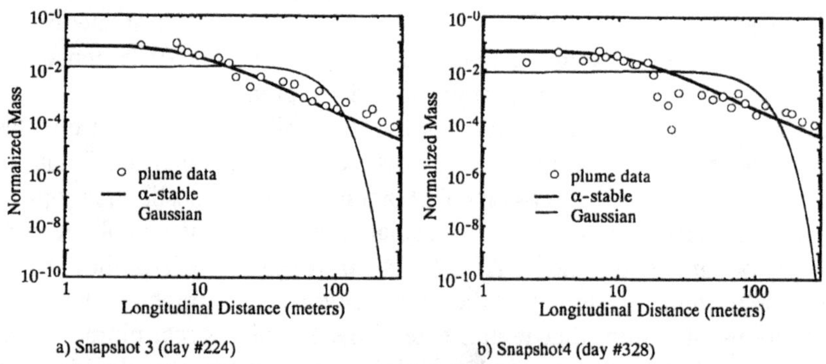

Figure 6. Log–log plots of the MADE-2 tritium plume. Analytic solutions of the ADE and the fractional ADE were gained by numerical integration of (A.15).

4.5. BEHAVIOR OF PLUME PEAK VERSUS MEAN

The rate at which any two portions of a density separate is well known. A density $f(x)$ of a scaled random variable grows by $a^{-1}f(x/a) = f(x)$. The fractional ADE density is invariant according to $\sigma^{-1}f(x/\sigma) = f(x)$, where $\sigma = (|\cos(\pi\alpha/2)|\mathcal{D}t)^{1/\alpha}$. This includes the ADE as well. The distance between the peak and the shift vt (the mean for $\alpha > 1$) at any time grows like $A(\cos(\pi\alpha/2)\mathcal{D}t)^{1/\alpha}$, where A is some constant that depends on the skewness and order α. If the skewness is zero, so is A. For a positively skewed plume, the peak lies behind the mean, so A is negative, since the distance is always smaller than vt. The distance travelled by the peak (x_p) is merely $x_p = vt + A(\cos(\pi\alpha/2)\mathcal{D}t)^{1/\alpha}$. The velocity of the peak is the time derivative:

$$v_p = v + \frac{A}{\alpha}(\mathcal{D}\cos(\pi\alpha/2))^{1/\alpha}t^{1/\alpha-1}. \tag{15}$$

Combining constants into $B = A\alpha^{-1}(\mathcal{D}\cos(\pi\alpha/2))^{1/\alpha}$ gives $v_p = v + Bt^{1/\alpha-1}$. For $1 < \alpha < 2$, the velocity of the peak (or any concentration percentile) approaches the mean velocity, as observed by Taylor (1953). For α very near unity, the velocity approaches the constant $v \pm B$ for any part of the plume, which is required for piston (wave equation) flow. For $\alpha < 1$, the apparent velocities of all parts of the plume are equal for vanishingly small time and diverge at later time (as does the theoretical plume centroid).

We are not aware of a handy expression for the mode location of an α-stable; however, one can generate a standard density (Appendix A) and get an accurate estimate for the value of A. Numerical estimators and plots of the mode location are also given by Fofack and Nolan (1998) and Nolan (1998). A maximally skewed density for $\alpha = 1.1$ has $A \approx -6$ (Nolan, 1998). The distance between the peak and the measured mean is very sensitive near $\alpha = 1$. For example, $\alpha = 1.14$ has $A \approx -4.5$. Therefore, inaccurate estimation of the parameters v, $\mathcal{D}$ and α could lead to a prediction that the plume peak is moving upstream. A corollary is that such an infeasible scenario would narrow the possible range of values of one of the parameters, given good estimates of the other two.

At the MADE site, the separation of the peak and mean positions is evident for later snapshots (Figure 7). For this plot we adjusted the values of $v = 0.12$ m/d and $\mathcal{D} = 0.14\,\mathrm{m}^{1.1}$/d predicted from the K distribution and J/η to better match the observed mean and peak locations. Our *a posteriori* estimates of $v = 0.11$ m/d and $\mathcal{D} = 0.12\,\mathrm{m}^{1.1}$/d based on this plot (Figure 7) are still very close to the initial estimates that were based only on the hydraulic properties of the aquifer. A thorough 'fitting' of $\mathcal{D}$ would still depend on an *a posteriori* best-fit value of α. The early time location of the peak concentration is inflated by the large (several meters in the x-direction) size of the initial tritium pulse. Also, the downstream edge of the plume was not adequately sampled during snapshot 5 at 440 days, so the mean was not calculated. The procedure of calculating peak location is complicated for values of $\alpha < 1$, since one must also estimate the plume's measured mean, which depends heavily on truncation from a finite sampling array. We have not performed this detailed sensitivity analysis. For illustrative purposes, we plot the behavior of the peak for $\alpha = 0.95$ using a best fit value $A(\mathcal{D}\cos(\pi\alpha/2))^{1/\alpha} \approx 0.016$ (Figure 7). A value of $\alpha = 0.95$ also explains the separation of the peak and measured plume mean. This would imply that the distance between readily identifiable portions of

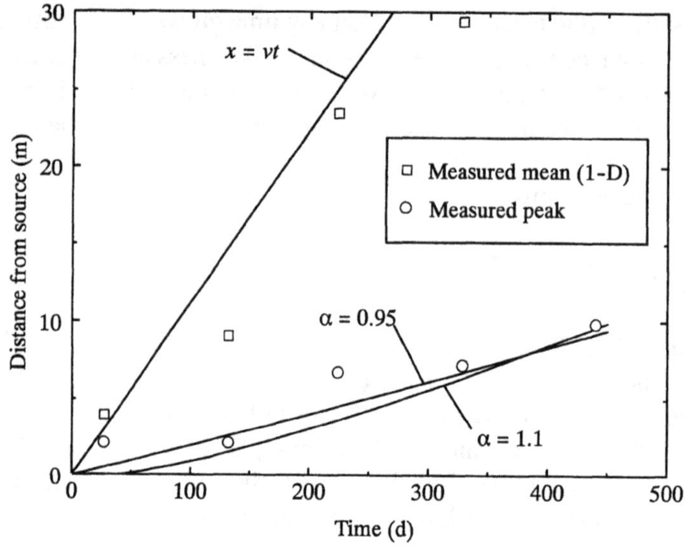

Figure 7. Locations of the center of mass and peak concentration for the positively skewed MADE site tritium plume (symbols) and theoretical estimates for two values of α (0.95 and 1.1). For $\alpha = 1.1$, the plotted mean velocity is $v = 0.11$ m/d and $\mathcal{D} = 0.12\ \mathrm{m}^{1.1}$/d. Measured center of plume mass calculated by interpolation of 1D values plotted in Figures 4–6.

the plume (i.e., the peak and the measured mean) is diverging with increasing time. The confidence in the slope of the peak location is not high, so we do not consider this a diagnostic tool for estimation of α.

5. Discussion

Skewed and/or heavy-tailed plumes do not necessarily imply a heavy-tailed velocity distribution. Non-stationary velocity fields that arise from variable recharge or boundary conditions such as aquifer thinning can preferentially stretch certain portions of a plume. We have not analyzed the aquifer boundary conditions in detail. Nor have we investigated the applicability of more capable solutions than a local ADE affords. In particular, can the following methods be implemented and produce accurate predictions of the plume? (1) high-variance decomposition methods (Serrano, 1995); (2) nonlocal, second-order equations (e.g., Deng *et al.*, 1993); (3) multi-rate mobile/immobile models (e.g., Haggerty and Gorelick, 1995); or (4) numerical implementations using the most recent non-Gaussian characterizations of the K field (Liu and Molz, 1997a,b).

At the MADE site, the very high degree of K heterogeneity seems to support an assumption of a power-law density and a simple analytic model. The assumption of negligible J/η fluctuations should be most valid when the variability of K is greatest. Therefore, the accuracy of the fractional ADE without any knowledge of plume history should be greatest at highly heterogeneous sites. We make these assertions without proof that a power-law K distribution gives rise to power-law velocity field with the same exponent. The skewness of the MADE plume may have arisen due to its initial conditions. Since the plume was placed in low-permeability material, the propensity for faster transitions is much higher for quite some time. Had the slug been placed in a high permeability channel, the opposite might be expected, although recent work by Schumer *et al.* (1999) indicates that natural α-stable plumes should be maximally skewed ($\beta = \pm 1$) based on the K distribution alone.

The parameters α and $\mathcal{D}$ come from a 2-parameter K distribution function, the assumption of a characteristic transition time (τ), and (J/η). The parameter $\mathcal{D}$ in a second-order equation comes from the variance of ln(K), the functional form of the autocorrelation (either a characteristic length or a simple function of separation), (J/η), and the local transverse dispersion coefficient. To model a skewed plume with any order equation, one or more additional parameters are needed: (β) for the fractional ADE and a mass transfer rate coefficient and some apportionment of the porosity for mobile/immobile formulations (Haggerty and Gorelick, 1995). Thus, the fractional equation (7) requires the same amount of information or less to produce qualitatively similar predictions as local second-order formulations. The distinguishing characteristic of the fractional solution is the prediction of at least one power-law tail in time and/or space (Benson, 1998).

Tests at mildly heterogeneous sites (Cape Cod) and even within a highly homogeneous sandbox show some of the hallmarks of α-stable plume development (Benson, 1998; Benson *et al.*, 1999b). This is not totally surprising, since random walks by conservative solutes without power-law temporal correlation should converge to some α-stable distribution, based on the GCLT (1). A Gaussian stable plume is a subset of these. The power-law K distribution must have an upper cutoff, hence a finite variance. Since the K distribution reports average velocity in finite volumes, one must assume that actual particle velocity is truncated by a much higher and unknowable upper bound. Mantegna and Stanley (1995) show that truncated α-stable motions stay similarly α-stable for very long times before quickly changing to a Gaussian, particularly when α is much lower than two. Painter (1997) observes this point and gives evidence that a plume might have to travel huge distances before integrating enough of these cutoff transitions. These studies lend support for a fractional ADE formulation for many conservative plumes. The fractional approach has also been shown to work well in unsaturated porous media (Pachepsky, 1998; Pachepsky *et al.*, 1999), where self-generating channels foster similar long-range particle velocity dependence.

6. Conclusions

In some approaches to modeling field-scale transport, such as small perturbation of a stochastic second-order ADE, the dispersive growth process is still fundamentally Gaussian; super-diffusive behavior is forced by scale-dependent macrodispersion parameters. If the actual growth process is stable but non-Gaussian, then such methods cannot accurately simulate dispersive growth at any scale. This is because the Gaussian fundamental solution cannot reproduce the heavy tails or scale correctly with time. The non-Fickian behavior of the fractional ADE is obtained by matching the order of differentiation of the dispersive derivative to the exponent of the growth process. The fundamental solution to the fractional ADE produces non-Gaussian, heavy-tailed densities (plumes) that simultaneously predict super-Fickian transport and the heavy breakthrough curve tails that are ubiquitously found at field sites and often attributed to channeling, dead-end pores, nonlinear adsorption, and/or zero permeability zones.

There are no restrictions on the fractional-order theory in terms of the variance of $\log(K)$ – it is based on an infinite variance velocity model. The theory is relatively simple to apply: the parameters are constants that can be estimated from early-time plume data or the underlying K data. In contrast, the early plume behavior gives no indication of the late-time macrodispersivity used in a second-order equation. At the MADE site, the fractional ADE gave better predictions with a very low information requirements: a simple equation reproduces very complex behavior.

Using the K data as a surrogate for velocity, the fractional equation parameters were accurately estimated at the MADE site. Our analysis of the K increments at the MADE site using robust tail estimators indicates a power-law of order $\alpha = 1.1$. The power-law tail velocity field imparts instantaneous solute displacements that are power-law tailed as well. These displacements converge to an α-stable plume that is modeled by a fractional dispersion term. This is reflected in the measurements of the MADE-2 tritium plume. Finally, the plume is positively and maximally skewed, which is simply handled by fractional derivatives and the α-stable fundamental solutions.

Acknowledgements

D.B. received support from the U.S. Dept. of Energy – Basic Energy Sciences grant #DE-FG03-98ER14885. R.S. was supported by the Sulo and Aileen Maki Fellowship from the Desert Research Institute. We thank Chunmiao Zheng at the University of Alabama and the Electric Power Research Institute (Palo Alto) for providing the MADE data.

Appendix A. Derivation of the 1D Fractional ADE

Derivation of a Fokker–Planck equation (FPE) starts with a simple mathematical statement of how a random measure changes state from one moment to the next, after some event has occurred. In this case, we are interested in the probability that a particle has moved from location x_1 to x_3 in the time t_1 to t_3, or

$p(x_3 - x_1, t_3 - t_1)$. This probability is referred to as the transition density, which is conditional on the initial position (x_1) and the time interval. The particle must move through an intermediate location x_2, so this probability can be found by summing over all possible intermediate points x_2. The probability of making both transitions (x_1 to x_2 to x_3) is the product of the single-transition probabilities, giving the Chapman–Kolmogorov equation (Sheshadri and West, 1982; Zaslavsky, 1994):

$$p(x_3 - x_1, t_3 - t_1) = \int p(x_3 - x_2, t_3 - t_2)p(x_2 - x_1, t_2 - t_1)\,\mathrm{d}x_2. \tag{A.1}$$

The particle's initial condition of being at position $x_1 = 0$ at time $t_1 = 0$ is placed into the density to give the probability of a particle's position at any time after moving for time (t):

$$p(x - x_0, t) = \int p(x - x_2, t - t_2)p(x_2 - x_0, t_2)\,\mathrm{d}x_2. \tag{A.2}$$

Note that, in general, the wandering particle is described by a joint space and time density (e.g., Shlesinger *et al.*, 1982; Klafter *et al.*, 1987; Berkowitz and Scher, 1995, 1998) that may be impossible to measure in a field situation. One can simplify by using the conditional space density for a specific time interval (Sheshadri and West, 1982; Zaslavsky, 1994; Benson, 1998; Meerschaert *et al.*, 1999) if one assumes that spatial dependence is much heavier-tailed than temporal correlation. This leads to the Markov assumption that the movement of a particle (a possibly very distant movement) is independent of past movements. We choose an equation that tracks the location of a particle at regular intervals of Δt, so the transition density $p(x - x_2|\Delta t)$ is now specific to (conditioned on) regular intervals, since taking the limit as Δt goes to zero leads to a differential equation, while the former, in general, does not (Benson, 1998). The 'propagator', or Green function, is the density describing the position of a particle at any time (t) having started from $x_0 = 0$ and is given the shorter notation $p(x - x_0|t - t_0) = P(x, t)$. This leads to the density of the particle position based on the present position and the transition density:

$$P(x, t + \Delta t) = \int p(x - \zeta|\Delta t)P(\zeta, t)\,\mathrm{d}\zeta. \tag{A.3}$$

Naturally, this is a convolution, since the density of a sum of two random variables (i.e., the distance traveled by a random walker from time zero to t and then again from time t to $t + \Delta t$) is the convolution of the densities of the summands (e.g., Ross, 1988). A requirement on the relative size of the total time (t) versus the transition time (Δt) is sometimes needed. For any general transition density, Δt must be much smaller than the total time, and Δt is called a *mixing time*. This argument is required so that an appropriate limit theorem such as the GCLT (1) can be invoked as a good approximation of the integration of a large number of transitions into a limit distribution. If the transitions are iid *stable* variables, then no restrictions are placed on the transition time and the convolution is always satisfied. For example, iid Gaussian transitions result immediately in a Gaussian propagator.

By taking infinitesimal values of Δt in (A.3), we will know the change in $P(x,t)$ over a very short time period, leading to a differential equation. One should expect that a particle that travels along fractal paths and/or requires power-law times to complete individual trajectories (e.g., Compte, 1996) will have different limiting behavior than a typical Gaussian process. For example, Hughes *et al.* (1981) show that fractal motions on a lattice have a heavy-tailed transition density. The integer time derivative of $P(x,t)$ is represented:

$$\frac{\partial P(x,t)}{\partial t} = \lim_{\Delta t \to 0} \frac{1}{\Delta t} \left(\int p(x-\zeta \,|\, \Delta t)\, \mathrm{d}\zeta - P(x,t) \right). \tag{A.4}$$

The convolution is simplified by taking Fourier transforms:

$$\frac{\mathrm{d}}{\mathrm{d}t} \hat{P}(k,t) = \lim_{\Delta t \to 0} \frac{\hat{P}(k,t)}{\Delta t} (\hat{p}(k|\Delta t) - 1), \tag{A.5}$$

where $\hat{p}(k|\Delta t)$ is the Fourier transform of the instantaneous particle transition density. Merely identifying a functional form for the instantaneous transition density and placing it in this equation gives an expression for the probability density of the particle at any place and time. Included in this functional form is the relationship of the density to time. A linear relationship is typically assumed, implying that a particle will travel twice as far in twice the time. Other relationships require a more complicated treatment with respect to time (e.g., fractional time derivatives, see Zaslavsky, 1994; Saichev and Zaslavsky, 1997; Benson, 1998). For nonreactive tracers, we shall assume this linear scaling. To illustrate, select a finite-variance transition density based on the cumulant expansion:

$$\hat{p}(k|\Delta t) = 1 + A_1(\mathrm{i}k) + \tfrac{1}{2} A_2(\mathrm{i}k)^2 + A_n(\mathrm{i}k)^n + \cdots \tag{A.6}$$

where $A_1 = -v\Delta t + \sigma(\Delta t)$, $A_2 = \sigma^2 \Delta t + \sigma(\Delta t)$, and $A_n = \sigma(\Delta t)$ for $n > 2$. The terms $\sigma(\Delta t)$ contain all trailing terms with powers of Δt greater than unity. The term σ^2 must have units of L^2/T so that the standard deviation of the jump size is L^2. This requirement provides the linear link between the travel variance (of a particle or plume) and time. These coefficients define the requirements for the transition density that generates the classical ADE – it must have a finite-variance and higher-order moments that essentially disappear as Δt becomes very small. The third moment describes skewness, so highly skewed transitions (finite-variance or not) may not be well modeled by the ADE until time becomes very large and the transition time Δt become proportionally small. Placing (A.6) into (A.5) and taking the limit, we have

$$\frac{\mathrm{d}}{\mathrm{d}t} \hat{P}(k,t) = -v(\mathrm{i}k)\hat{P}(k,t) + (\mathrm{i}k)^2 \frac{\sigma^2}{2} \hat{P}(k,t) \tag{A.7}$$

which is easily inverted using the property of Fourier transforms $\mathcal{F}^{-1}[(\mathrm{i}k)^n \hat{g}(k)] = \mathrm{d}^n/\mathrm{d}x^n g(x)$:

$$\frac{\partial}{\partial t}P(x,t) = -v\frac{\partial}{\partial x}P(x,t) + \frac{\sigma^2}{2}\frac{\partial^2}{\partial x^2}P(x,t) \tag{A.8}$$

One can develop a parallel theory for the heavy-tailed (infinite variance) case, just as a GCLT (1) arose from the classic CLT. The Fourier transform of the Lévy density can be written:

$$\hat{f}(k) = \exp[q\mathcal{C}(-\mathrm{i}k)^\alpha + p\mathcal{C}(\mathrm{i}k)^\alpha - \mu \mathrm{i}k], \quad \alpha \neq 1 \tag{A.9}$$

where the relative weight of coefficients $p + q = 1$ describes the skewness, and $\mathcal{C}$ is the scale (the spread similar to the variance in the Gaussian). The transition density scales linearly with the transition interval Δt for $\Delta t \to 0$. This stable transition density has the Fourier transform:

$$\hat{p}(k|\Delta t) = \exp[q\mathcal{D}\Delta t(-\mathrm{i}k)^\alpha + p\mathcal{D}\Delta t(\mathrm{i}k)^\alpha - v\Delta t(\mathrm{i}k)], \quad \alpha \neq 1 \tag{A.10}$$

with an expansion of $\hat{p}(k|\Delta t) = 1 - v\Delta t(\mathrm{i}k) + q\mathcal{D}\Delta t(-\mathrm{i}k)^\alpha + p\mathcal{D}\Delta t(\mathrm{i}k)^\alpha + \sigma(\Delta t)$. This units of $\mathcal{D}$ are $\mathrm{L}^\alpha\mathrm{T}^{-1}$. This density has an infinite variance for $\alpha < 2$. Placing the expansion into (A.5) and taking the limit:

$$\frac{\mathrm{d}}{\mathrm{d}t}\hat{P}(k,t) = \hat{P}(k,t)[-v(\mathrm{i}k) + q\mathcal{D}(-\mathrm{i}k)^\alpha + p\mathcal{D}(\mathrm{i}k)^\alpha]. \tag{A.11}$$

One can easily solve this eigenvalue ODE and find that it is the equation of a 1D Lévy motion (see Samorodnitsky and Taqqu, 1994). But this equation admits a remarkable form, based on the Fourier transform of a fractional derivative. The fractional derivative can be defined (Samko *et al.*, 1993; Debnath, 1995; Benson *et al.*, 1999a) by extending the well-known action of Fourier transform on integer derivatives $\mathcal{F}^{-1}[(\mathrm{i}k)^n \hat{f}(k)] = (\mathrm{d}^n/\mathrm{d}x^n)f(x)$ to rational order: $\mathcal{F}^{-1}[(\mathrm{i}k)^\alpha \hat{f}(k)] = (\mathrm{d}^\alpha/\mathrm{d}x^\alpha)f(x)$. In the forward sense,

$$\mathcal{F}\left[\frac{\mathrm{d}^\alpha}{\mathrm{d}x^\alpha}f(x)\right] = (\mathrm{i}k)^\alpha \hat{f}(k).$$

If the derivative is taken with respect to $-x$, we substitute $-k$ for k:

$$\mathcal{F}\left[\frac{\mathrm{d}^\alpha f(x)}{\mathrm{d}(-x)^\alpha}f(x)\right] = (-\mathrm{i}k)^\alpha \hat{f}(k).$$

By inverse transform (Samko *et al.*, 1993; Benson *et al.*, 1999a), one finds that the fractional derivative has the 1D representation (note the change of sign inside the integral):

$$\frac{\mathrm{d}^q f(x)}{\mathrm{d}(\pm x)^q} = \frac{(\pm 1)^n}{\Gamma(n-q)}\frac{\mathrm{d}^n}{\mathrm{d}x^n}\int_0^\infty \xi^{n-q-1} f(x \mp \xi)\,\mathrm{d}\xi. \tag{A.12}$$

The real-space inverse transform for the propagator (A.11) is now

$$\begin{aligned}\frac{\partial}{\partial t}P(x,t) &= -v\frac{\partial}{\partial x}P(x,t)+p\mathcal{D}\frac{\partial^\alpha}{\partial x^\alpha}P(x,t)+\\ &\quad +q\mathcal{D}\frac{\partial^\alpha}{\partial(-x)^\alpha}P(x,t),\end{aligned} \tag{A.13}$$

or equivalently (Benson, 1998):

$$\frac{\partial P}{\partial t}=-v\frac{\partial P}{\partial x}+(1+\beta)\frac{\mathcal{D}}{2}\frac{\partial^\alpha P}{\partial x^\alpha}+(1-\beta)\frac{\mathcal{D}}{2}\frac{\partial^\alpha P}{\partial(-x)^\alpha}, \tag{A.14}$$

where the dimensions of $\mathcal{D}$ (the dispersion coefficient) are $L^\alpha T^{-1}$. This is valid for light- or heavy-tailed particle excursions in 1D. We have introduced the description 'fractional ADE' with the understanding that only the dispersion term is correctly described by a fractional derivative. For a large number of independent solute 'particles', the probability propagator is replaced by ensemble concentration $C(x,t)$ (cf., Fürth, 1956; Bhattacharya and Gupta, 1990), and the general governing equation for the expected concentration solute movement (the 1D fractional ADE) simplifies to (7). For Gaussian or other light-tailed random motions, $\alpha = 2$ and the classical ADE is recovered.

A number of methods can be used to generate the α-stable density solutions (9). Denote the density $f_{\alpha\beta\sigma\delta}(x)$ where α is the index of stability, β is the skewness, $\sigma = [|\cos(\pi\alpha/2)|\mathcal{D}t]^{1/\alpha}$ is the scale, and $\delta = vt$ is the shift. A standard density has $\sigma = 1$ and $\delta = 0$. A general density is related to the standard by $f_{\alpha\beta\sigma\delta}(x) = \sigma^{-1}f_{\alpha\beta10}((x-\delta)/\sigma)$. Given v and $\mathcal{D}$, one can specify any x and t, and find the standard density for the argument $(x-vt)/[|\cos(\pi\alpha/2)|\mathcal{D}t]^{1/\alpha}$. The canonical forms of the density assume many different definitions of a skewness parameter. Our choice of p is related to Samorodnitsky and Taqqu's (1994) common skewness parameter (β) by $\beta = 2p - 1$. A standard density can be found by integrating the expression (Benson, 1998):

$$f_{\alpha\beta10}(x)=\frac{|cx|^{1/(\alpha-1)}\alpha c}{2|1-\alpha|}\int_{-\theta}^{1}U_\alpha(\phi,\theta)\exp(-|cx|^{\alpha/(\alpha-1)}U(\phi,\theta))\,\mathrm{d}\phi, \tag{A.15}$$

where

$$c=[1+(\beta\tan(\pi\alpha/2))^2]^{-1/2\alpha}, \qquad \theta=\frac{2}{\pi\alpha}\tan^{-1}(\beta\tan(\pi\alpha/2)),$$

$$U_\alpha(\phi,\theta)=\left(\frac{\sin\frac{1}{2}\pi\alpha(\phi+\theta)}{\cos\frac{1}{2}\pi\phi}\right)^{\alpha/(1-\alpha)},$$

or by Feller's (1971) series for $1 < \alpha \leqslant 2$:

$$f_{\alpha\gamma10}(x)=\frac{1}{\pi x}\sum_{k=0}^{\infty}\frac{\Gamma(k\alpha^{-1}+1)}{k!}(-x)^k\sin\frac{\pi k}{2\alpha}(\gamma-\alpha), \tag{A.16}$$

where Feller's skewness parameter γ is obtained from β by Samorodnitsky and Taqqu (1994):

$$\gamma = \frac{2}{\pi} \arctan[\beta \tan(\pi(\alpha - 2)/2)]. \tag{A.17}$$

For symmetric densities, setting $\beta = \gamma = 0$ in (A.16) yields a formula that converges rapidly even for large arguments. The symmetric standard density for $1 < \alpha \leqslant 2$ is given by the series:

$$f_{\alpha 010}(x) = \frac{1}{\pi} \sum_{k=0}^{\infty} \frac{(-1)^k}{(2k+1)!} \Gamma\left(\frac{2k+1}{\alpha} + 1\right) x^{2k}. \tag{A.18}$$

Appendix B. Estimation of Power-Law Content

Several numerical and graphical routines are typically used to estimate the 'power-law' content, if any, in data sets. Here we discuss the basis for each estimator and the pitfalls encountered when using each on different types of data. Each estimator is applied to the MADE K data in raw and incremental form, which has serious effect on the stability of some estimators.

The family of stable distributions is parameterized by the tail thickness $0 < \alpha \leqslant 2$, the skewness $-1 \leqslant \beta \leqslant 1$, the location $-\infty < \mu < \infty$, and the scale $\sigma > 0$ (Equation (3)). There are several established methods for stable parameter estimation, which can be used to fit a stable probability distribution to a data set. McCulloch (1986) tabulates quantiles of a stable distribution as a function of the parameters, and uses these tables to interpolate the best-fitting parameter values. These rough estimates can also be used to initialize the optimization for maximum likelihood estimation (MLE). Although the stable density cannot be written in closed form, Nolan (1997) has developed fast numerical approximations that allow practical computation of the stable MLE on a personal computer. A computer program to compute both the McCulloch quantile estimators and the MLE, including diagnostics to check for distributional fit, is available at http://www.cas.american.edu/~jpnolan/.

Anderson and Meerschaert (1998) review parameter estimation methods for heavy tails and apply them to a river flow time series. For data sets that are known to be stable, the MLE is the best parameter estimation method. Robust estimators are more appropriate when the exact probability distribution of the data is unknown. The most popular robust estimator was introduced by Hill (1975). Hill's estimator is the conditional MLE for a Pareto distribution $P[X > x] = Wx^{-\alpha}$ based on the largest order statistics. This model is appropriate because the upper tail of a stable distribution is approximately Pareto with the same index (α) (e.g., Samorodnitsky and Taqqu, 1994). A recent extension of Hill's estimator by Aban and Meerschaert (1999) fits a shifted Pareto $P[X > x] = W(x - s)^{-\alpha}$ using the same method. Since Hill's original estimator is not shift-invariant, this new method is more reliable.

Table II. Estimates of K and ΔK tail parameter α using four estimators. Due to the robust methodology of the Meerschaert and Scheffler (M&S) method, a value of $\alpha = 1.1$ is thought to be reliable.

Estimator (type)	McCulloch (quantile stable)	Nolan (MLE stable)	Hill (MLE Pareto)	M&S (asymptotics)
Raw K	0.8	0.4	3.0	1.1
K increments	0.7	0.8	1.1	1.1

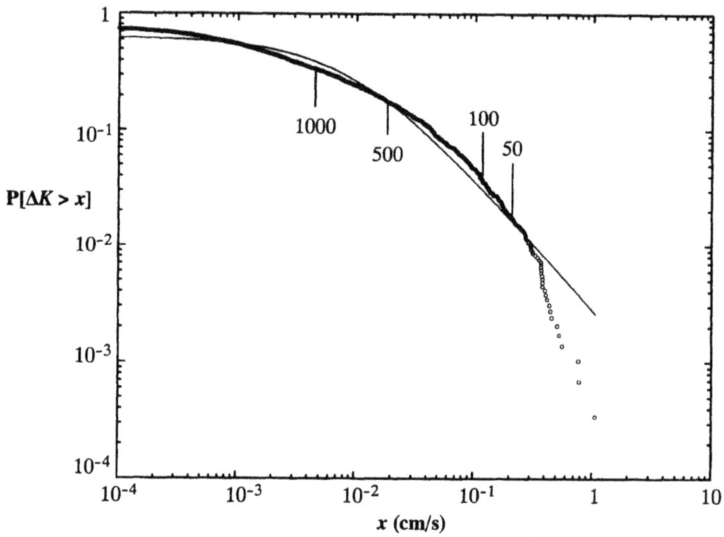

Figure 8. Mandelbrot plot of measured K increments at 15 cm lag ($|\Delta K|$) and shifted Pareto $P[|\Delta K| > x] = 0.0028(x + 0.01)^{-1.1}$ (solid line). Numbers on plot indicate rank of the datum.

Hosking and Wallis (1987) suggest using the MLE for the generalized Pareto distribution to fit hydrologic data. This two parameter model is equivalent to a Pareto with a specific shift that makes X range over the positive reals. McCulloch (1997) shows that Hill's estimator can yield misleading results when applied to stable data. Meerschaert and Scheffler (1998) suggest another robust estimator based on the asymptotics of the sum. This estimator performs about as well as Hill's estimator in most practical situations, and it performs best in exactly those situations (stable or nearly stable data with alpha near two) in which Hill's estimator is most likely to fail.

For the untransformed MADE K data the McCulloch quantile estimator gives $\alpha = 0.8$ and $\beta = 1$ (totally skewed to the right). Nolan's MLE gives $\alpha = 0.4$ and $\beta = 1.0$, indicating very heavy tails. A probability plot indicates significant deviation from the best fitting stable model, indicating that the data are probably not drawn from a stable distribution. This is a situation in which robust estimators are more appropriate. The generalized Pareto MLE yields $\alpha = 0.5$ but once again a probability plot indicates that there is not a good fit to this particular distribution. Hill's estimator uses the largest r observations out of $n = 2915$ for this data set. The resulting parameter estimates vary with r, for example we get $\alpha = 2.3$ at $r = 50$ and $\alpha = 1.0$ at $r = 500$, and in general, α steadily decreases with r, indicating a possible distributional misspecification. For the Pareto model $y = P[X > x] = Wx^{-\alpha}$ leads to $\ln(y) = \ln(W) - \alpha \ln(x)$. If we sort the data $X_1 \ldots X_n$ in descending order to obtain the order statistics $X(1) \geqslant \cdots \geqslant X(n)$, then we can estimate $y = i/n$ and $x = X(i)$. A log–log plot of i/n versus $X(i)$ should show a straight line with slope $-\alpha$ for Pareto data. This method was pioneered by Mandelbrot (1963) in his early work on heavy tails, and we will call this the Mandelbrot plot. Simple linear regression can be used to estimate (α) from the Mandelbrot plot, but this method typically performs poorly. More importantly, this graphical method can be used to test for fit to a Pareto model. The choice of r for Hill's estimator, based on the Pareto distribution, should be small enough that the Mandelbrot plot of the r largest order statistics appears linear. Since Hill's estimator is not shift-invariant, it is also reasonable to shift the data until the plot appears linear. This amounts to fitting a shifted Pareto. Visual inspection indicates that a shift of $s = -0.2$ cm/s (add 0.2 cm/s to each K observation) gives a Mandelbrot plot that appears linear, and Hill's estimator for the largest $r = 375$ observations yields $\alpha = 3$. The more exact shifted Hill's estimator (conditional MLE) yields $\alpha = 3.019$ at the optimal shift of $s = -0.1998$, so in this case, visual inspection is adequate. Overall, there is a wide discrepancy in α estimates (Table II), due to mismatches of estimation technique with the data distribution. Estimators like the MLE, which are tailored to a specific distributional form, are not robust with respect to deviations from that probability distribution. After rescaling the data (divide by the median), the robust estimator of Meerschaert and Scheffler yields $\alpha = 1.1$ and in this case we consider this our best parameter estimate.

To test the fits to a specific distribution, we also analyzed the variations in K. It is typical in econometrics to difference the data in this way (subtract one observation from the next) to unmask the effects of heavy tails. Theoretically speaking, the differences should exhibit the same tail thickness parameter (α) as the original data set (e.g., Davis and Resnick, 1985). For our data set, the McCulloch quantile method gives $\alpha = 0.7$ and $\beta = 1$ and Nolan's MLE gives $\alpha = 0.8$ and $\beta = 1$ for the differenced data, but once again, diagnostic plots indicate a significant deviation from a stable distribution. Meerschaert and Scheffler's robust estimator (again we rescale by the median) yields $\alpha = 1.1$, the same as for the raw data (Table I). Hill's estimator with $r = 150$ also gives $\alpha = 1.1$ and the Mandelbrot plot indicates that the fit to a shifted Pareto is good (Figure 8). In addition, Hill's estimator gives a scale coefficient $W = 0.0028\,(\text{cm/s})^{1.1}$. Since the conditional MLE of Aban and Meerschaert indicates that the optimal shift is near zero, the usual Hill's estimator should give good results. The generalized Pareto MLE is not reported here because we could not obtain convergence. Overall, it seems safe to say that the K data fit a heavy-tail model with $\alpha = 1.1$ indicating an infinite variance but a finite mean.

References

Aban, I. and Meerschaert, M.: 1999, Shifted Hill's estimator for heavy tails, Preprint.

Adams, E. E. and Gelhar, L. W.: 1992, Field study of dispersion in a heterogeneous aquifer, 2, Spatial moments analysis, *Water Resour. Res.* **28**(12), 3293–3307.

Anderson, P. and Meerschaert, M. M.: 1998, Modeling river flows with heavy tails, *Water Resour. Res.* **34**(9), 2271–2280.

Benson, D. A.: 1998, The fractional advection–dispersion equation: Development and application, Unpublished Ph.D. thesis, Univ. of Nevada, Reno, http://www.hydro.unr.edu/homepages/benson.

Benson, D. A., Wheatcraft, S. W. and Meerschaert, M. M.: 1999a, The fractional-order governing equation of Lévy motion, Preprint, http://www.hydro.unr.edu/homepages/benson.

Benson, D. A., Wheatcraft, S. W. and Meerschaert, M. M.: 1999b, Application of a fractional advection–dispersion equation, Preprint, http://www.hydro.unr.edu/homepages/benson.

Berkowitz, B. and Scher, H.: 1995, On characterization of anomalous dispersion in porous and fractured media, *Water Resour. Res.* **31**(6), 1461–1466.

Berkowitz, B. and Scher, H.: 1998, Theory of anomalous chemical transport in random fracture networks, *Phys. Rev. E* **57**(5), 5858–5869.

Bhattacharya, R. and Gupta, V. K.: 1990, Application of central limit theorems to solute transport in saturated porous media: from kinetic to field scales, Chapter IV, in: J. H. Cushman (ed.), *Dynamics of Fluids in Hierarchical Porous Media*, Academic Press.

Boggs, J. M. and Adams, E. E.: 1992, Field study of dispersion in a heterogeneous aquifer, 4; Investigation of adsorption and sampling bias, *Water Resour. Res.* **28**(12), 3325–3336.

Boggs, J. M., Beard, L. M., Long, S. E. and McGee, M. P.: 1993, Database for the second macrodispersion experiment (MADE-2), EPRI report TR-102072, Electric Power Res. Inst., Palo Alto, CA.

Brusseau, M.: 1992, Transport of rate-limited sorbing solutes in heterogeneous porous media: Application of a one-dimensional multifactor nonideality model to field data, *Water Resour. Res.* **28**(9), 2485–2497.

Compte, A.: 1996, Stochastic foundations of fractional dynamics, *Phys. Rev. E* **53**(4), 4191–4193.

Compte, A. and Caceres, M. O.: 1998, Fractional dynamics in random velocity fields, *Phys. Rev. Lett.* **81**, 3140–3143.

Crank, J.: 1975, *The Mathematics of Diffusion*, Oxford University Press, Oxford, Great Britain.

Dagan, G.: 1984, Solute transport in heterogeneous porous formations, *J. Fluid Mech.* **145**, 151–177.

Davis, R. and Resnick, S.: 1985, Limit theory for moving averages of random variables with regularly varying tail probabilities, *Ann. Probab.* **13**, 179–195.

Debnath, L.: 1995, *Integral Transforms and Their Applications*, CRC Press, New York.

Deng, F.-W., Cushman, J. H. and Delleur, J. W.: 1993, A fast Fourier transform stochastic analysis of the contaminant transport problem, *Water Resour. Res.* **29**(9), 3241–3247.

Einstein, A.: 1908, *Investigations on the Theory of the Brownian Movement*, translation by Dover Publications in 1956 of the original manuscript.

Feller, W.: 1971, *An Introduction to Probability Theory and Its Applications*, Volume II, 2nd ed., Wiley, New York.

Fofack, H. and Nolan, J.: 1998, Tail behavior, modes and other characteristics of stable distributions, Preprint, http://www.cas.american.edu/~jpnolan/.

Fogedby, H. C.: 1994, Lévy flights in random environments, *Phys. Rev. Lett.* **73**(19), 2517–2520.

Freyberg, D. L.: 1986, A natural gradient experiment on solute transport in a sandy aquifer, 2, Spatial moments and the advection and dispersion of nonreactive tracers, *Water Resour. Res.* **22**(13), 2031–2046.

Fürth, R.: 1956, Notes in: Einstein, A. E., *Investigations on the Theory of the Brownian Movement*, translation by Dover Publications.

Gelhar, L. W. and Axness, C. L.: 1983, Three-dimensional stochastic analysis of macrodispersion in aquifers, *Water Resour. Res.* **19**(1), 161–180.

Gnedenko, B. V. and Kolmogorov, A. N.: 1954, *Limit Distributions for Sums of Random Variables*, Addison-Wesley, Reading, Mass.

Gorenflo, R. and Mainardi, F.: 1998, Fractional calculus and stable probability distributions, *Arch. Mech* **50**(3), 377–388.

Grigolini, P., Rocco, A. and West, B. J.: 1999, Fractional calculus as a macroscopic manifestation of randomness, *Phys. Rev. E* **59**, 2603.

Haggerty, R. and Gorelick, S. M.: 1995, Multiple-rate mass transfer for modeling diffusion and surface reactions in media with pore-scale heterogeneity, *Water Resour. Res.* **31**(10), 2383–2400.

Hill, B.: 1975, A simple general approach to inference about the tail of a distribution, *Ann. Statist.* 1163–1173.

Hosking, J. and Wallis, J.: 1987, Parameter and quantile estimation for the generalized Pareto distribution, *Technometrics* **29**, 339–349.

Hughes, B. D., Shlesinger, M. F. and Montroll, E. W.: 1981, Random walks with self-similar clusters, *Proc. Natl. Acad. Sci. USA* **78**(6), 3287–3291.

Janicki, A. and Weron, A.: 1994, Can one see α-stable variables and processes?, *Stat. Sci.* **9**(1), 109–126.

Klafter, J., Blumen, A. and Shlesinger, M. F.: 1987, Stochastic pathway to anomalous diffusion, *Phys. Rev. A* **35**(7), 3081–3085.

LeBlanc, D. R., Garabedian, S. P., Hess, K. M., Gelhar, L. W., Quadri, R. D., Stollenwerk, K. G. and Wood, W. W.: 1991, Large-scale natural gradient tracer test in sand and gravel, Cape Cod, Massachusetts, 1, Experimental design and observed tracer movement, *Water Resour. Res.* **27**(5), 895–910.

Lévy, P.: 1937, *Théorie de L'addition des Variables Aléatoires*, Gauthier-Villars, Paris.

Liu, H. H. and Molz, F. J.: 1997a, Comment on 'Evidence for non-Gaussian scaling behavior in heterogeneous sedimentary formations' by Scott Painter, *Water Resour. Res.* **33**(4) 907–908.

Liu, H. H. and Molz, F. J.: 1997b, Multifractal analyses of hydraulic conductivity distributions, *Water Resour. Res.* **33**(11) 2483–2488.

Mandelbrot, B.: 1963, The variation of certain speculative prices, *J. Business* **36**, 394–419.

Mantegna, R. N. and Stanley, H. E.: 1995, Ultra-slow convergence to a Gaussian: the truncated Lévy flight, in: M. F. Shlesinger, G. M. Zaslavsky and U. Frisch (eds), *Lévy Flights and Related Topics in Physics*, Springer-Verlag, pp. 301–312.

McCulloch, J. H.: 1986, Simple consistent estimators of stable distribution parameters, *Comm. Statist. Simul. Comput.* **15**, 1109–1136.

McCulloch, J. H.: 1997, Measuring tail thickness to estimate the stable index alpha: A critique, *J. Business Econ. Statist.* **15**, 74–81.

Meerschaert, M.: 1986, Regular variation and domains of attraction in $\mathbb{R}^k$, *Stat. Prob. Lett.* **4**, 43–45.

Meerschaert, M. and Scheffler, H.-P.: 1998, A simple robust estimator for the thickness of heavy tails, *J. Stat. Plann. Inference* **71**(1–2), 19–34.

Meerschaert, M. M., Benson, D. A. and Bäumer, B.: 1999, Multidimensional advection and fractional dispersion, *Phys. Rev. E* **59**(5) 5026–5028.

Metzler, R., Klafter, J. and Sokolov, I. M.: 1998, Anomalous transport in external fields: Continuous time ransom walks and fractional diffusion equations extended, *Phys. Rev. E* **58**, 1621–1633.

Metzler, R., Barkai, E. and Klafter, J.: 1999, Deriving fractional Fokker–Planck equations from a generalized master equation, *Europhys. Lett.* **46**, 431–436.

Molz, F. J., Liu, H. H. and Szulga, J.: 1997, Fractional Brownian motion and fractional Gaussian noise in subsurface hydrology: A review, presentation of fundamental properties, and extensions, *Water Resour. Res.* **33**(10), 2273–2286.

Nolan, J.: 1997, Numerical calculation of stable densities and distribution functions: Heavy tails and highly volatile phenomena, *Comm. Statist. Stoch. Models* **13**, 759–774.

Nolan, J.: 1998, Parameterizations and modes of stable distributions, *Statist. Probab. Lett.* **38**(2), 187–195.

Oldham, K. B. and Spanier, J.: 1974, *The Fractional Calculus*, Academic Press, New York.

Pachepsky, Y. A.: 1998, Transport of water and chemicals in soils as in fractal media, *Agronomy Abstracts*, p. 202.

Pachepsky, Y. A., Benson, D. A. and Rawls, W.: 1999, Simulating scale-dependent solute transport in soils with the fractional advective-dispersive equation, Preprint.

Painter, S.: 1996a, Evidence for non-Gaussian scaling behavior in heterogeneous sedimentary formations, *Water Resour. Res.* **32**(5), 1183–1195.

Painter, S.: 1996b, Stochastic interpolation of aquifer properties using fractional Lévy motion, *Water Resour. Res.* **32**(5), 1323–1332.

Painter, S.: 1997, Reply to comment on 'Evidence for non-Gaussian scaling behavior in heterogeneous sedimentary formations' by H. H. Liu and F. J. Molz, *Water Resour. Res.* **33**(4) 909–910.

Rajaram, H. and Gelhar, L. W.: 1991, Three-dimensional spatial moments analysis of the Borden tracer test, *Water Resour. Res.* **27**(6), 1239–1251.

Rehfeldt, K. R., Boggs, J. M. and Gelhar, L. W.: 1992, Field study of dispersion in a heterogeneous aquifer. 3: Geostatistical analysis of hydraulic conductivity, *Water Resour. Res.* **28**(12), 3309–3324.

Rocco, A. and West, B. J.: 1999, *Physica A* **265**, 535.

Ross, S.: 1988, *A First Course in Probability*, 5th ed., Prentice Hall, NY.

Saichev, A. I. and Zaslavsky, G. M.: 1997, Fractional kinetic equations: solutions and applications, *Chaos* **7**(4), 753–764.

Samko, S. G., Kilbas, A. A. and Marichev, O. I.: 1993, *Fractional Integrals and Derivatives: Theory and Applications*, Gordon and Breach.

Samorodnitsky, G. and Taqqu, M. S.: 1994, *Stable Non-Gaussian Random Processes: Stochastic Models with Infinite Variance*, Chapman and Hall, New York.

Schumer, R., Benson, D., Meerschaert, M. and Wheatcraft, S.: 1999, A physical derivation of the fractional advection–dispersion equation, Preprint, http://www.hydro.unr.edu/homepages/benson.

Serrano, S. E.: 1995, Forecasting scale-dependent dispersion from spills in heterogeneous aquifers, *J. Hyd.* **169**, 151–169.

Sheshadri, V. and West, B. J.: 1982, Fractal dimensionality of Levy processes, *Proc. Natl. Acad. Sci.* **79**, 4501–4505.

Shlesinger, M. F., Klafter, J. and Wong, Y. M.: 1982, Random walks with infinite spatial and temporal moments, *J. Stat. Phys.* **27**(3), 499–512.

Sudicky, E. A.: 1986, A natural gradient experiment on solute transport in a sandy aquifer: Spatial variability of hydraulic conductivity and its role in the dispersion process, *Water Resour. Res.* **22**(13), 2069–2082.

Taylor, Sir, G. I.: 1953, Dispersion of soluble matter in solvent flowing slowly through a tube, *Proc. R. Soc. A., London* **219**, 186–203.

Zaslavsky, G. M.: 1994, Renormalization group theory of anomalous transport in systems with Hamiltonian chaos, *Chaos* **4**(1), 25–33.

Zheng, C. and Jiao, J. J.: 1998, Numerical simulation of tracer tests in heterogeneous aquifer, *J. Environ. Eng.* **124**(6), 510–516.

Transport in Porous Media **42:** 241–263, 2001.

The Role of Probabilistic Approaches to Transport Theory in Heterogeneous Media

BRIAN BERKOWITZ and HARVEY SCHER
Department of Environmental Sciences and Energy Research, Weizmann Institute of Science, Rehovot 76100, Israel

(Received: 10 March 1999; in final form: 28 December 1999)

Abstract. A physical picture of contaminant transport in highly heterogeneous porous media is presented. In any specific formation the associated governing transport equation is valid at any time and space scale. Furthermore, the advective and dispersive contributions are inextricably combined. The ensemble average of the basic transport equation is equivalent to a continuous time random walk (CTRW). The connection between the CTRW transport equation, in a limiting case and the familiar advection–dispersion equation (ADE) is derived. The CTRW theory is applied to the results of laboratory experiments, field observations, and simulations of random fracture networks. All of these results manifest dominant non-Gaussian features in the transport, over different scales, which are accounted for quantitatively by the theory. The key parameter β controlling the entire shape of the contaminant plume evolution and breakthrough curves is advanced as a more useful characterization of the transport than the dispersion tensor, which is based on moments of the plume. The role of probabilistic approaches, such as CTRW, is appraised in the context of the interplay of spatial scales and levels of uncertainty. We then discuss a hybrid approach, which uses knowledge of non-stationary aspects of a field site on a larger spatial scale (trends) with a probabilistic treatment of unresolved structure on a smaller scale (residues).

Key words: contaminant transport, heterogeneous media, continuous time, random walk, CTRW, non-Gaussian transport.

1. Introduction

Most natural aquifers contain heterogeneities with statistically complex morphologies. Steady water flow through highly heterogeneous aquifers gives rise to a strongly varying velocity field with multiscale coherence lengths. The key problem we address is how to characterize tracer movement in such a field. This problem presents an immediate difficulty. As there are different levels of uncertainty in our characterization of these heterogeneities we need to learn how to best deal with what we do and do not know. One promising approach involves the intensive efforts now being devoted to measurement and delineation of aquifer properties, usually the hydraulic conductivity, at as high a resolution as possible. The idea is to resolve the hydraulic conductivity (or velocity) field at a sufficiently high level, say $\sim 10\,m^3$ blocks in a formation extending over hundreds of meters or more. One can then predict contaminant migration in the formation by application of a

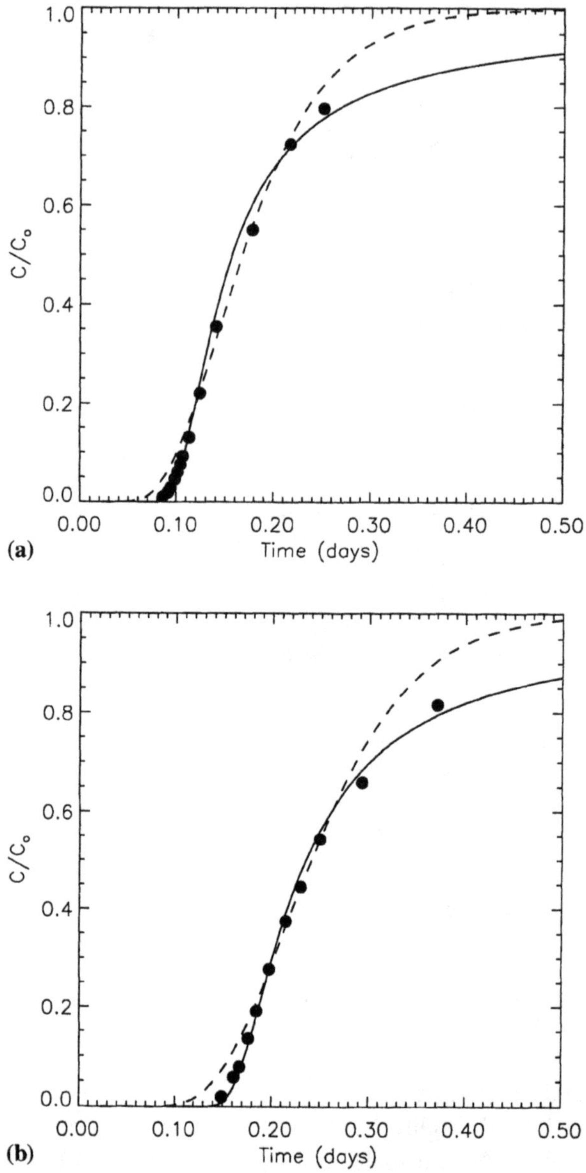

Figure 1. Measured concentrations (filled circles), at three distances from the tracer source, in a (two-dimensional) laboratory flow cell (data from Silliman and Simpson, 1987). The flow cell, 2.13 m in length, was packed with a uniform coarse sand interspersed with randomly-placed, fine sand lenses. The dashed lines show solutions of the ADE using 'best' estimated values of dispersivity (which vary) at each distance from the inlet. The solid lines show FPTD ('anomalous transport') solutions (see Sections 2.3 and 3.1) , with $\beta = 0.87 \pm 0.01$. Distance, L, from tracer source: (a) $L = 0.91$ m, (b) $L = 1.37$ m, and (c) $L = 1.83$ m (after Berkowitz *et al.*, 2000).

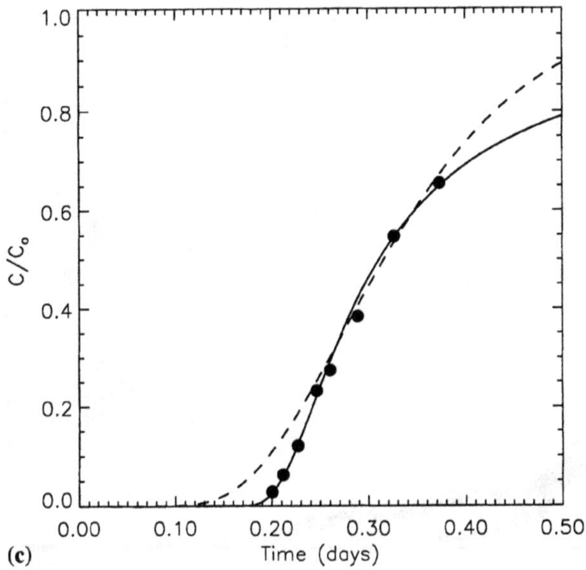

(c)

Figure 1. *(continued)*

numerical code that incorporates an advection–dispersion transport model at the scale of these blocks (e.g., Eggleston and Rojstaczer, 1998; LaBolle *et al.*, 1998).

However, we must ask ourselves whether all aspects of this approach are correct. The critical assumption is that if one can capture the larger scale 'trends', the advection–dispersion equation (ADE) can be applied at the resolution of smaller scale blocks to deal with the 'residues'. In contrast, as demonstrated frequently in the literature, the ADE does not adequately capture contaminant migration even in many 'homogeneous' systems, evidently because heterogeneities which cannot be ignored are present at all scales. These residues can cause transport in these regions to be non-Fickian. Throughout this paper we shall use the term 'anomalous' to describe such transport behavior. Notable examples of anomalous transport in stationary heterogeneous systems are shown in Figures 1 and 2: studies of laboratory flow cells containing porous media (Silliman and Simpson, 1987), and numerical studies of random fracture networks (Berkowitz and Scher, 1998), respectively. These results are discussed further in Section 3.

The inevitable decision of what equations to use, and on what spatial scale, must be a pragmatic one (e.g., Gelhar, 1993). On the one hand we should not use an ensemble approach to characterized, non-stationary structures, but on the other hand we should not 'homogenize' a domain containing features below our level of resolution. Hence, a probabilistic approach incorporating the unresolved heterogeneities should be integrated with the different scales of resolved aquifer features. We first develop our basic physical picture for a specific formation in the next section. The ensemble average of the resulting transport equation is then shown to be equivalent to a continuous random walk (CTRW) and is the basis of

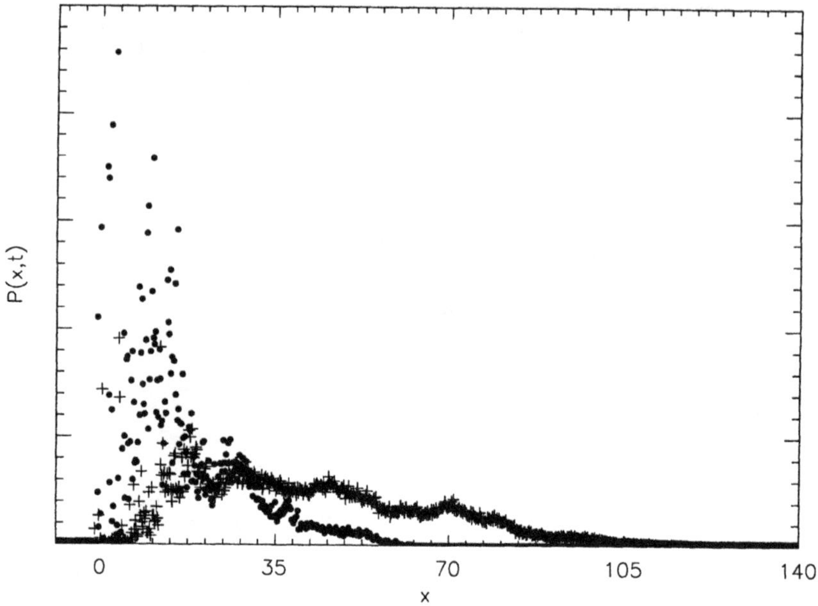

Figure 2. Profile of the spatial distribution of tracer at two relative times ($t = 1$, points; $t = 2.5$, crosses), determined from numerical simulations of particle transport in a two-dimensional, random fracture network (after Berkowitz and Scher, 1997). The concentration profile, $P(x, t)$, is the vertical average (along the y axis) of $P(\mathbf{s}, t)$, in arbitrary units; particle injection point is at $x = 0$.

our probabilistic approach. The connection between the transport equation of this approach, in a limiting case, and the familiar advection–dispersion equation (ADE) is derived. We subsequently demonstrate the significance of our transport model in the analysis of laboratory, field and simulation studies of a variety of heterogeneous media. We then return to the question of how to use an ensemble averaged equation when we have knowledge of important non-stationary features of a field site. The complete answer to this question is work in progress.

2. The Physical Framework of the Transport Equation

The picture of contaminant motion in a strongly varying velocity field can be envisioned as each particle executing a series of steps or transitions between locales where a change in velocity, $\mathbf{v}$, occurs, as shown schematically in Figure 3. The transport is controlled by the coherence lengths of the velocity field and the scattering among different velocity values. The lengths of the arrows shown in Figure 3 are a measure of these coherence lengths, while the arrow widths are a measure of the local concentration of particles. Note that particles in high and moderate velocity paths interact sporadically with low velocity regions. This occurrence has a strong influence on the possible non-Fickian behavior of the transport. We em-

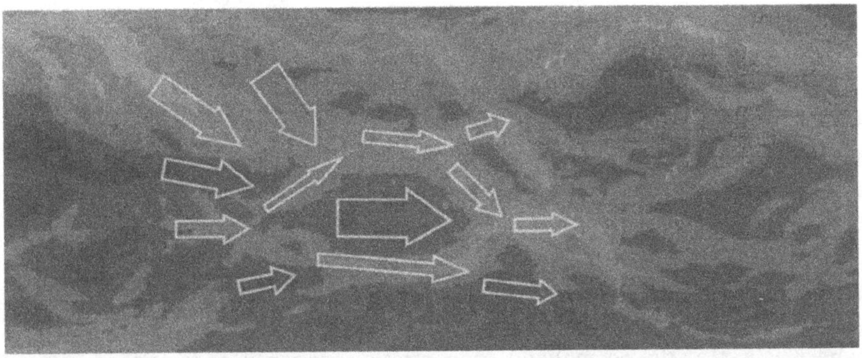

Figure 3. Schematic illustration of contaminant transport in heterogeneous porous media. The underlying velocity field map is representative of flow in such media (adapted from Grindrod and Impey, 1993); lighter scale represents regions of higher velocity. The white arrows indicate a conceptual mapping of contaminant flow paths onto a mixture of discrete paths with varying lengths and ranges of transition times (after Berkowitz *et al.*, 2000).

phasize that the presence of a large range of velocities is necessary but not sufficient to produce anomalous transport. As discussed by Berkowitz and Scher (1995), the encounter of the particles with this range of velocities is crucial. Anomalous transport can arise if the encounter-range relationship produces a wide spread of different sequences in the flow paths for the migrating particles.

A transport equation that can enumerate all these possible paths and cover the motion from continuous to discrete over a range of spatial and temporal scales is

$$\frac{\partial C(\mathbf{s}, t)}{\partial t} = -C(\mathbf{s}, t) \sum_{\mathbf{s}'} w(\mathbf{s}, \mathbf{s}') + \sum_{\mathbf{s}'} w(\mathbf{s}', \mathbf{s}) C(\mathbf{s}', t) \tag{1}$$

the 'Master Equation' (Oppenheim *et al.*, 1977; Shlesinger, 1996) for $C(\mathbf{s}, t)$, the particle concentration at point $\mathbf{s}$ and time t, where $w(\mathbf{s}, \mathbf{s}')$ is the transition rate from $\mathbf{s}$ to $\mathbf{s}'$. The transition rates describe the effects of the velocity field on the particle motion; the transport equation in (1) does not separate these effects into advective and dispersive features. The determination of $w(\mathbf{s}, \mathbf{s}')$ involves a detailed knowledge of the system. In relatively homogeneous regions the $C(\mathbf{s}, t)$ will be slowly varying over some length scale s_o and (1) in this region can be reduced to the familiar ADE equation (this procedure will be shown below for the ensemble-averaged version of (1)) with a local average velocity $\mathbf{v}$ and dispersivity $\mathcal{D}$. However, in regions of heightened heterogeneity, such as tightly interspersed permeability layers, $C(\mathbf{s}, t)$ will not necessarily vary slowly on the same scale s_o, and/or the volume needed for averaging can change dramatically from point to point. Conversely, if one fixes the volume to a practical pixel size (e.g., $10\,\mathrm{m}^3$) the use of a local average $\mathbf{v}$ and $\mathcal{D}$ in each volume can be quite limited. We will return to this issue in a broader context later. It essentially involves the degrees of uncertainty and its associated spatial scales. We start, at first, with systems containing an approximately stationary distribution of heterogeneities, which one assumes can be treated with an ensemble average.

The ensemble average of (1) can be shown (Klafter and Silbey, 1980) to be of the form

$$\frac{\partial P(\mathbf{s},t)}{\partial t} = -\sum_{\mathbf{s}'}\int_0^t \phi(\mathbf{s}'-\mathbf{s},t-t')P(\mathbf{s},t')\,\mathrm{d}t' + \\ +\sum_{\mathbf{s}'}\int_0^t \phi(\mathbf{s}-\mathbf{s}',t-t')P(\mathbf{s}',t')\,\mathrm{d}t', \tag{2}$$

where $P(\mathbf{s},t)$ is the normalized concentration, and $\phi(\mathbf{s},t)$ is defined below in (7). The form of (2) is a 'generalized master equation' (GME) which, in contrast to (1), is non-local in time and the transition rates are stationary (that is, depend only on the difference $\mathbf{s}-\mathbf{s}'$) and are time-dependent. It is straightforward to show (Kenkre *et al.*, 1973; Shlesinger, 1974), using the Laplace transform, that the GME is equivalent to a CTRW

$$R(\mathbf{s},t) = \sum_{\mathbf{s}'}\int_0^t \psi(\mathbf{s}-\mathbf{s}',t-t')R(\mathbf{s}',t')\,\mathrm{d}t', \tag{3}$$

where $R(\mathbf{s},t)$ is the probability per time for a walker to just arrive at site $\mathbf{s}$ at time t, and $\psi(\mathbf{s},t)$ is the probability rate for a displacement $\mathbf{s}$ with a difference of arrival times of t. The correspondence between (2) and (3) is

$$P(\mathbf{s},t) = \int_0^t \Psi(t-t')R(\mathbf{s},t')\,\mathrm{d}t', \tag{4}$$

where

$$\Psi(t) = 1 - \int_0^t \psi(t')\,\mathrm{d}t' \tag{5}$$

is the probability for a walker to remain on a site,

$$\psi(t) \equiv \sum_{\mathbf{s}} \psi(\mathbf{s},t) \tag{6}$$

and

$$\tilde{\phi}(\mathbf{s},u) = \frac{u\tilde{\psi}(\mathbf{s},u)}{1-\tilde{\psi}(u)}, \tag{7}$$

where the Laplace transform of a function $f(t)$ is denoted by $\tilde{f}(u)$. The CTRW accounts naturally for the cumulative effects of a sequence of transitions. The challenge is to map the important aspects of the particle motion in the medium onto a $\psi(\mathbf{s},t)$. The identification of $\psi(\mathbf{s},t)$ lies at the heart of the CTRW formulation. The CTRW approach allows a determination of the evolution of the particle distribution (plume), $P(\mathbf{s}, t)$, for a general $\psi(\mathbf{s}, t)$; there is no a priori need to consider only the moments of $P(\mathbf{s}, t)$. As we discuss below, a $\psi(\mathbf{s}, t)$ with an algebraic tail (cf. (15)) for large time leads to the description of anomalous transport (e.g., non-Gaussian plumes). Once $\psi(\mathbf{s}, t)$ is defined, (3)–(5), which are in the form of a convolution in space and time, can be solved by use of Fourier and Laplace transforms.

One way to separate advection and dispersion and also show the link between our physical picture and the basis for the ADE is to make series expansions of $P(\mathbf{s}, t)$ and $\tilde{\psi}(\mathbf{s}, u)$. The first approximation is based on assuming that $P(\mathbf{s}, t)$ varies sufficiently slowly in space that

$$P(\mathbf{s}', t) \approx P(\mathbf{s}, t) + (\mathbf{s}' - \mathbf{s}) \cdot \nabla P(\mathbf{s}, t) + \tfrac{1}{2}(\mathbf{s}' - \mathbf{s})(\mathbf{s}' - \mathbf{s}) : \nabla\nabla(\mathbf{s}, t) \tag{8}$$

with the dyadic symbol : denoting a tensor product. Inserting (8) in (2) yields

$$\begin{aligned} \frac{\partial P(\mathbf{s}, t)}{\partial t} &= \int_0^t \mathrm{d}t' \sum_{\mathbf{s}'} \phi(\mathbf{s} - \mathbf{s}', t - t')(\mathbf{s}' - \mathbf{s}) \cdot \nabla P(\mathbf{s}, t') + \\ &+ \int_0^t \mathrm{d}t' \sum_{\mathbf{s}'} \phi(\mathbf{s} - \mathbf{s}', t - t')\frac{1}{2}(\mathbf{s}' - \mathbf{s})(\mathbf{s}' - \mathbf{s}) : \nabla\nabla P(\mathbf{s}, t'). \end{aligned} \tag{9}$$

This equation has the form of an ADE generalized to non-local time responses as a result of the ensemble average.

The next step is a crucial one in distinguishing between normal and anomalous transport. If $\psi(\mathbf{s}, t)$ has both a finite first and second moment in t the transport is normal and one can expand $\tilde{\psi}(\mathbf{s}, u)$ as (Scher and Montroll, 1975)

$$\begin{aligned} \tilde{\psi}(\mathbf{s}, u) &\cong p_1(\mathbf{s}) - p_2(\mathbf{s})u + \cdots \quad \text{and} \\ \tilde{\psi}(u) &= \sum_{\mathbf{s}} \tilde{\psi}(\mathbf{s}, u) \cong 1 - \bar{t}u + \cdots, \end{aligned} \tag{10}$$

with $\sum_{\mathbf{s}} p_1(\mathbf{s}) = 1$ the normalization of $\psi(\mathbf{s}, t)$ and $\sum_{\mathbf{s}} p_2(\mathbf{s}) = \bar{t}$, the temporal moment of $\psi(\mathbf{s}, t)$. The functions $p_i(\mathbf{s})$ are asymmetric due to the bias in the velocity field; $p_1(\mathbf{s})$ is the probability to make a step of displacement $\mathbf{s}$. One now inserts (10) into (7) and retains only the leading term (independent of u). The inverse Laplace transform of the result is inserted into (9) to yield the ADE

$$\frac{\partial P(\mathbf{s}, t)}{\partial t} = -\mathbf{v} \cdot \nabla P(\mathbf{s}, t) + \hat{\mathbf{D}} : \nabla\nabla P(\mathbf{s}, t), \tag{11}$$

where the effective velocity $\mathbf{v}$ is equal to the first spatial moment of $p_1(\mathbf{s})$, $\bar{\mathbf{s}}$, the mean displacement for a single transition, divided by the mean transition time $\bar{t}$, and the dispersion tensor $\hat{\mathbf{D}} \equiv D_{ij}$ is the second spatial moment divided by $\bar{t}$, which can be written as

$$\mathbf{v} = \sum_{\mathbf{s}} p_1(\mathbf{s})\mathbf{s}/\bar{t} \equiv \bar{\mathbf{s}}/\bar{t}, \tag{12}$$

$$D_{ij} = v\frac{1}{2}\sum_{\mathbf{s}} p_1(\mathbf{s})s_i s_j/\bar{s}, \tag{13}$$

where $v = |\mathbf{v}|$ and $\bar{s} = |\bar{\mathbf{s}}|$. Thus, our underlying physical picture of advective-driven dispersion reduces to the familiar ADE when one can assume smooth spatial variation of $P(\mathbf{s}, t)$ and finite first and second temporal moments of $\psi(\mathbf{s}, t)$.

If these latter moments do not exist (Scher and Montroll, 1975), that is

$$\tilde{\psi}(\mathbf{s}, u) \cong p_1'(\mathbf{s}) - p_2'(\mathbf{s})u^{\beta}, \tag{14}$$

for $u \to 0$, where $0 < \beta < 1$ (which corresponds to the time dependence of $\psi(\mathbf{s}, t)$ shown in (15)), then the transport is anomalous as we shall document. Inserting (14) into (7) and then into the Laplace transform of (9), then inverting the Laplace transform, one can obtain a form similar to (11) with the time derivative replaced by the fractional derivative $\partial^{\beta} P(\mathbf{s}, t)/\partial t^{\beta}$, where $\mathcal{L}\{\partial^{\beta} f(t)/\partial t^{\beta}\} \equiv u^{\beta}\tilde{f}(u)$ (Hilfer and Anton, 1995; $\mathcal{L}$ denotes a Laplace transform). Thus we can define a fractional ADE that is an asymptotic special case of the general CTRW equations which obtains for the limiting case of (14). (We stress that this definition of 'asymptotic' does not correspond to the limit of complete self-averaging, in sharp contrast to the usual definition of the term used in stochastic hydrology (e.g., in order to define macrodispersion coefficients)). Similarly, if one cannot assume the expansion in (8), and makes an expansion analogous to (14) in the Fourier transform of $\psi(\mathbf{s}, t)$, that is, infinite spatial moments (Shlesinger *et al.*, 1982), it is easily shown (Compte, 1996, 1997), that one can derive a fractional ADE with fractional spatial derivatives; again, as a special limiting case of the CTRW. Generally, the fractional derivative $\partial^{\beta} f(t)/\partial t^{\beta}$ is essentially a notation for the convolution of $f(t)$ with $t^{-1-\beta}$ (cf. (15)).

We stress in this paper a particular physical picture, however it is important to emphasize that the general CTRW formalism can encompass a large number of physical processes with the appropriately determined $\psi(\mathbf{s}, t)$, for example, multiple-trapping, multiple-rate models (e.g., Scher *et al.*, 1991; Hatano and Hatano, 1998; Haggerty and Gorelick, 1995) and dispersion in stratified formations (Matheron and de Marsily, 1980; Zumofen *et al.*, 1991). We continue now with the physical origin of (14) and its quantitative consequences.

2.1. MOMENT CHARACTERIZATION OF ANOMALOUS TRANSPORT

The origin of (14) derives from a long tail behavior for $\psi(\mathbf{s}, t)$ (Scher and Montroll, 1975)

$$\psi(\mathbf{s}, t) \to t^{-1-\beta} \quad \text{as } t \to \infty, \quad \text{for } 0 < \beta < 1. \tag{15}$$

The asymptotic form of $\psi(\mathbf{s}, t)$ at large time (i.e., or $u \to 0$ for $\tilde{\psi}(\mathbf{s}, u)$; see discussion following (14)) determines the time dependence of the mean position $\bar{\ell}(t)$ and standard deviation $\bar{\sigma}(t)$ of $P(\mathbf{s}, t)$. In the presence of a pressure gradient (or 'bias'),

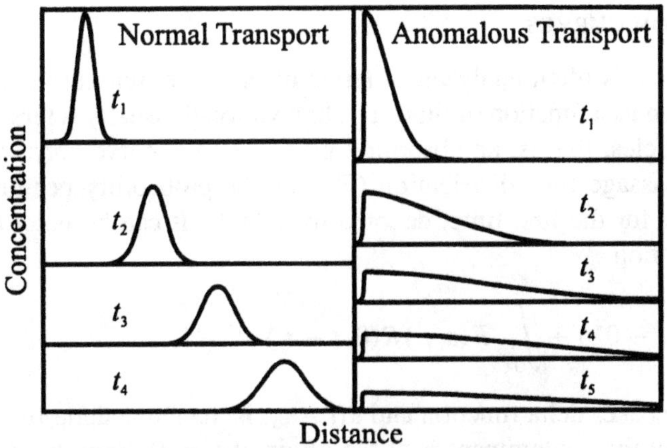

Figure 4. Schematic illustration of the contrast between anomalous and Gaussian transport, in terms of concentration profiles, $P(x, t)$ (adapted from Scher *et al.*, 1991; after Berkowitz *et al.*, 2000).

and for (15), it can be shown (Scher and Montroll, 1975, p. 2466; Shlesinger, 1974, p. 423) that

$$\bar{\ell}(t) \sim t^{\beta} \tag{16}$$

and

$$\bar{\sigma}(t) \sim t^{\beta}. \tag{17}$$

The unusual time dependence of $\bar{\ell}(t)$ and $\bar{\sigma}(t)$, which results from the infinite temporal mean of $\psi(\mathbf{s}, t)$, that is, does not fulfill the conditions of the central limit theorem, is the hallmark of the non-Gaussian propagation of $P(\mathbf{s}, t)$. This behavior is in sharp contrast to Gaussian models where $\bar{\ell}(t) \sim t$, $\bar{\sigma}(t) \sim t^{1/2}$ (as an outcome of the central limit theorem) and the position of the peak of the distribution coincides with $\bar{\ell}(t)$. Note that in Gaussian transport, $\bar{\ell}(t)/\bar{\sigma}(t) \sim t^{1/2}$; an important distinguishing feature of anomalous transport is that $\bar{\ell}(t)/\bar{\sigma}(t) \sim$ constant. The contrasting behavior between Gaussian and anomalous transport is shown in Figure 4. Anomalous transport is characterized by a concentration peak which moves much more slowly than the Gaussian, with a longer forward advance of particles. The relative shapes of the anomalous transport curves, and the rate of advance of the peak, vary strongly as a function of β. Thus the parameter β effectively quantifies the contaminant dispersion. This parameter will be discussed below. Hence, the crucial considerations for the appearance of anomalous transport in a specified scale of a heterogeneous medium are the physical criteria for the long tail behavior (15) and its (time) range of applicability.

2.2. BREAKTHROUGH CURVES

Contaminant migration is often analyzed in terms of breakthrough curves which quantify concentration as a function of time. The breakthrough usually refers to the plane of exiting particles, that is, an absorbing plane, and the curve corresponds therefore to a first passage time distribution (FPTD), the probability per time to reach site **s** at time t for the first time, denoted by $F(\mathbf{s}, t)$. It can be determined from the implicit relation

$$R(\mathbf{s}, t) = \delta_{\mathbf{s},0}\delta(t - 0^{+}) + \int_0^t F(\mathbf{s}, t')R(0, t - t')\, \mathrm{d}t', \tag{18}$$

where $\delta_{i,j}$ is the Kronecker delta function and $\delta(t - t_o)$ is the Dirac delta function, indicating in (18) that the contaminant is at the origin at $t = 0$. Equation (18) is a convolution relation between $R(\mathbf{s}, t)$ and $F(\mathbf{s}, t)$. The meaning of (18) is that a particle might have visited s at an earlier time t' for the first time and returned an arbitrary number of times, in the remaining $t - t'$, described by $R(0, t - t')$. We solve (18) using Laplace transforms and obtain

$$F(\mathbf{s}, t) = \mathcal{L}^{-1}\{\tilde{R}(\mathbf{s}, u)/\tilde{R}(0, u)\} \quad \text{for } \mathbf{s} \neq 0. \tag{19}$$

For anomalous transport the asymptotic regime (15) is the important one (see also section 3.1), and we evaluate (19) in this regime. A similar problem was considered by Montroll and Scher (1973), where it was shown (for $0 < \beta < 1$) that

$$f(L, \tau) \equiv \langle F(\mathbf{s}, t)\rangle = \mathcal{L}^{-1}\{\exp(-bu^{\beta})\}, \tag{20}$$

where $b \equiv L/\bar{s}$, L is the location of an absorbing plane, τ is a suitably defined dimensionless time and $f(L, \tau)$ is the average of (19) over the plane. The evaluation of (20) follows closely Appendix C in Scher and Montroll (1975).

The $\mathcal{L}^{-1}$ expression in (20) can be evaluated in the form of an infinite series; the function $\exp(-bu^{\beta})$ is expanded in a power series and integrated term by term to derive

$$f(L, \tau) = \tau^{-1} \sum_{j=0}^{\infty} \frac{(-b/\tau^{\beta})^j}{\Gamma(j+1)\Gamma(-j\beta)}, \tag{21}$$

where $\Gamma(x)$ is the gamma function and $1/\Gamma(-j\beta) = 0$ for integer values of $j\beta$. The series in (21) represents an entire function (the exponential is an example of such a function) and is easily computed for most of the range of b/τ^{β} of interest. For $b/\tau^{\beta} \gg 1$ the series in (21), however, is not useful and one can evaluate (20) as an asymptotic series in $\omega \equiv (\tau^{\beta}/\beta b)^{1/1-\beta}$ with the saddle-point method (Copson, 1965); the lowest order term is ($\omega \ll 1$)

$$f(L, \tau) \cong \exp\{-(1 - \beta)/(\beta\omega)\}/[2\pi(1 - \beta)\omega]^{1/2}\tau. \tag{22}$$

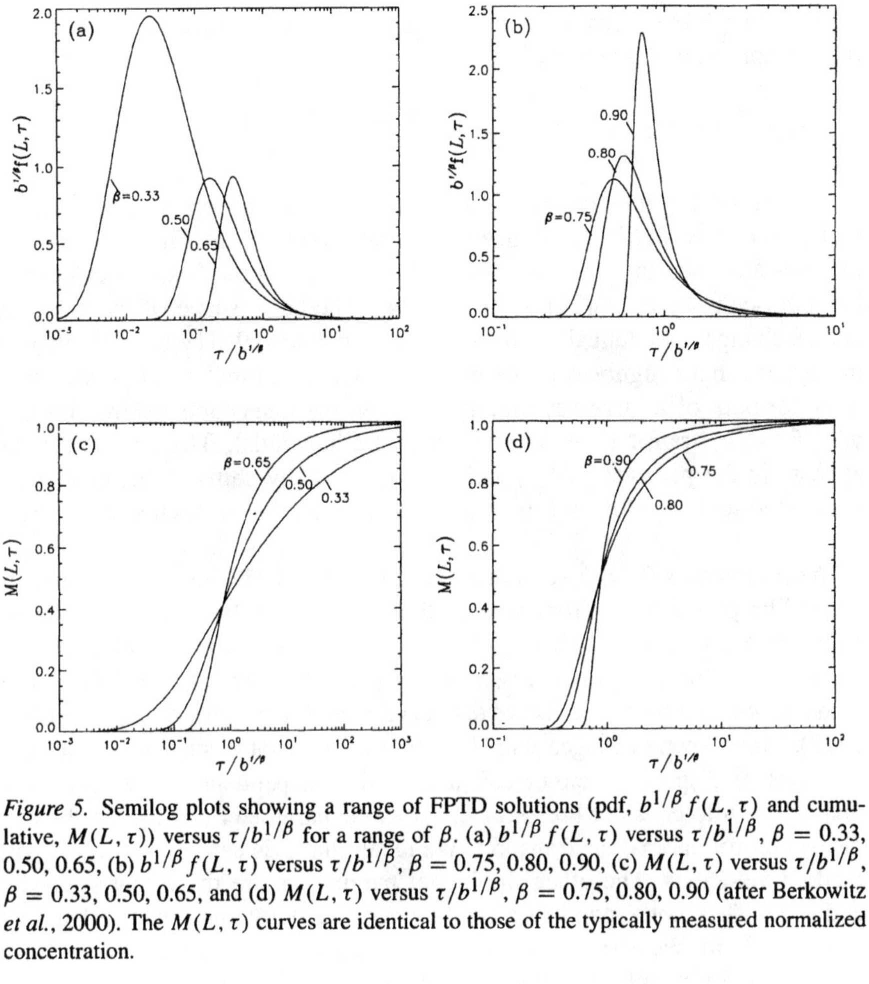

Figure 5. Semilog plots showing a range of FPTD solutions (pdf, $b^{1/\beta} f(L, \tau)$ and cumulative, $M(L, \tau)$) versus $\tau/b^{1/\beta}$ for a range of β. (a) $b^{1/\beta} f(L, \tau)$ versus $\tau/b^{1/\beta}$, $\beta = 0.33$, 0.50, 0.65, (b) $b^{1/\beta} f(L, \tau)$ versus $\tau/b^{1/\beta}$, $\beta = 0.75$, 0.80, 0.90, (c) $M(L, \tau)$ versus $\tau/b^{1/\beta}$, $\beta = 0.33$, 0.50, 0.65, and (d) $M(L, \tau)$ versus $\tau/b^{1/\beta}$, $\beta = 0.75$, 0.80, 0.90 (after Berkowitz *et al.*, 2000). The $M(L, \tau)$ curves are identical to those of the typically measured normalized concentration.

Note: Results equivalent to (21), (22) can also be obtained from the recent exact evaluation (Metzler *et al.*, 1994, 1998) of (20) in terms of Fox's H-function:

$$f(L, \tau) = \frac{1}{\beta t} H_{1,1}^{1,0}\left[\frac{b^{1/\beta}}{t} \middle| \begin{matrix} (0, 1) \\ (0, 1/\beta) \end{matrix} \right]. \tag{23}$$

In Figure 5, we show semilog plots of $f(L, \tau)$ and

$$M(L, \tau) \equiv \int_0^{\tau} f(L, \tau')\, \mathrm{d}\tau', \tag{24}$$

versus $\tau/b^{1/\beta}$ for a range of β. The function $M(L, \tau)$ is the accumulated mass reaching L due to a pulse of solute at the origin at $\tau = 0$ (normalized to a total mass of unity). In the case of a uniform concentration C_o injection starting at $\tau = 0$, the concentration C reaching L is

$$C(L, \tau) \equiv C_o \int_0^{\tau} f(L, \tau - \tau')\, d\tau'. \tag{25}$$

A change of the integration variable, $\tau - \tau' \rightarrow \tau'$ reproduces (24), that is, $C(L, \tau)/C_o = M(L, \tau)$. Hence the curves $M(L, \tau)$ in Figure 5 are those of the typically measured normalized concentration and the basis of calculating the breakthrough curves is (21), (22) and (24). The basic shape of the breakthrough curves changes dramatically from $\beta = 0.33$ to $\beta = 0.9$. The breakthrough curves in Figure 5 have highly variable basic features as a function of β with the common attribute of a large asymmetry between the dispersion before and after the $\tau/b^{1/\beta} \approx 1$ 'transit time' (that is, long trailing tails). The spatial dependence appears in the parameter $b^{1/\beta}$; a change in L rigidly shifts $M(L, \tau)$ along the x-axis in the semilog plots in Figure 5. We note in all the curves that $M(L, \tau) = 1/2$ corresponds to $\tau/b^{1/\beta} \approx 1$.

As β increases ($0 < \beta < 1$) there is a dramatic 'stiffening' of the concentration front. The plume has a front steeper than a Gaussian but yet with a trailing tail softer than a Gaussian. Also for $\beta > 0.5$ the peak of the spatial distribution of the particles (recall (4) and Figure 4), $P(x, t)$ (the average of $P(\mathrm{s}, t)$ over the plane orthogonal to the x-axis), is displaced from the origin (McLean and Ausman, 1977). These shape changes with β cannot be associated simply with a change in an effective dispersion, either constant or time-dependent. We will argue below that it is best to consider the exponent β as the parameter controlling the effective dispersion for anomalous transport. We recall that β is determined by the range of random velocities. The other significant parameters are those that scale distance and time. They correspond to a characteristic velocity and intrinsic length scales of the medium. We assume values for these parameters based on the experimental conditions, for example, average fluid velocity (cf. Section 3.1).

3. Comparison of the Theory with Experiments and Simulations

The purpose of this section to review recent applications of the theory, described in the previous section, to a variety of experiments. All the cases treated assume stationary heterogeneity. We use the generally good agreement with the experimental results as a basis for advancing a more complete description of transport in field sites with known non-stationary features in the last section.

3.1. FPTD ANALYSIS OF LABORATORY MEASUREMENTS

We have examined a tracer migration experiment in a heterogeneous sandbox model (Berkowitz *et al.*, 2000) as a critical application of the FPTD solutions. Silliman and Simpson (1987) constructed a flow cell packed with a uniform coarse sand.

Small rectangular regions consisting of a fine sand, with a hydraulic conductivity about one order of magnitude lower than the coarse sand, were interspersed randomly during the packing process. Chloride was used as a tracer, which was introduced along the inlet face of the flow cell as a step concentration boundary condition ($C = C_o$). The flow cell contained electrode columns at increasing distances from the inlet. Using electrical conductivity measurements, breakthrough curves delineating tracer advance were obtained at each column distance. Silliman and Simpson (1987) argued that the breakthrough curve data presented in their paper (see Figure 1) are indicative of the so-called 'scale effect', and incompatible with the classical ADE solutions (cf. Figure 8b in Silliman and Simpson, 1987).

Our recent analysis (Berkowitz *et al.*, 2000) shows that the FPTD solution fits the range of measured data remarkably well (Figure 1), with a β value which is invariant over the spatial/temporal scale of the experiment, and with simple estimates of $\bar{s}$ and v. These latter parameters translate the dimensionless curves (as shown in Figure 5) to dimensional units compatible with the actual experiment. The choice of β is straightforward, and is made by trial and error fitting with the data sets. The fits of the theoretical curves to the data shown in Figure 1 were thus obtained using values of $\beta \approx 0.87 \pm 0.01$. In contrast, the dispersivities used in the ADE solutions in Figure 1 were determined individually at each electrode column, representing 'best fit' estimates for the data (Silliman and Simpson, 1987). While the solutions are fit to the data in the concentration range around $C/C_o = 0.5$, deviations clearly occur outside this range. We note that imposition of a single, overall dispersivity value, which is intrinsic to the theory underlying the ADE, leads to even greater discrepancy between the ADE solution and the data.

The curves shown in Figure 1 provide a striking contrast between behaviors of the ADE (Gaussian) and CTRW (non-Gaussian) solutions. In Figure 1(a), the difference in the two fits occurs at early times. We note that the electrode distance in Figure 1(a) is about 12 times the length of the low permeability heterogeneities. At the larger electrode distances in Figures 1(b) and 1(c), there is a clearer difference between the fits for the developing plume. The largest measured concentration range is in Figure 1(b) – the CTRW theory accounts for both the early and late time breakthroughs. The evolving differences between the CTRW and ADE predictions become more substantial for the fully developed plume. These results demonstrate that anomalous transport behavior, quantifiable by CTRW theory, can exist in uniformly heterogeneous porous media on a scale much larger than the size of the heterogeneities. Of course, if the relative size of the heterogeneities is small at extremely large length scales (which may be impractical), the effect of the low velocities can diminish (that is, the relative sizes of the low permeability regions can decrease faster than the drop-off in the velocity in these regions), and the behavior can become Gaussian-like.

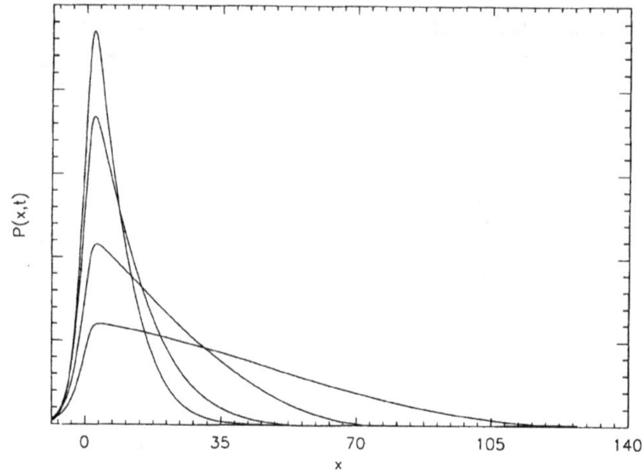

Figure 6. Profiles of the spatial distribution of tracer, $P(x, t)$, determined from the CTRW solution for $\beta = 0.5$, at four relative times $t = 1, 2.5, 10$ and 37.5 (after Berkowitz and Scher, 1997).

3.2. CTRW ANALYSIS OF FIELD MEASUREMENTS AND NUMERICAL SIMULATIONS

In our studies, we have developed a combined analytical-numerical evaluation of the (two-dimensional) profile of the particle spatial distribution, $P(x, t)$, for the case $\beta = 0.5$ (Berkowitz and Scher, 1997). These results can be extended to arbitrary β. The approach was first applied to consideration of Monte Carlo transport simulations in a series of synthetic random fracture networks (Berkowitz and Scher, 1997). The functional form of $\psi(\mathbf{s}, t)$ was determined from the distribution of velocities in the fracture segments (Berkowitz and Scher, 1997). The spatial and temporal behavior of $P(x, t)$ is shown in Figure 6 (compare also Figure 4). These curves are characterized by a peak that lies close to the origin, with tails that spread in response to the flow field. As time progresses, the distribution becomes increasingly uniform in space. Further detailed discussion of the character of these curves appears in Section 4.

The shapes of the spreading pulse $P(x, t)$ shown in Figure 6 are qualitatively the same as those found from the particle tracking simulations in these fracture networks, as shown in Figure 2. Despite the statistical noise in these results due to vertical averaging and a relatively small number of realizations, the CTRW theory reproduces the development of highly skewed particle plumes. Significantly, the CTRW solution captures the variation of both the movement of the center of mass of the particles and the standard deviation of particle location around this center of mass. For these simulations, transport was anomalous with the mean and standard deviation of the particle position scaling as t^β (recall (16) and (17)), with $\beta \approx$ 0.7–0.8.

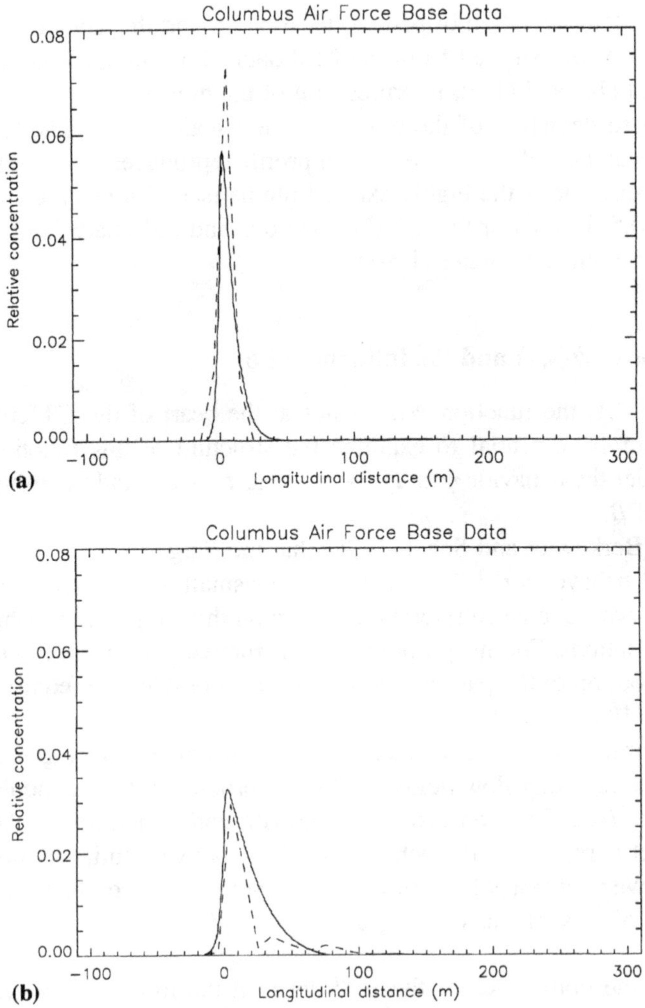

Figure 7. Comparison of the advance of the measured tracer plume (dashed lines) at the Columbus Air Force Base site (Figures 7(a) and 7(e) in Adams and Gelhar, 1992) and the CTRW solutions (solid lines), for $\beta = 0.5$. (a) $t = 49$ days (b) $t = 370$ days (after Berkowitz and Scher, 1998).

We have also used these solutions (Berkowitz and Scher, 1998) to suggest the occurrence of this same type of anomalous transport in natural, heterogeneous porous media, by interpreting field data from a large-scale field study performed at the Columbus Air Force Base (Adams and Gelhar, 1992). At this site, a highly heterogeneous alluvial aquifer, bromide was injected as a pulse and traced over a 20-month period by sampling from an extensive three-dimensional well network. The tracer plume that evolved was remarkably asymmetric, and could not be described by classical Gaussian models (Adams and Gelhar, 1992). Significantly, the CTRW captured the variation of both the movement of the center of mass of the particles and the standard deviation of particle location around this center of mass in the direction of flow, determined from the field data. Here, in agreement with CTRW theory (recall (16) and (17)), interpretation of the moments suggested that the mean and standard deviation of the plume position scale as $t^{0.6}$. The CTRW predictions for the evolution of the concentration profile reproduced very satisfactorily the observed behavior of the highly skewed plume (see Figure 7), using the solution with $\beta = 0.5$. Details of the CTRW solutions and full analysis of these data are given in Berkowitz and Scher (1998).

4. On the Behavior of $\psi(\mathbf{s}, t)$ and the Influence of β

As noted in Section 2.1, the function $\psi(\mathbf{s}, t)$ lies at the heart of the CTRW formulation. It is, therefore, essential to examine the structure of this function: in particular, we consider the behavior $\psi(\mathbf{s}, t) \to t^{-1-\beta}$ as $t \to \infty$, and the influence and determination of β.

We have found (Berkowitz and Scher, 1998) that the 'large time' $t \to \infty$ for which $\psi(\mathbf{s}, t)$ could behave as $t^{-1-\beta}$ is in fact, rather small – of the order of 10 transitions. As such, we can concentrate our attention on this 'large time' behavior for most practical situations. For this purpose, it is instructive to also work in terms of the velocity distribution of the particle transitions, and consider the (equivalent) behavior $\Phi(v) \to v^{1+\beta}$ as $v \to 0$.

One functional form of $\Phi(v)$, determined from analysis of computer-generated fracture networks and resulting flow fields, contains a product of both exponential and power-law terms (see Equation (16) in Berkowitz and Scher, 1998). While there is a scale effect here, in that the behavior of $\Phi(v)$ is exponential in most of the velocity regime, we note that at large times the low velocity part of the flow field dominates. In terms of $\psi(\mathbf{s}, t)$, substituting $v = s/t$ yields (15) with coefficients depending on s.

Why, then, does the dominance of the $v^{1+\beta}$ term in the low velocity regime lead to anomalous transport behavior? The slow decrease of v means that there is a higher probability that the particles encounter low velocity transitions (and we stress that not all particles must encounter low velocity transitions in order to produce anomalous transport), so that the low velocity regime cannot be ignored. It is the rate of drop-off of the velocity distribution in the $v \to 0$ regime, controlled by β, and the unique interplay between long and short time transitions (low and high velocities) that can lead to anomalous transport (recall Figure 3).

In light of this discussion, and before discussing β further, it is natural to question the degree and form of heterogeneity in fractured and in porous media that give rise to anomalous transport. More specifically, we might ask what permeability and velocity fields produce such $\psi(\mathbf{s}, t)$, with the characteristic algebraic drop-off of the velocities in the low velocity range. Qualitatively, the $\Phi(v)$ determined from the fracture network simulations (Section 3.2) should be general if the disorder in the system heterogeneities is sufficiently large. In such a random flow field, there are many spatial regions of low flow and hence a slow drop-off in their relative number.

Clearly though, anomalous transport need not arise in all cases of tracer migration in heterogeneous media. As discussed in Section 3.1, Berkowitz *et al.* (2000) demonstrated the existence of anomalous transport in a uniformly heterogeneous porous medium, which was caused by the fluid and tracer encounter with the low flow regions. In contrast, using the same experimental setup, Silliman and Simpson (1987) conducted a series of parallel experiments in homogeneous media, and found that the breakthrough curves followed closely the shape predicted using a standard Fickian model with a constant dispersivity. Silliman *et al.* (1987) then examined transport behavior in a homogeneous sand interspersed with a random packing of vertically-oriented, impermeable plastic disks. While the flow field was disturbed so that fluid and tracer paths were diverted around these disks, thus increasing the 'tortuosity' of the flow lines, the overall breakthrough curves again exhibited Fickian behavior. We can conclude, therefore, that 'unusual' low velocity behavior arises when flow lines must distribute themselves through combinations of higher and lower conductivity regions.

More generally, whether anomalous or Fickian transport occurs is a function of several key aspects: (1) the structure of the heterogeneities (e.g., random, correlated, stratified); (2) the scale of the heterogeneities relative to the overall length scale of the flow domain; and (3) the relative influence of the temporal scales of the transport mechanisms (e.g., strength of advective transport relative to diffusion) and their interplay among the different heterogeneities (i.e., the length scale of transport versus the velocity distribution). We note parenthetically that non-Fickian transport also occurs in non-stationary systems, as well as in highly specialized cases such as perfectly stratified formations (which usually can be treated by a superposition of independent tracks of different velocities) or formations possessing evolving scales of heterogeneity. These cases require specialization of the present model or other theoretical frameworks. We shall address non-stationary systems in Section 5.

We observe also that many other stochastic theory treatments of macroscopic dispersion (e.g., the reviews in Dagan and Neuman, 1997) consider velocity fields in nonhomogeneous domains. In strong contrast to the current approach, however, we note that these 'heterogeneous' domains are generally limited to having small conductivity variances. Moreover, in these stochastic treatments, the velocity (or hydraulic conductivity) distributions are usually assumed to be characterized by their first two (finite) moments. As such, the key role of the ('heavy') tails of these distributions are not recognized or accounted for. This is especially so with regard to perturbation approaches, which simply truncate these tails. Application of the CTRW to the types of velocity distributions (transition probabilities) used in these other approaches will ultimately lead to Gaussian transport behavior (again, as a function of the length and temporal scales of the problem).

We focus now on the influence of β. We have shown that β is the key parameter that controls the effective dispersion behavior of migrating contaminants, and that the value of β is determined by the low velocity tail of the flow field distribution. Thus, β is a physically-based parameter which can be estimated on the basis of knowledge on the velocity (or hydraulic conductivity) field. (In this context, it is worth noting recent studies which suggest that hydraulic conductivity distributions may follow non-Gaussian, heavy-tailed distributions (e.g., Painter, 1996; Liu and Molz, 1997). Such distributions might lead directly to the $v^{1+\beta}$ behavior in the low velocity regime, and allow estimation of β). Alternatively, as done for estimation of the dispersivity coefficient in the usual ADE equation, β can be determined by direct fitting to measured contaminant breakthrough curves.

Remarkably, this parameter controls a very wide range of dispersion behaviors. In the range $0 < \beta < 1$, the first and second (temporal) moments of $\psi(\mathbf{s}, t)$ are infinite, and transport is anomalous (recall Figure 4). As β increases, it is clear that $\psi(\mathbf{s}, t)$ has a finite first moment, but still an infinite second moment, for $1 < \beta < 2$. In this range, the transport behavior is in transition from anomalous to Gaussian. And for $\beta > 2$, $\psi(\mathbf{s}, t)$ has finite first and second moments, and the transport is Gaussian (recall (10) and (11)).

In terms of concentration profiles, such as in Figures 2, 4 and 6, anomalous transport arises for $0 < \beta < 1$, and the peak of the concentration distribution lags behind the mean plume position. For $\beta \geqslant 1$, the peak is coincident with the mean position ($\bar{\ell}(t) \sim t$), but still does not vary as $\sqrt{t}$; rather $\bar{\sigma}(t) \sim t^{(3-\beta)/2}$ (Shlesinger, 1974). In this range of β, the propagating distribution resembles a Gaussian distribution, but with heavier tails. The transport tends to a Gaussian form for $\beta \geqslant 2$ (Montroll and Scher, 1973; McLean and Ausman, 1977), with $\bar{\ell}(t) \sim t$, $\bar{\sigma}(t) \sim t^{1/2}$.

To illustrate the change in transport behavior with increasing β, we compare the concentration profiles shown in Figure 6, calculated for $\beta = 0.5$, with the profiles shown in Figure 8, calculated for the case $\beta = 0.87$. This latter value of β corresponds to that found for the dispersion experiments in the laboratory flow cell, discussed in Section 3.1 (and Figure 5). The most notable effect is that as β increases (even in the range $0 < \beta < 1$), the peak begins to move more noticeably with time, and the character of the dispersion changes, becoming less skewed. This trend in the behavior of the concentration profiles continues for $\beta > 1$. Profiles for the case $\beta = 1.5$ (presented by Montroll and Scher, 1973) and using approximate solutions (provided by McLean and Ausman, 1977) demonstrate how the transport behavior continues to evolve. Although the propagating distribution begins to resemble a Gaussian distribution, distinct heavier tails remain. Hence, through the parameter β, the range of random velocities essentially controls the character of the transport.

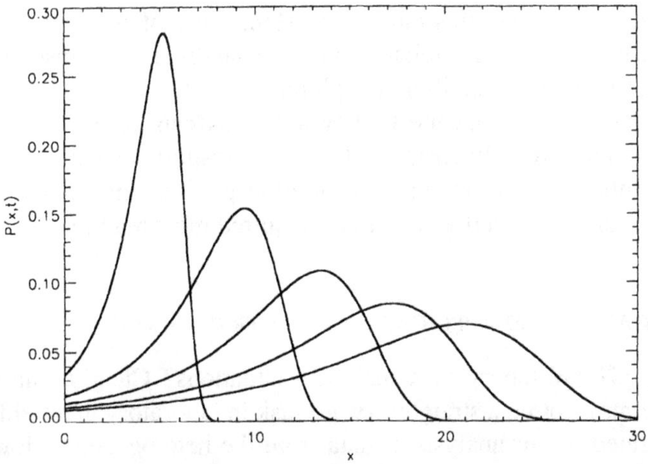

Figure 8. Profiles of the spatial distribution of tracer, $P(x, t)$, determined from the approximate form of the CTRW solution for $\beta = 0.87$, at the relative times $t = 5, 10, 15, 20$ and 25. The profile, $P(x, t)$, is the vertical average (along the y axis) of $P(\mathrm{s}, t)$, in arbitrary units; particle injection point is at $x = 0$. These approximate solutions assume no diffusion upstream of the injection point, which accounts for the shape of the profiles at this boundary.

5. Discussion and Concluding Remarks

5.1. CTRW THEORY

Complex distributions of sediments and high contrasts in hydraulic properties present in heterogeneous aquifers naturally give rise to a mixture of preferential pathways and stagnant regions, which can lead to anomalous transport behavior. We have introduced and applied a CTRW theory that qualitatively and quantitatively describes observed transport behaviors in porous media which have not been adequately explained by other existing theories. Simply stated, the CTRW accounts for a full, appropriately weighted, sampling of flow paths. This sampling includes paths containing statistically rare events. These latter events can have large effects on the plume distribution especially in pre-ergodic conditions, which are the ones prevailing in all the studies reviewed above. In contrast, the stochastic approaches such as surveyed in Dagan and Neuman (1997) essentially truncate these critical events and tend to focus on the ergodic behavior of a migrating contaminant plume.

The CTRW is equivalent to a GME transport equation, (2), which is the ensemble average of a ME, (1), description of particles making transitions in space and time in a random flow field. In a fracture network, these transitions are clearly defined by the fracture segments. In a heterogeneous porous medium, the definition of these transitions is more subtle; particles move lengths controlled by the coherence properties of the velocity field and the scattering features among different velocity values. Depending on the velocity distribution the theory can account for the range of transport behavior from anomalous to Gaussian. The key parameter quantifying the plume shape over this range is β. Hence, use of β to characterize the transport is more meaningful than reliance on the usual dispersion tensor, which is based on statistical moments describing the plume.

Finally, we emphasize here that the CTRW solutions can be developed and applied to one-, two- and three-dimensional flow and transport systems. Here, we have focused on solutions for averaged concentration profiles and breakthrough curves, because these are the data that tend to be acquired most readily.

5.2. LABORATORY AND FIELD APPLICATIONS OF CTRW THEORY

How can we apply CTRW theory in actual field situations? Clearly, analytical definition of $\psi(\mathbf{s}, t)$ may not be a straightforward task in laboratory and field situations. As demonstrated in our analysis of data from the heterogeneous flow cell (Berkowitz *et al.*, 2000), and from the Columbus field site (Berkowitz and Scher, 1998), acquisition of detailed measurements of the flow field structure which might permit precise evaluation of $\psi(\mathbf{s}, t)$ is not of prime importance. Rather, estimation of the 'dispersion parameter', β, is critical. If estimates of the temporal variation of the mean and standard deviation of the contaminant plume are available (as in the case of the Columbus field site), one can use estimates of β to determine if anomalous transport is in fact occurring, and if so, to estimate how the plume will continue to evolve. In other situations, as in the laboratory flow cell, only breakthrough curves are of interest. We need then only work with the 'large time' form of $\psi(\mathbf{s}, t)$ and the parameter β, determining breakthrough curves using the technique given in Sections 2.3 and 3.1. As noted in Section 4, the 'large times' (length and time scales) needed to apply this formulation are, in practical terms, rather short.

Even at the field scale, where the point at which 'large time' is reached may not be known (at least *a priori*), the question of the appropriate length and time scales is not problematic. In actual field situations, we need to measure the advancing plume (the cross-sectional breakthrough curve concentrations) at, say, two locations. Such an arrangement is convenient in design of an assessment/monitoring system, because we only require that families of observation wells be installed at two distances from the contaminant source. Measuring breakthrough curves at these locations provides sufficient information to use curve fitting to estimate β (as done in Berkowitz *et al.*, 2000) and to make predictions about the continuing plume evolution. If we have measurements at several distances, as well as several times, we can further constrain our estimates. The estimates of $\bar{s}$ and v can then also be derived from basic information on the medium heterogeneities and flow field.

We argue here that this approach has another distinct strength: rather than (as is usually done) trying to delineate a conductivity or velocity field, and then attempting to account for effects of, for example, diffusion, mixing or adsorption in a transport model, we work directly with concentration measurements which already reflect the history that the advancing particles have experienced. If we can determine β from the breakthrough curve we can calculate the plume. Generally, we can 'convert' FPTD solutions to $P(x, t)$ solutions, by combining many FPTD solutions at a set time and a series of distances from the inlet.

5.3. TOWARDS COMPREHENSIVE QUANTITATIVE TRANSPORT MODELS

It is clear that no single generic model can fully account for transport in heterogeneous aquifers, and that we must incorporate site-specific geological characterizations in our quantitative models. Discretizing heterogeneous formations into small-scale (high-resolution) blocks, with levels of discretization sufficient to resolve the larger-scale aquifer features, is a realistic modeling approach. However, as noted by Eggleston and Rojstaczer (1998) with regard to the Columbus site (discussed in Section 3.2), small-scale variations in the hydraulic conductivity within these small blocks strongly affect large-scale contaminant migration patterns, and large-scale stationarity conditions cannot be met (i.e., there is no unambiguous way to filter out large-scale trends from small-scale 'residues'). Indeed, our analysis of experimental data demonstrates that anomalous transport can exist in heterogeneous porous media even when the system size is much larger than the size of the heterogeneities and even with relatively small contrasts in hydraulic conductivity.

We suggest an approach which accounts for non-Fickian transport induced by the small-scale heterogeneities, and which also accounts for different, larger spatial scales. The underlying approach is the probabilistic CTRW, which can be applied and integrated with an accounting for interplay of spatial scales and levels of uncertainty. If only limited geological and hydraulic information are available for consideration of a large-scale field problem involving contaminant transport, then it is reasonable to use a 'black box' (ensemble average) approach, wherein the entire domain is treated as a single unit, with an approximately stationary distribution of heterogeneities. Consideration of the sandbox model and MADE data, discussed in Section 3, illustrates that the CTRW approach, in sharp contrast to other existing approaches, captures well the large-scale transport behavior.

On the other hand, if more detailed, site-specific information is available, then we can apply a hybrid approach which uses knowledge of non-stationary aspects on a larger spatial scale with a probabilistic treatment of unresolved structure (residues) on a smaller scale. We suggest that the CTRW analysis can account for transport behavior on relatively small scales at which we have little or no information. In relatively homogeneous porous media, definition of blocks at a certain (small) scale as homogeneous, with transport quantifiable by the ADE, seems reasonable. But in moderately and highly heterogeneous formations, we suggest the use of the CTRW framework to characterize transport in domains of unresolved small-scale features, in regions of relatively similar geological and hydraulic properties. In other words, the CTRW is applicable over a range of length scales as long as stationarity can be assumed. We can then consider groupings of stationary units on a larger scale, that is, nonstationary field-scale systems comprised of stationary units. Here the challenge, currently under investigation, is to 'integrate' transport behavior quantified in each of these stationary units, either analytically or numerically, in order to yield predictions for transport over the nonstationary, field-scale systems. One consequence of such an approach is that the definition of $\psi(\mathbf{s}, t)$, or more simply, the value of β, will vary in each of the different geological units into which the particles migrate. The complete solution then involves the treatment of the boundary conditions between these geological units analogous to the present use of the ADE in the context of large numerical models of complex fields (LaBolle *et al.*, 1998). Our development of a hybrid numerical model utilizing CTRW theory and important non-stationary features of a field site is work in progress.

Acknowledgements

The authors thank Ghislain de Marsily and an anonymous reviewer for comments which improved the manuscript, and Gennady Margolin for preparation of Figure 8. Brian Berkowitz thanks the European Commission (Contract No. ENV4-CT97-0456) for support. Harvey Scher thanks the Sussman Family Center for Environmental Sciences for partial support.

References

Adams, E. E. and Gelhar, L. W.: 1992, Field study of dispersion in a heterogeneous aquifer: 2. Spatial moment analysis, *Water Resour. Res.* **28**(12), 3293–3308.

Berkowitz, B. and Scher, H.: 1995, On characterization of anomalous dispersion in porous and fractured media, *Water Resour. Res.* **31**(6), 1461–1466.

Berkowitz, B. and Scher, H.: 1997, Anomalous transport in random fracture networks, *Phys. Rev. Lett.* **79**(20), 4038–4041.

Berkowitz, B. and Scher, H.: 1998, Theory of anomalous chemical transport in fracture networks, *Phys. Rev. E* **57**(5), 5858–5869.

Berkowitz, B., Scher, H. and Silliman, S. E.: 2000, Anomalous transport in laboratory-scale, heterogeneous porous media, *Water Resour. Res.* **36**(1), 149–158.

Compte, A.: 1996, Stochastic foundations of fractional dynamics, *Phys. Rev. E* **53**(4), 4191–4193.

Compte, A.: 1997, Continuous time random walks on moving fluids, *Phys. Rev. E* **55**(6), 6821–6831.

Copson, E. T.: 1965, *Asymptotic Expansions*, Cambridge University Press, Cambridge.

Dagan, G. and Neuman, S. P. (eds): 1997, *Subsurface Flow and Transport A Stochastic Approach*, Cambridge University Press, New York.

Eggleston, J. and Rojstaczer, S.: 1998, Identification of large-scale hydraulic conductivity trends and the influence of trends on contaminant transport, *Water Resour. Res.* **34**(9), 2155–2168.

Gelhar, L.W.: 1993, *Stochastic Subsurface Hydrology*, Prentice-Hall, Inc., Englewood Cliffs, NJ.

Grindrod, P. and Impey, M. D.: 1993, Channeling and Fickian dispersion in fractal simulated porous media, *Water Resour. Res.* **29**(12), 4077–4089.

Haggerty, R. and Gorelick, S. M.: 1998, Multiple-rate mass transfer for modeling diffusion and surface reactions in media with pore-scale heterogeneity, *Water Resour. Res.* **31**(10), 2383–2400.

Hatano, Y. and Hatano, N.: 1998, Dispersive transport of ions in column experiments: An explanation of long-tailed profiles, *Water Resour. Res.* **34**(5), 1027–1033.

Hilfer, R. and Anton, L.: 1995, Fractional master equations and fractal time random walks, *Phys. Rev. E* **551**(2), R848–R851.

Kenkre, V. M., Montroll, E. W. and Shlesinger, M. F.: 1973, Generalized master equations for continuous-time random walks, *J. Stat. Phys.* **9**(1), 45–50.

Klafter, J. and Silbey, R.: 1980, Derivation of continuous-time random-walk equations, *Phys. Rev. Lett.* **44**(2), 55–58.

Kinzelbach, W.: 1988, The random walk method in pollutant transport simulation, in: E. Custodio, A. Gurgui and J. P. Lobo Ferreria (eds), D. Reidel, Norwell, Mass., *Groundwater Flow and Quality Modelling, NATO ASI Ser., Ser. C Math and Phys. Sci.*, Vol. 224, pp. 227–246.

LaBolle, E. M., Quastel, J. and Fogg, G. F.: 1998, Diffusion theory for transport in porous media: Transition probability densities of diffusion processes corresponding to advection–dispersion equations, *Water Resour. Res.* **34**(7), 1685–1693 .

Liu, H. H. and Molz, F. J.: 1997, Multifractal analyses of hydraulic conductivity distributions, *Water Resour. Res.* **33**(11), 2483–2488.

Matheron, G. and de Marsily, G.: 1980, Is transport in porous media always diffusive? A counter example, *Water Resour. Res.* **16**(5), 901–917.

McLean, F. B. and Ausman, G. A., Jr.: 1977, Simple approximate solutions to continuous-time random-walk transport, *Phys. Rev. B* **15**(2), 1052–1061.

Metzler, R., Glöckle, W. G. and Nonnenmacher, T. F.: 1994, Fractional model equation for anomalous diffusion, *Physica A* **211**, 13–24.

Metzler, R., Klafter, J. and Sokolov, I. M.: 1998, Anomalous transport in external fields: Continuous time random walks and fractional diffusion equations extended, *Phys. Rev. E* **58**(2), 1621–1633.

Montroll, E. W. and Scher, H.: 1973, Random walks on lattices. IV. Continuous-time walks and influence of absorbing boundaries, *J. Stat. Phys.* **9**(2), 101–135.

Oppenheim, I., Shuler, K. E. and Weiss, G. H.: 1977, *The Master Equation*, MIT Press, Cambridge.

Painter, S.: 1996, Evidence for non-Gaussian scaling behavior in heterogeneous sedimentary formations, *Water Resour. Res.* **32**(5) 1183–1195.

Prickett, T. A., Naymik, T. G. and Lonnquist, C. G.: 1981, A random walk solute transport model for selected groundwater quality evaluations, *Bull. Ill. State Water Surv. 65*, Champaign.

Scher, H. and Lax, M.: 1973, Stochastic transport in a disordered solid, I. Theory, *Phys. Rev. B* **7**(10), 4491–4502.

Scher, H. and Montroll, E. W.: 1975, Anomalous transit-time dispersion in amorphous solids, *Phys. Rev. B* **12**(6), 2455–2477.

Scher, H., Shlesinger, H. F. and Bendler, J. T.: 1991, Time-scale invariance in transport and relaxation, *Physics Today* January, 26–34.

Shlesinger, M. F.: 1974, Asymptotic solutions of continuous-time random walks, *J. Stat. Phys.* **10**(5), 421–434.

Shlesinger, M. F.: 1996, Random Processes, in *Encyclopedia of Applied Physics*, Vol. 16, VCH Publishers, Inc., New York.

Shlesinger, M. F., Klafter, J. and Wong, Y. M.: 1982, Random walks with infinite spatial and temporal moments, *J. Stat. Phys.* **27**(3), 499–512.

Silliman, S. E. and Simpson, E. S.: 1987, Laboratory evidence of the scale effect in dispersion of solutes in porous media, *Water Resour. Res.* **23**(8), 1667–1673.

Silliman, S. E., Konikow, L. F. and Voss, C. I.: 1987, Laboratory investigation of longitudinal dispersion in anisotropic porous media, *Water Resour. Res.* **23**(11), 2145–2151.

Uffink, G. J. M.: 1985, A random walk method for the simulation of macrodispersion in a stratified aquifer, in: *Relation of Groundwater Quality and Quantity, IAHS Publ. 146*, Int. Assoc. of Hydro. Sci., Gentbrugge, Belgium, pp. 103–114.

Zumofen, G., Klafter, J. and Blumen, A.: 1991, Trapping aspects in enhanced diffusion, *J. Stat. Phys.* **65**(5/6), 991–1013.

CPSIA information can be obtained
at www.ICGtesting.com
Printed in the USA
LVHW051136180623
750102LV00005B/308

9 780792 367796